2007 QUEENS COLLEGE
Mathematics Competition

Problem Solving
Through
Recreational Mathematics

Bonnie Averbach

and

Orin Chein

DOVER PUBLICATIONS, INC.
Mineola, New York

Bibliographical Note

This Dover edition, first published in 2000, reproduces entirely the text of *Mathematics: Problem Solving Through Recreational Mathematics,* originally published in 1980 by W. H. Freeman and Company, San Francisco. A substantial new appendix on the subject of probability, with accompanying exercises, hints and solutions, and answers, as well as a publisher's note have been added to the present edition. In addition, the index, contents, and some of the prefatory material have been updated to incorporate the new material.

Library of Congress Cataloging-in-Publication Data

Averbach, Bonnie, 1933–
 [Mathematics]
 Problem solving through recreational mathematics / Bonnie Averbach and Orin Chein.
  p. cm.
 Originally published: Mathematics. San Francisco : W. H. Freeman, c1980.
 Includes bibliographical references and index.
 ISBN 0-486-40917-1 (pbk.)
  1. Mathematics. 2. Mathematical recreations. 3. Problem solving. I. Chein, Orin, 1943– II. Title.

QA39.2 .A896 2000
510—dc21

99-052198

Manufactured in the United States of America
Dover Publications, Inc., 31 East 2nd Street, Mineola, N.Y. 11501

Contents

☆ Indicates sections that may be omitted without loss of continuity.

Publisher's Note

This book is a reproduction of *Mathematics: Problem Solving through Recreational Mathematics* which was published by W. H. Freeman and Company in 1980. The original manuscript for that book contained a chapter on probability which was omitted from that edition. We still believe that the material in this chapter (which was Chapter 7 in the original manuscript) is worthy of publication, so in this new Dover edition it has been reinstated as Appendix C, found on page 336. This appendix includes all the detailed discussion, sample problems, exercises, hints and solutions, and selected answers that any of this book's other chapters include; we feel it makes the present edition an even greater aid to students and instructors than the original edition, which is now unavailable.

When *Mathematics: Problem Solving through Recreational Mathematics* was in print, it was accompanied by a solutions manual available to instructors using the book as a text for their courses. This manual is now available from the authors, at a nominal charge. Readers desiring a copy of the manual should contact Bonnie Averbach at the Department of Risk Management and Actuarial Science, Temple University, Philadelphia, PA 19122.

Preface

How can students be given a feeling for mathematics in a one or two semester college course? With most of the usual approaches—through history, culture, or applications—the student is an observer. He or she learns about what others have done, but does not learn to think independently. The student learns some techniques and practices them on problems that are essentially the same as those done in class. If applications are stressed, only highly simplified cases can usually be considered. For example, linear programming techniques are usually presented for functions of only two variables, a case that rarely comes up in practice. The result is that the student comes away with little of permanent benefit and unconvinced of the actual usefulness of the subject.

A few years ago, we decided to try a different approach. We said to the student: "You participate and be the mathematician. Take a problem and use anything and everything you know to solve it. Think about it; strain your mind and imagination; put it aside, if necessary; keep it in mind; come back to it. If you can solve it on your own, isn't the feeling great? If you can't solve it, maybe some mathematics (new to you) would be helpful to know; let's develop some and see."

What have we accomplished with this approach? As teachers we love teaching this course, because of the enthusiasm generated by the students. The students we taught learned that mathematics is not just numbers and manipulation—it's thinking; it's strategy. In fact, aspects of mathematics can be seen all around us and are useful for any kind of problem, in any kind of setting. Most important, our students have learned how to think critically. Several of our former students have returned to tell us how they have been able to apply techniques that they learned in our course to a variety of situations, ranging from student teaching to business.

Our approach is based on recreational mathematics—recreational problems, puzzles, and games. Why? First, because they are fun, and motivation is perhaps the central problem in teaching mathematics. Second, because, historically, many important mathematical concepts arose from problems that were recreational in origin. Recreational mathematics has proved ideal for introducing most of the topics usually covered in liberal arts mathematics courses. However, our emphasis is not on the mathematical results themselves, but rather on how these results can be used in thinking about problems and solving them. Problem solving is one of the most important skills a person can acquire. It is useful to the mathematician and is also of great importance in many other fields of endeavor and in life itself.

The major difficulty we encountered in developing a course based on recreational problem solving was the selection of a text. Although there are many books available that contain excellent selections of mathematical problems, puzzles, and games, none was completely suitable for our purposes. Very few of these books attempt to explain how to go about attacking a problem or try to present any formal mathematics. For this reason, we decided to write our own notes for the course—the basis for this book.

During the several years in which we have been teaching this course, the actual topics covered have varied from semester to semester. Since we have decided to include most of these topics in this book, there is more than enough material for a one semester course. Because of this abundance of material, an instructor using this book as a text for a course lasting one semester or less must select which chapters to cover. For this reason, we have made the chapters as independent of each other as possible.

The only dependencies are the following:

1. The ability to translate a verbal quantitative problem into equations is required for some of the problems and exercises in Chapter 4. This translation process is discussed briefly in Chapter 3, and drill in using it can be obtained from the Chapter 3 exercises. A student with a good background in algebra could omit Chapter 3 completely.

2. Some of the exercises in Chapter 5 may be solved most simply by applying the notion of congruence, which is discussed in Chapter 4. These exercises are designated by the symbol #.

3. Some of the games in Chapters 7 and 8 refer back to ideas presented in earlier chapters. In particular, a matchstick game is first introduced in Chapter 1; congruences (Chapter 4) are useful in analyzing some of the Chapter 7 matchstick games; the binary system (Chapter 5) is needed for a discussion of Nim (7) and is also of interest with regard to the Tower of Brahma (8); the two solutions

of the colored cubes problem (8) refer to unique factorization (4) and graph theory (6).

The chapters are otherwise independent, although we do recommend beginning with Chapter 1.

Each chapter of the book contains several types of problems with different purposes. The Sample Problems are to whet the reader's appetite and motivate the discussions of the chapter (these are solved in the text). The Practice Problems are to firm up the reader's grasp of the mathematical ideas presented. The Exercises are to challenge the reader and help develop problem solving ability. Exercises that are unusually difficult or lengthy are marked by ★ (or, if they are very difficult, by ★★). The notation [A] after an Exercise or Practice Problem number indicates that the answer to the problem is given in the Answers to Selected Problems in the back of the book; [H] indicates that a hint is given in the Hints and Solutions; and [S] indicates that the problem is solved completely in the Hints and Solutions. In each chapter, a number of exercises are solved completely. These solutions augment the discussion of the chapter and occasionally present some techniques that will be helpful with other problems.

The sample problems and exercises in this book are, for the most part, problems we have accumulated over the years. These are mostly our variations of problems given us by students and colleagues or that have appeared in columns, journals, and the many puzzle books in recreational mathematics. Some of these problems may be found, in some form or another, in more than one source; many are now considered to be classics. As a result, although we originally intended to trace the true origin of each problem, we found this a difficult, if not impossible, task. For this reason and with a few exceptions, we decided to give references only for problems that we have taken essentially verbatim. However, we want to draw your attention to the many excellent books on recreational mathematics. These have been an inspiration to us and to our students, and we hope they will inspire others as well. A selected list of these may be found in the Bibliography on page 376. In addition, when we are aware of the creator of a problem or game, we have so indicated in parentheses. However, our knowledge in this matter is very incomplete, and we apologize for any mistakes or oversights.

May 1980   Bonnie Averbach
                                                            Orin Chein

To the Reader

Problem solving can be a stimulating and exciting pastime. Sometimes one can see the method of solution immediately. At other times, a problem can lead to hours of frustration. But the satisfaction that comes with the eventual solution of a difficult problem is, in itself, a reward that makes all the effort worthwhile.

In working with this book, think of yourself as the mathematician, the problem solver. Near the beginning of each chapter, you will find some sample problems. Try to solve the first of these before reading the chapter. If you are successful, think about how you approached the problem—specifically what you did to solve it. If you are unsuccessful, think about what further information would be helpful to you, what kind of mathematical considerations might be useful.

As you read the chapter, keep the original sample problem in mind. If you reach a point at which you think you might be able to solve the original problem, stop reading and try it again. Then read on to see our discussion of the solution. (If you have been able to solve the problem, you may find that your solution is better—simpler or more elegant—than ours. However, continue reading our discussion anyway. We have usually chosen our approach for a reason, and, in any case, it is often instructive to see different approaches to the same problem.)

After you have solved (or read about the solution of) the first sample problem, try the second one. If you get stuck, read on. Continue this approach throughout the chapter.

Do not be discouraged if you cannot solve all, or even any, of the sample problems in a given chapter. Each problem is there to try to lead you to an idea or a mathematical area that may be new to you. Even when you are unable to solve a problem, if you have really thought about it, then you have learned something. You may have partial success. You may realize where the difficulties lie. But, even if you made no progress at all with the problem, when the solution is finally shown to you, you will better appreciate the power of the techniques and mathematical theory used.

When you have finished reading a chapter, try as many of the exercises as you can. We do not expect you to attempt all the problems immediately (leave some for your future enjoyment), but you only learn to be a good problem solver by solving problems.

Although most of the exercises in each chapter relate to the basic mathematical tools or strategies presented in the chapter, you will find that being able to solve one problem does not mean that you will be able to solve them all—each problem presents its own challenge.

Analyze each exercise you try, as you did the sample problems. If you have been successful with a problem, try to take note of your approach. What were the crucial steps? If you have been unsuccessful after a reasonable effort, don't give up. Put the problem aside for a while, but keep it in mind; sleep on it. Good ideas often come at the strangest times.

If you are reading this book as part of a group or class, become involved in any class discussions. Active participation is far more beneficial than passive attendance. Ask questions; create your own problems; share your ideas with others; try to provide helpful criticism of the thought processes of others; and seek criticism of your own thinking.

The more you get involved with the problems in this book, the more benefits you will derive from the book. If you approach the problems conscientiously, we think you will find that you will not only become a better problem solver, but your general thinking processes, as they apply to all fields, will benefit from the experience.

B.A.
O.C.

Acknowledgments

Like most authors, we are indebted to many people who have assisted us in a variety of ways from the time our idea for this book was first conceived until final publication.

For their advice, encouragement, and many helpful suggestions, we would like to thank David Blauer, Isidor Chein, Martin Gardner, Nancy Hagelgans, William Karjane, Archibald Krenzel, Robert Morse, Leon Steinberg, William Traylor, William Wisdom, and David Zitarelli, as well as the reviewers who read our manuscript at various stages during its preparation and the editors who worked with us during the production process.

We would also like to thank Raelaine Ballou for her continual assistance in tracking down some of the many sources we used in preparing our manuscript, Susan Graham for drawing preliminary sketches for many of our cartoon ideas, John Johnson for embellishing these ideas and bringing them to their delightful final form, and Nadia Kravchenko for her excellent typing of the entire manuscript.

For their constant encouragement and patience throughout the years we were writing this book, we extend our love and thanks to the members of our families—Rachelle, Wayne, Debra, Gary, Robert, Stephen, Jake, and Alex Averbach, and Carrie, Adam, and Jason Chein.

The origins of many mathematical problems are lost in the mists of antiquity; thus our attempts to locate original sources for our problems were not very successful. We decided to settle for identifying those sources from which we have taken problems essentially verbatim. Each such problem in the text is followed by a reference number, which refers to a source appearing in the Bibliography. In particular, we would like to acknowledge the following publishers and authors for granting us permission to reprint or adapt some of their material.

Exercises 1.19 and 1.22 are reprinted with the permission of

Al B. Perlman. Problem 4.4 and Exercises 5.9 and 7.22 have been adapted from *Mathematical Puzzles for Beginners and Enthusiasts* by Geoffrey Mott-Smith (copyright © 1946 by Copeland and Lamm, Inc., copyright renewed in 1974; published by Dover Publications, Inc.), with the permission of Copeland and Lamm, Inc. Problem 8.2 and its solution and Exercise 8.9 have been adapted from *Polyominoes* by Solomon W. Golomb (copyright © 1965 by Solomon W. Golomb), with the permission of Charles Scribner's Sons and Allen & Unwin, Ltd.

Dover Publications has granted us permission to use problems from several of the books in their fine series on mathematical recreations. In particular, we have taken problems from the Dover books by Friedland, Kraitchik, Phillips, and Read listed in our Bibliography, and also from *Mathematical Bafflers* by Angela Dunn, some of whose material was published earlier by Litton Industries in their *Problematical Recreations* series.

Several journals have also been helpful: *Technology Review*, with its "Puzzle Corner" edited by Allan Gottlieb; *Arithmetic Teacher*; *Mathematics Magazine*; and *Sphinx: Revue Mensuelle des Questions Récréatives*, edited by M. Kraitchik and M. Pigeolet.

The American Mathematical Society, Emerson Books, Inc., the Longman Group Limited, and McGraw-Hill, Inc. also granted permission for use of problems.

Finally, we wish to thank Phillip Matthew Swift for his work in graphics and word processing to prepare the new chapter on probability (Appendix C) for this edition.

B.A.
O.C.

1

Following the Clues

An enjoyable and interesting way of studying problem solving in general is to consider recreational problems in particular.

Recreational problems and games have been a source of amusement and interest for hundreds of years. Although the origins of many problems can be traced back to ancient times, probably the first important collection of problems is a *Greek Anthology* attributed to Metrodorus (c. 500). Many other collections followed. Among these, *Problèmes plaisans et delectables, qui se font par les nombres,* by Claude-Gaspar Bachet de Méziriac (1612), is the first noteworthy collection to appear in print.

It was not until the late nineteenth and early twentieth centuries, though, that recreational mathematics acquired great popularity. Not only were many books on the subject published at that time, but the newspaper columns of Sam Loyd in America and H. E. Dudeney in England helped to bring recreational puzzles to the attention of the general public. The decades since have seen a continuation of this trend: The newspaper columns of Hubert Phillips ("Caliban") in England, the Belgian journal *Sphinx : Revenue mensuelle des questions récréatives*, the *American Mathematical Monthly*, and many books continue to supply a wide selection of recreational problems.

Current developments may be followed in the *Journal of Recreational Mathematics*, in the books of Martin Gardner, and in his columns in *Scientific American*.

Although some mathematical background is required to solve certain problems, many problems of recreational mathematics can be solved

1

with little more than fundamental reasoning skills, a power of concentration, perseverance, and a touch of imagination and ingenuity. The techniques of solution can be used to help solve problems in all fields.

We will investigate some of these techniques by considering the following problems.

SAMPLE PROBLEMS

Problem 1.1

An elimination boxing tournament was organized. There were 114 participants and so there were 57 matches in the first round of the tournament. In the second round, the 57 fighters remaining were paired, resulting in 28 matches; one fighter received a bye (that is, did not have to fight in that round). The 29 fighters remaining were then paired, and so on.

(a) How many matches in all were required to determine a winner of the tournament?

(b) How many matches would be required if n people participated in the tournament (where n represents a fixed but unspecified whole number)?

Problem 1.2

Ms. X, Ms. Y, and Ms. Z—an American woman, an Englishwoman, and a Frenchwoman, but not necessarily in that order, were seated around a circular table, playing a game of Hearts. Each passed three cards to the person on her right. Ms. Y passed three hearts to the American. Ms. X passed the queen of spades and two diamonds to the person who passed her cards to the Frenchwoman.

Who was the American? The Englishwoman? The Frenchwoman?

Problem 1.3

Armand Alloway, Basil Bennington, Col. Carlton Cunningham, Durwood Dunstan, and Everett Elmsby, Esq. are the five senior members of the Devonshire Polo Club. Each owns a pony that is named after the wife of one of the others.

Mr. Alloway's pony is named Georgette; Col. Cunningham owns Jasmine; and Mr. Elmsby owns Inez. Francine, owned by Mr. Dunstan, is named after Alloway's wife. Georgette's husband owns the pony that is named after Mr. Bennington's wife. Helene Cunningham is the only wife who knows how to ride a horse.

Who is Jasmine's husband? Who owns Helene?

Problem 1.4

Messrs. Baker, Dyer, Farmer, Glover, and Hosier are seated around a circular table, playing poker. Each gentleman is the namesake of the profession of one of the others. The dyer is seated two places to the left of Mr. Hosier. The baker sits two places to Mr. Baker's right. The farmer is seated to the left of Mr. Farmer. Mr Dyer is on the glover's right.

What is the name of the dyer? ([51],* Problem 36)

Problem 1.5

Two players, A and B, take turns in the following game. There is a pile of six matchsticks. At a turn, a player must take one or two sticks from the remaining pile. The player who takes the last stick wins. Player A makes the first move and each player always makes the best possible move.

Who wins this game?

Problem 1.6

Figure 1.1 pictures a railroad track on which are found a locomotive and two railroad cars. Car B at the right has just been filled with coal from the conveyor belt; car A at the left is empty. The tunnel is large enough to accommodate either car but not the locomotive, L. Furthermore, each car is longer than the tunnel so that, when in the tunnel, it is accessible to the locomotive from either side.

Using the locomotive to push or pull the cars, how can the two cars be made to switch places and the locomotive end up between them?

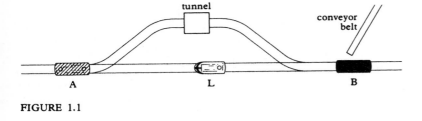

FIGURE 1.1

---

* References appear in the bibliography at the end of the book (page 336).

Before reading further, try to solve Problem 1.1 (page 2).

Solution of Problem 1.1

Did you approach part (a) of the problem as follows, by counting matches?

1st round:	57 matches, 57 fighters remain.
2nd round:	28 matches, 29 fighters remain.
3rd round:	14 matches, 15 fighters remain.
4th round:	7 matches, 8 fighters remain.
5th round:	4 matches, 4 fighters remain.
6th round:	2 matches, 2 fighters remain.
7th round:	1 match, 1 winner remains.

Thus, $57 + 28 + 14 + 7 + 4 + 2 + 1 = 113$ matches must be held.

If you used this approach, you have found a correct solution for part (a).

Does this method carry over to give a solution for part (b)? Not readily.

Is there another way in which you can approach the problem? Can you adopt another point of view?

Consider part (a) again. Since there could be only one winner, 113 fighters had to be eliminated; to eliminate 113 fighters requires 113 matches.

Although this second approach seems simpler, both methods of solving part (a) are correct. The advantage of the second approach is that it does carry over to part (b): Since $n - 1$ fighters must be eliminated, $n - 1$ matches are required.

What can we learn from this example? Often, there is more than one way to look at a problem. Sometimes several ways will lead to a solution; other times, only one. Even when several approaches to a problem do succeed, one approach may be simpler or more satisfying than another. (As you gain more experience in problem solving, you may find several different solutions to a problem. You can then compare these and select the one that seems the most elegant to you.)

In this sense, solving a problem is like finding the way through a maze or labyrinth. There may be several paths that lead to the goal, but one path may be shorter than the others.

On the other hand, there may be only one correct path and the other paths may lead to blind alleys. Similarly, many approaches to a particular problem may lead to a dead end.

At the entrance of a labyrinth, there is usually no way to know which is the correct path to take. We try one and, if we reach a blind alley or seem to be going around in circles, we return to the beginning and try another.* Similarly, there are no hard and fast rules that tell us how to approach every problem. If we seem to be getting nowhere with one approach, then we try another. Since there are only a finite number of possible paths in a labyrinth, eventually, by trial and error, we will hit upon the right one. Unfortunately, however, not all the paths in solving a problem are clearly visible. Sometimes they are, but often success as a problem solver depends on the ability to discover the "hidden paths."

Fortunately, however, there are important aspects of solving problems that don't apply in traversing a labyrinth. When we are in a labyrinth, we can see only the paths immediately before us but have no idea where they will lead. When we attack a problem, on the other hand, we usually have an overview of the entire arena—that is, we can look at the problem from above rather than within. This overview can give us different perspectives from which we can gain added insight.

In addition, we may be able to draw on our experience. We may recognize similarities with other problems. If one problem reminds us of another that we have already solved, we may try the techniques of the original solution as our first approach to the new problem. In this sense, the more experience we have in problem solving, the more successful we are likely to be in attacking new problems.

Although there are no hard and fast rules applicable to all problems, there are some basic techniques for starting to find a solution. Before we discuss some of these techniques, you should attempt some of the sample problems again. As you work on them, try to be aware of your thinking processes. What techniques prove successful? Where do you encounter difficulties?

Try Problems 1.2 and 1.3 (page 2).

Now that you have thought about and possibly solved these problems, we will use them as models to illustrate some techniques.

In Problem 1.2, we are given the following information:

1. Each person passed to her right.

2. Ms. Y passed to the American. (The fact that she passed three hearts is irrelevant.)

Solution of Problem 1.2

---

* There are systematic ways of attacking a labyrinth. See, for example, [13], pp. 127–137.

3. Ms. X passed to the person who passed to the Frenchwoman.

How can we use this information?

A diagram will help. Since Ms. Y passed to the American, she is not the person who passed to the Frenchwoman (Figure 1.2). Ms. X passed to the person who passed to the Frenchwoman (Figure 1.3).

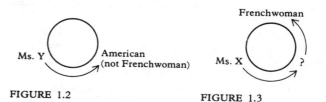

FIGURE 1.2   FIGURE 1.3

Did Ms. X pass to Ms. Y? Clearly not, when you compare Figures 1.2 and 1.3.

Therefore, Ms. X passed to Ms. Z (Figure 1.4), and Ms. Y is the Frenchwoman (Figure 1.5).

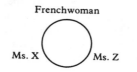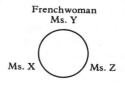

FIGURE 1.4   FIGURE 1.5

Combining these with Figure 1.2, we find that Ms. X is the American (Figure 1.6), which leaves Ms. Z to be the Englishwoman (Figure 1.7).

FIGURE 1.6   FIGURE 1.7

The solution of this problem was relatively easy. All we had to do was list the pertinent information that was given and draw some straightforward conclusions. With the aid of several diagrams, out came the desired identifications.

Before we try to make a list of the steps we followed in attacking this problem, let us consider Problem 1.3 (page 2).

We begin by listing the information given in the statement of the problem:

Solution of Problem 1.3

1. Mr. Alloway's pony is Georgette.

2. Col. Cunningham's pony is Jasmine.

3. Mr. Elmsby's pony is Inez.

4. Mr. Dunstan's pony is Francine.

5. Francine is married to Mr. Alloway.

6. Helene is married to Col. Cunningham.

7. Georgette's husband owns the pony that is named after Mr. Bennington's wife.

8. Each man's pony is named after somebody else's wife, and not after his own wife.

Here, a chart with three columns is helpful—one column each for the name of the gentleman, the name of his horse, and the name of his wife. By entering the given information along with the number of the pertinent clue, we obtain Figure 1.8.

gentleman	horse	wife
Mr. Alloway	Georgette (1)	Francine (5)
Col. Cunningham	Jasmine (2)	Helene (6)
Mr. Elmsby	Inez (3)	
Mr. Dunstan	Francine (4)	
Mr. X	Y	Georgette (7)
Mr. Bennington		Y (7)

FIGURE 1.8

Note that the only way in which clue 7 could be entered in the chart was to introduce symbols such as X and Y to represent people who have not yet been identified. (This is similar to the technique of algebra in which letters are used to represent unknown number quantities.) In Figure 1.8, Mr. X represents Georgette's husband and Y represents the name of Mr. Bennington's wife, which is also the name of Mr. X's horse.

We now make several observations:

Mr. Bennington's horse must be Helene because the other four names are already assigned in the horse column.

Mr. X cannot be Col. Cunningham, as the latter is married to Helene whereas Mr. X is married to Georgette.

Hence, since Col. Cunningham owns Jasmine and Mr. X owns Y, Y cannot be Jasmine.

Nor can Y be Francine or Helene, since Y is married to Mr. Bennington, and Francine and Helene are not.

Y cannot be Georgette, since Georgette's husband owns Y, according to clue 8.

The only possibility remaining is that Y is Inez. But then Mr. X must be Mr. Elmsby (since he owns Inez).

We can now consolidate our chart to obtain Figure 1.9. By the process of elimination, we have discovered that Jasmine is married to Durwood Dunstan.

gentleman	horse	wife
Mr. Alloway	Georgette	Francine
Col. Cunningham	Jasmine	Helene
Mr. Elmsby	Inez	Georgette
Mr. Dunstan	Francine	Jasmine
Mr. Bennington	Helene	Inez

FIGURE 1.9

Alternative Approach to Problem 1.3

Instead of the chart used above, we could have used a table of the type shown in Figure 1.10. The letters A, B, C, D, E are the first letters of the gentlemen's last names, and the letters F, G, H, I, J are the initials of the wives' first names. When we discover that a particular gentleman is married to a particular lady, we mark ✓ where his row meets her column; if we know they are not married, we place an X. Once a check appears, then the other boxes of that row and column must contain X's, as each man is married to only one woman and each woman is married to only one man. With this in mind, we can deduce Figure 1.11 from clues 5 and 6.

Also, from clue 3, we know that E is not married to I and, from clue 7, that B is not married to G. (Otherwise, Mr. B's pony would be named after Mr. B's wife, contrary to clue 8.) We obtain Figure 1.12. But now we seem to be stuck. It appears that we have not made full use of clue 7; but how can we use it?

Since, in Figure 1.12, there are only two possible alternatives for Georgette's husband, let us see what happens when we assume one of them to be true.

Let us first marry Georgette off to Mr. Dunstan. Then, from clue 7, Mr. Dunstan owns the pony that is named after Mr. Bennington's wife. But we know that Mr. Dunstan's pony is Francine (clue 4) and that Francine is not Mr. Bennington's wife (clue 5). We thus reach a contradiction, and so this marriage does not work, and perhaps never could have:

Georgette is not married to Mr. Dunstan.

Note that when an assumption logically leads to a contradiction (that is, something is shown to be true that has previously been seen to be false, or vice versa), then the assumption must be incorrect. The item assumed to be true must therefore be false.

We enter this new information in our chart, and get Figure 1.13.

Observe that there is now only one possibility for Georgette's husband—Mr. Elmsby.

We may, therefore, put a check in the E–G box in the chart and then put an X in the E–J box.

Using clue 7 again, Mr. Elmsby owns the pony named after Mr. Bennington's wife. But Mr. Elmsby owns Inez. It follows that:

Mr. Bennington is married to Inez.

Entering this information in the chart, we arrive at Figure 1.14.

We now put X's in the B–J and D–I boxes; and, hence, the D–J box must receive a check.

Thus, as we found before,

Jasmine is married to Mr. Dunstan.

wives

	F	G	H	I	J
A					
B					
C					
D					
E					

FIGURE 1.10

	F	G	H	I	J
A	✓	X	X	X	X
B	X		X		
C	X	X	✓	X	X
D	X		X		
E	X		X		

FIGURE 1.11

	F	G	H	I	J
A	✓	X	X	X	X
B	X	X	X		
C	X	X	✓	X	X
D	X		X		
E	X		X	X	

FIGURE 1.12

	F	G	H	I	J
A	✓	X	X	X	X
B	X	X	X		
C	X	X	✓	X	X
D	X	X	X		
E	X		X	X	

FIGURE 1.13

	F	G	H	I	J
A	✓	X	X	X	X
B	X	X	X	✓	
C	X	X	✓	X	X
D	X	X	X		
E	X	✓	X	X	X

FIGURE 1.14

As a final step in the solution, regardless of which approach was used, the original problem should be checked to verify that the conclusions reached are consistent with the given facts.

WHICH CHART OR DIAGRAM TO CHOOSE

In our two approaches to the solution of Problem 1.3, a specific chart or table helped us to envision the solution. There are other useful visual aids. For example, in a table we could head the rows with the horses' names and the columns with the women's names. We would then place a check in, say, the F-row, J-column if Francine is the name of the horse that belongs to the man who is married to Jasmine. Or, we could plot the men against their horses; this time, instead of checks, we use circles with the wives' initials, as in Figure 1.15.

This figure tells us that Mr. Alloway is married to Francine and owns Georgette, Mr. Bennington does not own Inez, and Mr. Dunstan owns Francine (but we do not yet know to whom he is married).

Another diagram we could use is Figure 1.16.

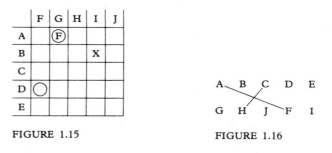

	F	G	H	I	J
A		Ⓕ			
B				X	
C					
D	◯				
E					

FIGURE 1.15

```
A   B   C   D   E
     \ /
      X
     / \
G   H   J   F   I
```

FIGURE 1.16

Here the letters G, H, J, F, I stand for both the wives' and the horses' initials. We indicate that a particular horse belongs to a particular owner by writing the horse's initial below that of its owner. The lines connect husbands' and wives' initials.

In general, it is difficult, if not impossible, to know in advance exactly what type of visual aid will be most helpful. Sometimes, none is needed; other times, more than one is necessary. You might start out with one aid and realize that another is better. The best guideline seems to be to try to use the chart or other aid that will incorporate as much of the given information as possible, in the least complicated way possible.

PRESENTING A SOLUTION

When you are solving a problem for yourself, you may be able to obtain a solution without doing any writing other than a few checks or X's or letters in a table. However, if you wish to present your solution to others, exhibiting a completed chart is unconvincing. They will want an explanation of the method of solution. This is best accomplished by listing with each step or ✓ or X the justification for it. You can do this by numbering the steps or clues and using these numbers for reference or as explanation of an entry, as we did in Figure 1.8. This is a good idea even when you do not intend to present your solution to anyone else, as it makes it easier for you to retrace your steps in the event that you find a mistake or want to check your work. (An illustration of the technique is coming up soon.)

It is also important to be critical of your own work before you present a solution to someone else. If you are not fully convinced by your argument, it is not likely that anyone else will be convinced.

SOME STEPS IN PROBLEM SOLVING

We are now ready to list some good techniques to follow in attacking a problem.

1. As a first step, it is important to understand exactly what information is given and what is to be found. It is also important to decide what information is relevant. Sometimes, restating the problem will make these tasks easier. Often, it is useful either to make a list of the given information or to use a chart, diagram, or other visual aid.

2. After all the given information is listed or charted, add to it all the information that you can logically deduce from what is given.

3. Continue to add to the list deductively, until either the desired information is obtained or an impasse is reached—that is, no further conclusions seem possible but the desired information has not yet been obtained. (Before you decide that an impasse has been reached, it is often a good idea to run through all the given clues once again, bearing in mind what has been found so far. Sometimes this leads to additional conclusions that had been overlooked.)

4. In the event of an impasse, choose some aspect of the problem for which only a finite (preferably small) number of nonoverlapping alternatives are possible. Arbitrarily assume any one of these alternatives to be true; add it temporarily to the list or chart of facts that are known; and see what conclusions you can logically draw. If a contradiction is reached, then conclude that the alternative assumed is, in fact, not correct and eliminate it as a possible alternative. Also, add the fact that the alternative is not correct to the list of things that are known. This fact may now be used in further arguments.

What happens, though, if an assumption does not lead to a contradiction? It may, for example, lead to an answer. In this case, you cannot yet conclude that this is the complete solution to the problem. *Each alternative assumption must also be investigated: Every possible case must be considered.* If all these other assumptions lead either to contradictions or to the same answer found originally, then you can conclude that this answer is the only solution to the problem. However, if some alternatives lead to different answers, it follows that either the problem has more than one solution—and this should be noted—or that the problem is not well posed. (If the wording of the problem indicates that there should be a unique solution and if, in fact, there can be more than one solution, then the problem is not well posed.) If a problem does have more than one solution, the truly inquisitive mind would not be satisfied until all solutions are at least categorized, if not worked out in detail.

Since you must always return eventually to the point at which you make an assumption—either because the assumption leads to a contradiction or to consider other alternatives—it is a good idea to enter that assumption in your visual aid with a pencil, preferably using a different color from the one you started with. Use the same pencil to note down all conclusions that are based on the particular assumption. Then, when you do return to the point of making the assumption, you will easily be able to eliminate all information based on it, leaving only what was known before the assumption was made.

It is possible that after you make an assumption, you still obtain neither an answer nor a contradiction—you still face an impasse. In this case, you may have to make a secondary assumption. We will consider this possibility in more detail in the solution of Problem 1.4.

5. Each solution you obtain should be shown to satisfy completely the conditions of the original problem. Sometimes an answer appears when you use part (but not all) of the given information. It is possible that this "answer" contradicts some of the other conditions. This is the reason a final check is important, especially if more than one answer has been obtained.

Now try to apply these techniques to Problem 1.4 (page 3).

The clues are:

1. Each gentleman is the namesake of the profession of one of the others.

2. The dyer is seated two places to the left of Mr. Hosier.

3. The baker sits two places to Mr. Baker's right.

4. The farmer is seated on Mr. Farmer's left.

5. Mr. Dyer is on the glover's right.

Solution of Problem 1.4

A diagrammatic aid helpful in the solution of this problem is just a simple picture: the dyer is seated somewhere. Call the chair in which he is seated chair 1 and number the other chairs in the clockwise direction. By clue 2, Mr. Hosier is in chair 4 (Figure 1.17).

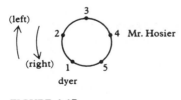

FIGURE 1.17

How can we use the other information? We can chart the numbers of the seats that could possibly be occupied by each of the people in clues 3, 4, and 5 (Figure 1.18).

By clue 3

Mr. B	baker
1	4
2	5
~~3~~	~~1~~	dyer is in 1
~~4~~	~~2~~	Mr. H is in 4
5	3

By clue 4

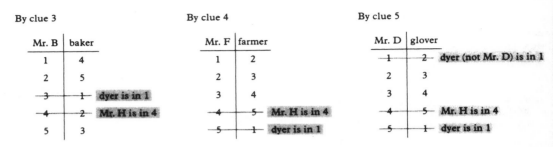

Mr. F	farmer
1	2
2	3
3	4
~~4~~	~~5~~	Mr. H is in 4
~~5~~	~~1~~	dyer is in 1

By clue 5

Mr. D	glover
~~1~~	~~2~~	dyer (not Mr. D) is in 1
2	3
3	4
~~4~~	~~5~~	Mr. H is in 4
~~5~~	~~1~~	dyer is in 1

FIGURE 1.18

There are only two possibilities left for Mr. Dyer: seat 2 or seat 3. Let's investigate each of these alternatives.

Case I Assume he is in seat 3, which means that the glover is in seat 4. In this case, we can eliminate some more seating possibilities (Figure 1.19).

In case I:

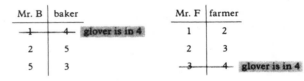

Mr. B	baker
~~1~~	~~4~~	glover is in 4
2	5
5	3

Mr. F	farmer
1	2
2	3
~~3~~	~~4~~	glover is in 4

FIGURE 1.19

We have not yet arrived at a solution or a contradiction. However, at this point, we see that there are only two possibilities for Mr. Baker: He is in seat 2 or in seat 5. We consider these as subcases, that is, as alternative **secondary assumptions**.

Case Ia Assume that Mr. Baker is in seat 2.

Then, the baker is in seat 5, and Mr. Farmer must be in seat 1 (by Figure 1.19, since Mr. Baker is assumed to be in seat 2), with the farmer in seat 2. This gives us Figure 1.20.

By the process of elimination, we can complete the diagram, obtaining Figure 1.21.

Thus Mr. Farmer is the dyer.

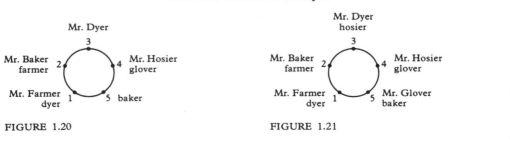

FIGURE 1.20

FIGURE 1.21

Does this complete the solution of the problem? Not yet. Our conclusion has been based on two assumptions. We must still investigate the alternatives. We begin with the alternative to our secondary assumption.

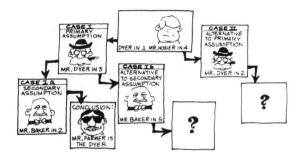

Case Ib Assume that Mr. Baker is in seat 5. (Remember that we are still working under the primary assumption that Mr. Dyer is in seat 3.)

Then, the baker must be in seat 3, and the farmer must therefore be in seat 2, with Mr. Farmer in seat 1. This gives the seating arrangement shown in Figure 1.22.

Again by the process of elimination, we can complete the diagram to obtain Figure 1.23.

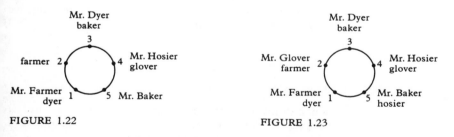

FIGURE 1.22   FIGURE 1.23

Note that although the seating arrangements in Figures 1.21 and 1.23 are not identical, both yield the same answer to the original question—Mr. Farmer is the dyer.

We are still not yet satisfied. We must return to consider the alternative to our original assumption.

Case II Assume that Mr. Dyer is in seat 2.

It follows that the glover is in seat 3 (see Figure 1.18). Again we can use this information to eliminate some of the possibilities from Figure 1.18. Specifically, Mr. Baker must be in seat 1, with the baker in seat 4 (see Figure 1.24).

Since the dyer is also in seat 1, this seems to give us a different answer—that Mr. Baker is the dyer. However, we must follow through to make sure that this situation is consistent with the given information.

It turns out that it is not, as there is now no seat for Mr. Farmer (see Figure 1.25).

In case II:

Mr. B	baker
1	4
2	5 Mr. Dyer is in 2
5	3 glover is in 3

Mr. F	farmer
1	2 Mr. Baker is in 1
2	3 Mr. Dyer is in 2
3	4 the baker is in 4

FIGURE 1.24   FIGURE 1.25

We have reached a contradiction. Hence, Mr. Dyer cannot be in seat 2 and must be in seat 3.

This completes the solution of the problem. We have found that two of the possible seating arrangements (Figures 1.21 and 1.23) are consistent with the given information. In both, the dyer is Mr. Farmer; and so that is the answer. (Note that we could also conclude that the glover is Mr. Hosier, since that is the case in both diagrams; but we cannot identify the occupations of the other three gentlemen.)

TREE DIAGRAMS

In the solution of Problem 1.4 and the second solution of Problem 1.3, we reached a point at which we resorted to making assumptions— that is, considering cases—successively testing the truth of each of several possible alternatives. However, there are many problems in which this technique is unnecessary—the solution may be deduced directly from the given information, without analysis by cases.

On the other hand, in a problem in which we have to assume something, even after we make that assumption we are not always able to proceed directly to a solution; we may still have an impasse. We may then have to consider subcases and make a secondary assumption, possibly followed by further assumptions. Whenever we do so, however, we must remember to consider each possible alternative to every assumption we make. This procedure, of considering cases and subcases, is sometimes referred to as **case analysis**.

To help keep track of these assumptions or cases and their alternatives, a diagram can be helpful. For example, in Problem 1.4, after we placed the dyer in seat 1 and Mr. Hosier in seat 4, we saw in Figure 1.18 that there were two possibilities (2 and 3) for the seat occupied by Mr. Dyer, three possibilities (1, 2, and 5) for the seat occupied by Mr. Baker, and three possibilities (1, 2, and 3) for the seat occupied by Mr. Farmer. We can indicate this diagrammatically, as in Figure 1.26.

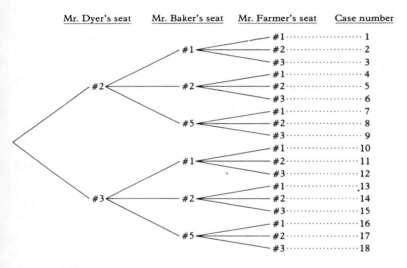

FIGURE 1.26

This figure is a **tree diagram**, so called because it branches out as a tree does. The points at which the branching takes place are called **nodes** or **branch points**. Each line segment connecting two nodes is called a **branch**, and a sequence of branches connecting the starting point of the tree to a terminal point is called a **path**. Each path indicates a case to be considered; the whole tree indicates all possible cases.

The tree in Figure 1.26 indicates all the possible alternatives for the seats in which Mr. Dyer, Mr. Baker, and Mr. Farmer could be seated. We read the cases in the tree in the following manner:

Case 1   Mr. Dyer is in seat 2, Mr. Baker in seat 1, and Mr. Farmer in seat 1. (This case cannot actually occur.)

Case 2   Mr. Dyer is in seat 2, Mr. Baker in seat 1, and Mr. Farmer in seat 2. (This case cannot occur either.)

Case 3   Mr. Dyer is in seat 2, Mr. Baker in seat 1, and Mr. Farmer in seat 3.

Case 4   Mr. Dyer is in seat 2, Mr. Baker in seat 2 (again an impossibility), and Mr. Farmer in seat 1.

Etc.

Obviously, many of the branches in the tree do not represent alternatives that can actually occur. Cases in which two of the three people occupy the same seat are eliminated immediately. In the remaining cases, the locations of Mr. Dyer, Mr. Baker, and Mr. Farmer determine the locations of the glover, the baker, and the farmer respectively. Figure 1.27 shows this information. In five of the seven cases, we again find two different people occupying the same seat. (For each path leading to a contradiction, the people occupying the same seat are indicated at the right.)

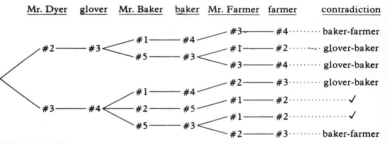

FIGURE 1.27

The two cases that remain correspond to the two solutions that we found when we originally solved the problem. In both, Mr. Farmer is the dyer.

An alternative tree diagram that more closely describes the procedure we actually followed in solving Problem 1.4 is shown in Figure 1.28.

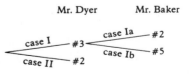

FIGURE 1.28

THE MULTIPLICATION PRINCIPLE

Many times we are interested in knowing the number of cases that could possibly arise in a problem, that is, how many paths there would be in a tree diagram. (Note that this number is the same as the number of terminal branches.)

In the original tree diagram (Figure 1.26) for Problem 1.4, there are two branches emanating from the starting points. Each of these splits into three branches, each of which in turn splits into three others.

How many terminal branches are there altogether?

The answer lies in the **Multiplication Principle**. The principle says that, if a tree has m primary branches, each of which splits into n secondary branches, then there are mn different paths through the tree (Figure 1.29).

More generally, if a task can be broken up into two steps, the first of which can be done in m ways and the second of which can be done in n ways (regardless of how the first step is done), then there are mn different ways of completing the task.

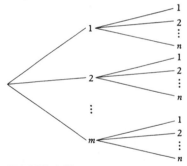

FIGURE 1.29

This principle generalizes to tasks that may be broken into more than two steps. Thus, the tree in Figure 1.26 has three sets of branch points. At the first or starting point, the tree splits into two branches; at the second, each branch splits into three; and at the third, each branch again splits into three. Hence, there are $2 \cdot 3 \cdot 3 = 18$ terminal branches or 18 different paths.

Note that the Multiplication Principle cannot be applied if the number of secondary branches depends on which primary branch is chosen. For example, the tree diagram in Figure 1.30 has two primary branches (a_1 and a_2). If a_1 is chosen, then there are only two possible secondary branches (b_1 and b_2); but if a_2 is selected, then there are three secondary branches (c_1, c_2, and c_3). Thus, the Multiplication Principle does not apply in this diagram.

The tree diagram approach can also be used to solve Problem 1.5 (page 3).

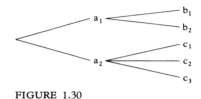

FIGURE 1.30

If you have not yet solved Problem 1.5 (page 3), try it now.

Solution of Problem 1.5

At the beginning of the game, A has two possible first moves. This is indicated in Figure 1.31. The upper branch indicates the possibility that A takes one stick (leaving five); the lower branch shows the case in which A takes two (leaving four).

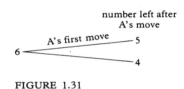

FIGURE 1.31

We now extend the diagram to show all possible moves by A and B (Figure 1.32). Subscript notation in the figures shows whose move it is and which move; for example, A_2 means A's second move.

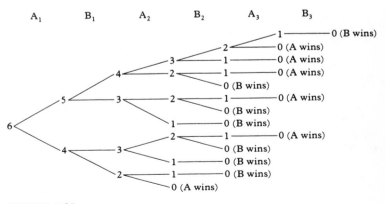

FIGURE 1.32

Note that there are 13 different ways in which the game could be played. Each of these is included as a path from start to finish in the diagram, and each must be considered in examining the possible outcome of the game. For example, Figure 1.33 highlights the case in which A first takes two matchsticks, then B takes one, then A takes one, and then B takes two.

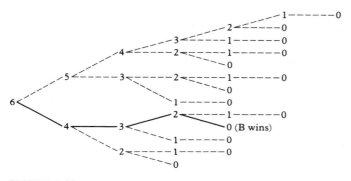

FIGURE 1.33

Note also that, in the diagram of all the game possibilities (Figure 1.32), some of the branches lead to a win for A while others result in B winning. Does this mean that we cannot analyze the game—that is, determine who should be the winner? Not at all. It's true that if both players play at random, either of them could win; but we have assumed that the players are intelligent and play to the best of their abilities. Under this assumption, we can show that one of the two players (the second one, B, in this case) can be sure of winning. How can we say this in light of the fact that the tree diagram does not always lead to the same answer?

The explanation is simple. We must bear in mind that a player has a choice at each turn. This choice determines the branch that is followed. Suppose B adopts the following strategy: Always take the opposite of what A just took. (We will ignore for now the question of how this strategy was discovered.) Then the game can be represented by the tree in Figure 1.34. In all cases, B wins. (Note that many

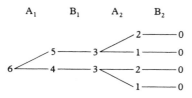

FIGURE 1.34

branches of the original tree have been eliminated because B avoids them.) Thus, we can say that B can always win the game by sticking to the given strategy—no matter what A does, B has a move that will eventually lead to a win.

This also shows that A has no winning strategy. No matter what move A makes, there will always be a branch of the tree leading to a win for B, namely, the branch chosen by B following the above strategy.

SIMPLIFICATION

So far we have discussed some basic steps for attacking a general reasoning problem. There are other techniques that are often helpful. One of the most important is to simplify the problem. This may be done in many different ways.

One way is to *consider special cases of the problem*. For example, if we want to analyze the matchstick game for all possible values of n (where n is the number of sticks we start off with), we might first try to solve it for specific values of n, say $n = 1$, $n = 2$, $n = 3$, ..., $n = 6$, and then try to generalize for all n.

A second way of simplifying the problem is to *reduce it to a previously solved problem*. For example, suppose we want to analyze the case where the matchstick game starts with seven sticks. Instead of drawing a tree diagram, we observe that the first player, by taking one stick, can reduce the game to a six-stick game in which he goes second. Because we know from before that this game will result in a win for the second player, we have proven that the first player can force a win in the seven-stick game.

Now try Problem 1.6 (page 3).

Solution of Problem 1.6

Another way of simplifying a problem is to *break it into parts*. Consider Problem 1.6. Part of the problem is to move car A to the right, and a second part is to move car B to the left. We simplify the problem by momentarily shunting car B out of the picture and directing our attention to the following question: How can the locomotive and car A switch places in Figure 1.35?

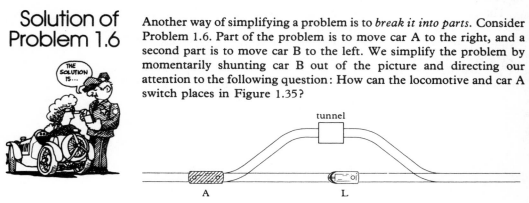

FIGURE 1.35

The solution of this problem is comparatively easy:

1. L pulls A to the right (Figure 1.36).

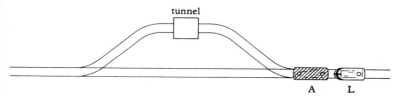

FIGURE 1.36

2. L pushes A to the tunnel (Figure 1.37).

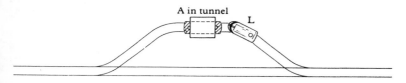

FIGURE 1.37

3. L goes back to the main track and then moves to the left (**Figure 1.38**).

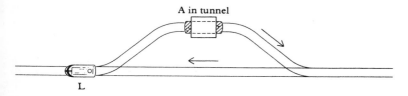

FIGURE 1.38

4. L picks up A at the tunnel and pulls A back to the main track on the left (Figure 1.39).

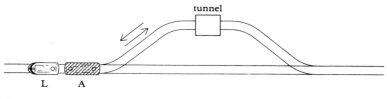

FIGURE 1.39

5. L pushes A to the right and comes back to A's starting position (Figure 1.40).

FIGURE 1.40

Returning to the original problem, we ask how simplifying it can help us. If we realize that the size of the locomotive plays no role in the solution of the simplified problem, then the answer is not difficult. Suppose we first attach L to car B and consider it as one unit LB—as if it were the locomotive. Then by following the procedure above, we can move A to the right of LB, not just to the right of L. This leaves the position shown in Figure 1.41.

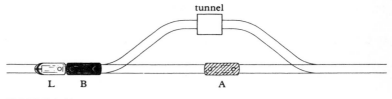

FIGURE 1.41

If LB now pushes A all the way to the right, we are left with Figure 1.42.

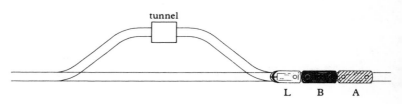

FIGURE 1.42

All that remains for us to do is to switch L and B, which is essentially the simplified problem revisited. Specifically:

1. L pulls B to the left.

2. L pushes B into the tunnel (Figure 1.43).

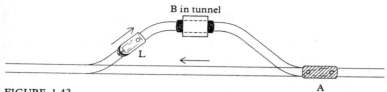

FIGURE 1.43

3. L returns to the main track and moves to the right (Figure 1.44).

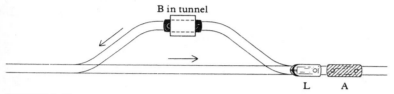

FIGURE 1.44

4. L picks up B at the tunnel and pulls B back to the main track at the right (Figure 1.45).

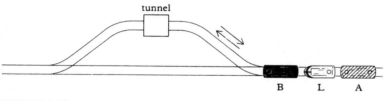

FIGURE 1.45

This completes the problem.

Problem 1.6 is one of a variety of "shunting problems." Among the exercises in this chapter is an additional problem of this type to further illustrate the idea of simplification. Several other shunting problems will be found in Chapter 9.

THE CHAPTER IN RETROSPECT

General reasoning is the basis of mathematics and logic as well as problem solving in general. This chapter has introduced you to some principles and techniques used in general reasoning. We have, in an informal and intuitive way, discussed the basic ideas of making assumptions, looking at alternatives, reaching a contradiction, using visual aids, and other approaches. Not all the techniques mentioned here will be applicable to all problems. In fact, for some problems, none of these will help. However, as you gain more experience with problem solving, you will develop other techniques.

If you find that you are not making any progress in trying to solve a problem, it is frequently a good idea to leave the problem for a while and return to it later, possibly with a fresh attack.

In fact, if you have difficulty with many of the problems, then, before you attack each one, make a list of all the different approaches you can think of for that problem. This way, if you should get stuck, you will already have another point to view from.

Remember that there are sometimes many ways to solve a problem, and any method that is logically sound is correct.

The remaining chapters of this book are devoted to specific types of problems and introduce some techniques that will be helpful in solving them. In particular, Chapter 2 takes a somewhat more formal approach to the reasoning process by considering some principles of symbolic logic.

Exercises

1.1 H A ⋆

There were five fine ladies from Carruther
Who named their pets after each other.
From the following clues,
Can you carefully choose
The pet which belongs to Sue's mother?

Toni Taylor owns a hog;
Belle Bradkowski owns a frog;
Janet Jackson owns a crow;
The garter snake is owned by Jo;

Sue's the name they call the frog;
And "Here Jo, here Jo" brings the hog;
The name by which they call the pony
Is the name of the woman whose pet is Toni;
The final clue, which I'll now tell,
Is that Sue's mother's pet is Belle.

1.2 Adam, Robert, Clifton, Stephen, and Brent are the five starters on the Doylestown Dribblers basketball team.

Two are left handed and three right handed.

Two are over 6 feet tall and three are under 6 feet.

Adam and Clifton are of the same handedness, whereas Stephen and Brent use different hands.

Robert and Brent are of the same height

---

* H at the beginning of a problem means you will find a hint about how to solve it in the Hints and Solutions section (page 340); S means the full solution is given there. A means that the answer is given in the section called Answers to Selected Problems (page 381).

range, while Clifton and Stephen are in different height ranges.

The man who plays center is over 6 feet and is left handed.

Who is he?

1.3. In the fall, the members of the small and very exclusive Lawnsand Garden Club began to plan the following season's garden arrangements. After several meetings, the five members of the club (Mr. Isaac Iris, Ms. Rita Rose, Madame Anastasia Azalea, Dr. Frederick Forsythia, and Sir Horace Holly) decided that each should send a plant to one of the others.

The five plants sent corresponded, in some order, to the names of these five people.

Each of the five received exactly one plant; in no case did the receiver, or the sender, have the same name as the plant.

Ms. Rose sent a holly to Dr. Forsythia.

The recipient of the plant sent by the doctor sent a rose.

The flower lover with the same name as the plant sent by Madame Azalea received a forsythia from the namesake of the plant that Madame Azalea received.

Who sent what to whom?

1.4. [S] [A] Allen, Bill, Chuck, Ed, Harry, Jerry, Mike, Paul, and Sam have formed a baseball team. The following facts are true: (a) Allen does not like the catcher. (b) Ed's sister is engaged to the second baseman. (c) The center fielder is taller than the right fielder. (d) Harry and the third baseman live in the same building. (e) Paul and Chuck each won $20 from the pitcher at poker. (f) Ed and the outfielders play cards during their free time. (g) The pitcher's wife is the third baseman's sister. (h) All the battery and infield except Chuck, Harry, and Allen are shorter than Sam. (i) Paul, Allen, and the shortstop lost $100 each at the race track. (j) The second baseman beat Paul, Harry, Bill, and the catcher at billiards. (k) Sam is in the process of getting a divorce. (l) The catcher and the third baseman each have two legitimate children. (m) Ed, Paul, Jerry, the right fielder, and the center fielder are bachelors; the others are married. (n) The shortstop, the third baseman, and Bill all attended the fight. (o) Mike is the shortest player on the team.

Determine the position of each player on the baseball team.

Note: On a baseball team there are three outfielders (right fielder, center fielder, and left fielder), four infielders (first baseman, second baseman, third baseman, and shortstop), and the battery (pitcher and catcher). ([27], Jan. 1976, p. 64)

1.5. [H] Six players—Pietrovich, Cavelli, St. Jacques, Smith, Lord Bottomly, and Fernandez—are competing in a chess tournament over a period of five days. Each player plays each of the others once. Three matches are played simultaneously during each of the five days. The first day, Cavelli beat Pietrovich after 36 moves. The second day, Cavelli was again victorious when St. Jacques failed to complete 40 moves within the required time limit. The third day had the most exciting match of all when St. Jacques declared that he would checkmate Lord Bottomly in 8 moves and succeeded in doing so. On the fourth day, Pietrovich defeated Smith.

Who played against Fernandez on the fifth day?

1.6. [H] [A] Joe Flannery had a week off from his trucker's job and intended to spend all nine days of his vacation (Saturday through the following Sunday) sleeping late. But his plans were foiled by some of the people who work in his neighborhood.

On Saturday, his first morning off, Joe was wakened by the doorbell; it was a salesman of magazine subscriptions who was "working his way through college."

On Sunday, the barking of the neighbor's dog as the paper boy tried to deliver the paper abruptly ended Joe's sleep.

On Monday, he was again wakened by the persistent salesman but was able to fall asleep again, only to be disturbed by the garbagemen.

In fact, the salesman, the garbagemen, and the neighbor's dog that barked at the paper boy combined to wake Joe at least once each day of his vacation, with one exception.

The salesman woke him on Wednesday; the garbagemen on the second Saturday; the dog on Wednesday and the final Sunday.

No one of the three noisemakers was quiet for three consecutive days; but yet, no pair of them made noise on more than one day during Joe's vacation—for example, the salesman and the garbagemen both woke Joe on Monday but did not both wake him up on any other day.

On which day of his vacation was Joe able to sleep late?

★* 1.7. ⬛H ⬛A Five schools competed for the gold medal in the finals of the Greater Philadelphia Scholastic Track Meet. They were Central, Franklin, Lincoln, Mastbaum, and Olney. The five events in the finals were:

---

* A star before an exercise means that you may find it harder to solve than some other problems. Two stars mean that the problem is a real challenge.

the high jump, shot put, 100-yard dash, pole vault and one-mile relay. In each event, the school placing first received five points; the one placing second, four points; the one placing third, three points; and so on. Thus, the one placing last received one point. At the end of the competition, the points of each school were totaled; these totals determined the final ranking.

1. Central won with a total of 24 points.

2. Otto O'Hara of Olney won the high jump hands down (and feet up), while Oliver Oats, also of Olney, came in third in the pole vault.

3. Lincoln had the same number of points in each of four events.

If each school had only one entry in each event, and if there were no ties, and if the schools ended up being ranked in the same order as the alphabetical order of their names, then how did Freddy Farkle of Franklin place in the high jump?

★ 1.8. ⬛H ⬛A One afternoon, having nothing to do, Bob, Carol, Ted, and Alice decided to have a tiddlywinks tournament. They each played one game against each of the others. Each game started with ten tiddlywinks on the table and ended when all ten had been flipped into a cup. The players alternated turns and each scored one point for each tiddlywink that he or she successfully flipped into the cup.

Alice was the clear victor, winning all three of the games she played in and outscoring her opponents by a total of 22 points.

Carol only won one game, but also outscored her opponents.

Bob had a better won-lost-tie record than Ted did; in fact, Bob scored as many points against Ted as Ted scored in all three of his games combined.

No two matches ended in the same score.

What was the outcome and the score of each match?

1.9. [H] The Jones family was looking for a good vacation spot. It had to have a swimming pool, tennis courts, a golf course, and fine food. Most importantly, it should not be too expensive.

Harry's Hideaway, Shangri La, the Cromwell, and Paradise Lodge were the four resorts that the Joneses discovered which were within their price range. Unfortunately, none of the four had all the other desired features. However, the Joneses decided to select the one that met the most of their requirements.

1. Both Harry's Hideaway and the Cromwell had an Olympic-size swimming pool.

2. Both Shangri La and Paradise Lodge had an 18-hole golf course.

3. The Cromwell and Shangri La had indoor tennis courts.

4. Paradise Lodge is especially known for its good food.

5. Of the six possible pairs of requirements, every pair was met by exactly one resort. For example, only one resort had both a swimming pool and tennis courts, and another had both a swimming pool and a golf course.

Which resort did the Jones family choose?

★ **1.10.** [A] In order to select capable contestants for their quiz shows, some producers require prospective contestants to compete off camera. One show, "Fractured Funnybones," usually chooses four candidates and conducts a series of head to head matches between each pair of them. The candidate who performs the best is then chosen to appear on the show.

The game is played as follows: The contestants are shown a key word, and whoever first thinks of a joke using that word in the punch line gets one point. The first player to get five points wins.

When Manny Morris, Nancy Novokov, Patti Proctor, and Thomas Twickenham competed against each other, the results were as follows:

Patti won all three of her games and her opponents scored a total of only 2 points against her.

Nancy scored a total of 10 points and her opponents also scored 10.

Manny and Thomas both scored 8 points, but Manny allowed his opponents 15 whereas Tom's opponents scored only 14. In addition, Tom scored more points against Patti than Manny did.

What was the outcome and the score of each head to head match?

★ ★ **1.11.** [H] [A] Good poker players have four characteristics in common: They are familiar with the odds associated with card distribution; they know when it is wise to bluff; they have poker faces; and they are lucky.

Angel invited four of his friends to play poker one evening, around his big circular table. They were Babs, Cleo, Dot, and Edie. The following was also true:

Everyone was sitting next to someone who knew the odds, but four of the five were sitting next to someone who was not well versed in the probabilistic aspects of the game.

Four of the people were sitting next to wise bluffers, but three of them were sitting next to people who did not know when to bluff.

Four of the players were sitting next to people with good poker faces, but everyone was sitting next to someone who could not keep a straight face.

Exactly three of the people were sitting next to someone who was noted for good luck.

Each of the players had at least one of the desirable traits; but only one, the big winner of the night, had all four.

Babs knows the odds and knows when to bluff, but does not have a poker face and is not noted for luck.

Edie is not sitting next to anyone who knows when to bluff.

The person on Dot's right has a poker face.

The person on Cleo's left does not know when to bluff.

Angel knows the odds, but is not lucky.

Who was the big winner of the evening, what was the seating arrangement, and which traits did each of the five players have?

1.12. [A] Lou Lion, Ted Tiger, Eli Elephant, Gary Gazelle, Peter Panther, and Leopold Leopard all recently returned from a photographic safari in Africa. Each "hunter" was able to photograph only two animals, each of which happened to be the namesake of one of the hunter's colleagues.

Each animal was photographed by exactly two of the men.

Each hunter photographed at least one member of the cat family; and Eli and Gary accounted for all four cats between them.

Gary and Peter both filmed a leopard.

The namesakes of the animals that Eli captured on film both photographed a gazelle.

Ted did not photograph any of the animals that Peter did.

Gary and Ted had photographs of the same animal.

Who caught the lion on film?

1.13. [H] [A] Messrs. Doctor, Lawyer, Plumber, Gardener, and Butcher met on a cruise ship and discovered that they all lived within a 5 mile radius of one another. They also discovered that the occupation of each is the namesake of one of the others.

Taking an instant liking to one another, they exchanged phone numbers, intending to use each other's professional services when they returned home. Unfortunately, because they were all slightly drunk at the time, the phone numbers got all mixed up. Several weeks later, when the doctor's pipes began leaking, he tried to call the plumber but reached Mr. Butcher

instead. And when he tried to call the butcher, he inadvertently reached Mr. Lawyer. The number he had written down for the gardener turned out to belong to Mr. Gardener instead.

But the doctor was not the only one to make a mistake. Actually, each of the five men had written down the phone numbers correctly but had mismatched all of them (except his own, of course). Furthermore, no two attributed the same number to the same person.

The lawyer tried to call Mr. Plumber but reached Mr. Lawyer.

The butcher tried to call Mr. Plumber but reached Mr. Butcher.

The butcher also reached Mr. Gardener when he tried to call the plumber.

Whom did Mr. Butcher reach when he tried to call Mr. Gardener?

★ ★ **1.14.** [S] Three merry logicians were seated about a table talking, while eating pizza and drinking beer. Inadvertently, each smeared his or her face without realizing it. Suddenly they all looked at each other and began to laugh. Then one stopped laughing, for he realized that his own face was smeared.

How did he come to this conclusion? (Assume that anyone realizing that his or her own face was smeared would have stopped laughing.)

★ ★ **1.15.** [H] When the producers of the new

show, "The Secret Life of Sherlock Holmes," were casting for the part of the protagonist, three young actors applied for the job. They all gave exemplary readings and the producers were hard pressed to choose between them. The director did have a slight preference for Byron Bentley, who came closest to fitting the director's image of what the great Holmes should look like; so it was decided to determine how Bentley compared to Holmes in the reasoning department.

The director obtained five identical handkerchiefs from the wardrobe department and, in front of the three applicants, wrote "Sherlock Holmes" on three of the handkerchiefs and "Professor Moriarty" on the other two.

He then stood Bentley center stage and pinned one of the handkerchiefs on Bentley's back. Next, one of the other applicants was told to stand behind Bentley and a handkerchief was pinned on his back. The second applicant could see the name on Bentley's handkerchief but could not see the name on his own. The third applicant was placed behind the other two, so that he could see the names on their handkerchiefs but neither they nor he could see the name on the handkerchief that was placed on his back.

Bentley, of course, could not see any handkerchief.

"The first one of you who can deductively determine the name on his own back will get the part," the director announced.

After a few minutes of silence, Bentley correctly asserted that he was Sherlock Holmes.

How did he know?

"Elementary, my dear reader."

★ ★ **1.16.** [H] [A] On a very busy day, the four senior partners of the law firm Smith, Smith, Smith, and Smith had some sandwiches delivered from the corner delicatessen. One of the partners had ordered three salami sandwiches; one had ordered two salami and one bologna; one had ordered one salami and two bologna; and the fourth had ordered three bologna. Each knew that the four orders were different.

Shortly after they started eating, the owner of the delicatessen came running in to apologize for his new waitress, who had accidentally mixed up the orders so that each order was placed in the wrong bag.

Ann Smith said, "Oh. I have already eaten two salami sandwiches, so I know what the third sandwich in my bag must be."

"In that case," Bob Smith declared, "since I know what Ann ordered and I have already eaten one salami sandwich, I know what the other two sandwiches in my bag must be."

Carla Smith said, "I have not yet opened my bag but, based on what I have just heard, I must have received three bologna sandwiches."

Assuming that the delicatessen owner's statement implies that each of the four received the order intended for one of the others, what did John Smith (who at that moment was out of the room and had not yet started eating) order, and what did he receive?

★ ★ **1.17.** [S] [A] The professor was a guest lecturer at a logic course for senior executives of major oil companies. She selected six male students for this demonstration. The professor placed fifteen dimes and fifteen nickels in six tin cups such that each cup contained the same number of coins but a different amount of money. She made six labels showing correctly how much money each cup held, but attached to each cup an incorrect label. She explained the situation to the six students and gave a cup to each. She asked each man in turn to feel the size of as many coins as he wanted in his own cup and announce something true about them. The only evidence each man had was the size of the coins he felt, the incorrect label on his own cup, and the statements made by those who preceded him. The first man said, "I feel four coins which are not all the same size; I know that my fifth coin must be a dime." The second man said, "I feel four coins which are all the same size; I know that my fifth coin must be a nickel." The third man said, "I feel two coins, but I shall tell you nothing of their size; I know what my other three coins must be." The fourth man said, "I feel one coin; I know what the other four coins must be."

Determine how the remaining two cups were labeled and what the total value of the money in those two cups was. ([27], Feb. 1970, p. 82)

★ ★ **1.18.** [A] After the senior prom at Merriweather High, six friends went to their favorite greasy spoon restaurant, where they shared a booth. The group consisted of the senior class president, the head cheerleader, a player on the school volleyball team, a player on the basketball team, the class valedictorian, and the school principal's only child. Their names were Bobbie, Frankie, Gerry, Jo Jo, Ronnie, and Sal, not necessarily in that order.

Each of the six was in love with one of the others, but no two had crushes on the same person.

Bobbie was in love with the person sitting opposite her.

Frankie liked the cheerleader but was sitting opposite the valedictorian.

Gerry was sitting next to the cheerleader and was crazy about the class president.

Jo Jo, who was not the valedictorian, was sitting between the volleyball player and the class president.

Ronnie disliked the basketball player.

Sal, an orphan, was sitting against the wall and had a crush on the volleyball player.

The volleyball player sat opposite the principal's child.

Identify each person's claim to fame.

★ ★ **1.19.** [A] The annual mixed doubles table tennis tourney took place in Paddleford City in April. The five schools represented were: East, West, North, South, and Central. Each team had two members, one male and one female. The girls were Becky, Emily, Sylvia, Vicki, and Helen. The five boys were Irv, Lee, Max, Paul, and Ted.

According to tournament rules, a team is eliminated after three defeats, and whichever team survives after the four other teams have been eliminated is the winner. The tournament opens on a Monday evening, with three matches scheduled nightly. If play continues beyond Thursday, the Friday matches continue until a winner is determined.

This year's schedule is as follows:

Monday:
    1st match, East vs. West.
    2nd match, North vs. South.
    3rd match, Central vs. the winner of the first match.

Tuesday:
    1st match, the losers of the first two Monday matches.
    2nd match, the winners of Monday's last two matches.

3rd match, the winner of the evening's first match vs. whichever team hadn't played in either of the evening's preceding matches.

Wednesday:

1st match, the losers of Tuesday's first two matches.

2nd match, the winners of Tuesday's last two matches.

3rd match, the winner of the evening's first match vs. whichever team hadn't played either of the preceding matches.

Thursday:

1st match, the winners of Wednesday's last two matches.

2nd match, the losers of Wednesday's first two matches.

3rd match, the winners of the evening's two previous matches.

Friday:

???

If

1. Emily played in the second match on both Monday and Tuesday.

2. West won its only match of the tourney on Monday.

3. Max and his partner participated in five matches before being eliminated at the conclusion of the second contest on Thursday.

4. South played in the first match on both Tuesday and Wednesday.

5. Helen and her partner won two matches on Tuesday.

6. Irv and his partner won Wednesday's second match.

7. Vicki and her partner won Thursday's first match.

8. Paul and Sylvia were on opposite teams in Wednesday's third match.

9. Ted and his partner lost Thursday's third match.

10. Central was defeated in Thursday's first match.

11. The team that handed North its third defeat played in two more games, but did not win the championship.

Determine:

(a) The members of each team.

(b) The outcome of Thursday's and Friday's matches.

(c) This year's championship team. (Al B. Perlman, [50], Problem 4)

★ ★ 1.20. ⊞ Ⓐ At the end of five events of the decathlon at an intercollegiate track meet, Juan Jimenez, Hal Harris, Bill Boone, Michael Manners, and Tom Twofeathers were in the first five places (without ties), but not necessarily in that order. They each wore a number between 1 and 5 on their jerseys, but no two wore the same number.

As luck or skill would have it, they also were the first five finishers (again without ties) in the sixth event—the high jump.

The five schools they represented were: Holbrook University, Jupiter College, Marmaduke University, Mamaraneck Military Academy, and West Rochedale U. (again, not necessarily in that order).

Prior to the start of the sixth event, the competitor from Jupiter was in first place and the competitor from Holbrook was in fifth. The competitor wearing #1 was in third, and the competitor wearing #5 was in fourth.

Boone's finishing position in the sixth event was one number lower (one place better) than his overall standing prior to that event; but Manners' position in the sixth event was one number higher than his prior standing. Twofeathers' finish in the high jump coincided with Harris' previous standing. Jimenez was the only competitor whose performance in the sixth event exactly matched his earlier performance.

The competitor who was in second after the first five events was the only competitor whose first initial coincided with the initial of the first name of his school.

The number worn by the jumper from Marmaduke was the same as his finishing position in the high jump; and the number worn by Harris was one greater than his standing after five events. Twofeathers wore a number that was one less than his prior standing but was equal to the prior standing of the competitor from Mamaraneck.

Identify each competitor with his school, his entry number, his standing after the first five events, and his position of finish in the high jump. (Adapted from [10], Problem 23)

★★ **1.21.** [H] [A] There are five houses in a row (east to west), each of a different color and inhabited by men of different nationalities, with different pets and preferences in beverages and cigarettes.

1. The Englishman lives in the red house.
2. The Spaniard owns the dog.
3. Coffee is drunk in the green house.
4. The Ukrainian drinks tea.
5. The green house is east of the ivory house and next to it.
6. The Old Gold smoker owns snails.
7. Kools are smoked in the yellow house.

8. Milk is drunk in the middle house.
9. The Norwegian lives in the most westerly house.
10. The man who smokes Chesterfields lives in the house next to the man with the fox.
11. Kools are smoked in the house next to the house where the horse is kept.
12. The Lucky Strike smoker drinks orange juice.
13. The Japanese smokes Parliaments.
14. The Norwegian lives next to the blue house.

Who drinks water? And who owns the zebra? ([27], Jan. 1971, p. 87)

★★ **1.22.** [H] [A] Big Data Computer Dating Service arranged a tour for ten of its members. Participating in this tour were five young men: Albert, Barney, Chuck, Danny, and Ernie; and five young women: Florence, Glenda, Helen, Inez, and Joan. From the tens of thousands of clients whose psyches had been programmed onto the electronic tape, they were the ones deemed to have the greatest compatibility potential with all five members of the opposite sex. One love they all had in common was theatergoing.

Big Data arranged for them to see five of the most popular shows on Broadway— "Kumquats and Kulaks," "The Loquacious Labradorian," "The Masticating Mahatma," "The Narcoleptic Nonconformist," and "The Omnipotent Ottoman." The theaters in which these stellar productions were housed were— not respectively—The Purgatory, The Quagmire, The Reptilian, The Sarcophagus, and The Thumbscrew.

The members of the Big Data tour attended each of the shows with a different date, and no more than one couple ever attended the same performance of the same show. Everybody went to the theater on Wednesday, Thursday, Friday, and Saturday evenings as well as to the traditional Saturday matinee.

But enough of the prologue! From the following facts, can you (1) link up each show with its theater and (2) deduce who saw what with whom and at which performance?

The facts are:

1. Though Glenda loved both "Mahatma" and the show at The Thumbscrew, neither thrilled her quite as much as the one she saw with Chuck, for this marked her very first visit to a real Broadway theater.

2. During an intermission, Glenda and Danny had an argument over which show had a more exciting last act—"Labradorian" or the one at The Reptilian.

3. As Helen and Danny were leaving the theater, they agreed that the show they had just seen wasn't as good as "Nonconformist" but was somewhat better than the one at The Quagmire.

4. In a period of some 26 hours, Barney had theater dates with Florence, Helen, and Inez (although not necessarily in that order).

5. Chuck's date at "Ottoman" had been Danny's date at The Reptilian.

6. As Ernie was leaving the theater with Florence, he remarked that of the five shows that he had seen that week, he enjoyed "Kumquats" the most, the one at The Thumbscrew the least, with the one they had just seen together falling somewhere in between.

7. Florence saw "Mahatma" and the show at The Sarcophagus on the same day.

8. The show at The Purgatory was the only one on the itinerary that consisted of one long act without any intermission. Inez saw it on the same night that Glenda and Barney were attending "Mahatma."

9. During what may well have been the most idyllic 26-hour period of her entire life, Joan saw "Mahatma" and the show at The Quagmire, and she also had a theater date with Albert (not necessarily in that order).

10. Albert's date at The Purgatory had previously attended "Kumquats" with Chuck.

11. The last show Glenda saw that week was "Nonconformist."

12. Neither of the people who attended a certain performance at The Thumbscrew together had as yet seen "The Narcoleptic Nonconformist."

13. During the intermission of "Ottoman," Chuck and Joan stepped outside for a breath of air, and they caught a glimpse of their friend Inez in the intermission crowd at The Thumbscrew, which was next door.

14. The night before Inez and Albert had their date, she had been with Danny, and he (Albert) saw "Ottoman."

15. Chuck attended The Thumbscrew with Helen.

16. The Quagmire is across the street from where "Kumquats" is playing.

17. At the same time that Joan and Albert had their date, Barney and Florence saw "Kumquats." (Al B. Perlman, [50], Problem 2)

1.23. [H] Figure 1.46 shows the same main

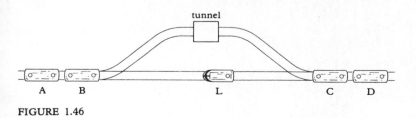

FIGURE 1.46

railroad track as shown earlier, but there are now four railroad cars, A, B, C, and D, and the locomotive, L, arranged in the order A B L C D.

(a) Using the side track, rearrange the cars in this order: C D L A B (Figure 1.47a).

(b) Starting again from the order A B L C D, rearrange them now in this order: D C L B A (Figure 1.47b).

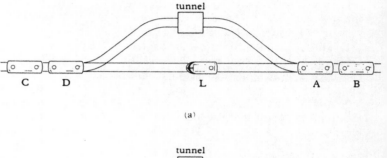

(a)

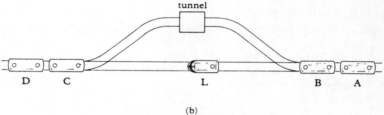

(b)

FIGURE 1.47

2

Solve It With Logic

"If it was so, it might be; and, if it were so, it would be; but as it isn't, it ain't. That's logic." says Tweedledee in Lewis Carroll's *Through the Looking Glass*.

What is logic? According to the *Britannica World Language Dictionary*, logic is a science—"the science of valid and accurate thinking." However, it is not only the scientist, the logician, or the mathematician who is interested in the principles of logic. We all use some form of logical reasoning regularly. In fact, we have already had to use it in our analysis of the problems of Chapter 1. Our approach, however, was informal. At times a little more formality can be helpful. For example, in elaborately stated problems (such as Exercise 2.24 in this chapter) the use of a formal or symbolic notation can greatly simplify the analysis.

In order to attack any problem, it is important to understand the precise meaning of the problem. This is also true in our everyday lives: In order for people to communicate efficiently, it is important for each one to understand precisely what the other means and what inferences can be drawn. If your mother says, "If I get an income tax refund, then I'll buy a new television," there should be no ambiguity about what she means. Does she mean that she will not get a new television if she does not get a tax refund? If, two months later, you see a new television in her house, can you conclude that she received a tax refund?

The rules of logic, which we study in this chapter, will tell us what she should mean and what conclusions we should be able to draw.

These rules were first formulated by Aristotle (c. 350 BCE). He observed that many logical arguments have the same form. For example, consider:

All students are bright people.	All cows are animals that love grass.
All bright people study logic.	All animals that love grass moo.
Therefore, all students study logic.	Therefore, all cows moo.

Both have the logical form of syllogism:

All A are B.

All B are C.

Therefore, all A are C.

Aristotle realized that the validity of an argument depends on its form and not on the specific words that are used. His ideas were expanded upon by many of the logicians and scholars in science who followed him. Although some symbols were used by Aristotle and his followers, it was not until the seventeenth century that the mathematician Leibnitz attempted to develop a complete algebra of logic. His efforts went unnoticed for almost two hundred years. Then new life was injected into the field when Boole, De Morgan, Frege, and others successfully introduced a symbolic notation that gave great precision and a new impetus to the subject.

From a recreational point of view, Lewis Carroll helped popularize logic by presenting many comically worded problems in his various works. These kinds of puzzles are still popular today and may be found in the works of H. Phillips [51], G. Summers [61] and [62], and C. R. Wylie [65], among others.

In this chapter, we study informally some basic aspects of symbolic logic. We will then see how this symbolism can be helpful in the solution of certain types of problems. We motivate our discussion by considering the following sample problems. Try to solve them before reading the rest of the chapter.

SAMPLE PROBLEMS

Problem 2.1

On Freshman Day, each new student had the option of being a Truthteller or a Liar for the day. The Truthtellers had to speak only the truth; the Liars would speak only lies. I came upon three freshmen, A, B and C, sitting on a step. I asked A whether he was a Truthteller or a Liar.

A answered with his back turned, so I could not hear what he said.

"What did he say?" I asked B.

B said, "A says he is a Truthteller."
C said, "B is lying."
Was C a Truthteller or a Liar?

Problem 2.2

Three siblings, Alice, Bob, and Carol, truthfully reported their grades
to their parents as follows:
    Alice: If I passed math, then so did Bob.
           I passed English if and only if Carol did.
    Bob: If I passed math, then so did Alice.
         Alice did not pass history.
    Carol: Either Alice passed history or I did not pass it.
        If Bob did not pass English, then neither did Alice.
    If each of the three passed at least one subject and each subject
was passed by at least one of the three, and if Carol did not pass the
same number of subjects as either of her siblings, which subjects did
they each pass?

Problem 2.3

Either Lucretia is forceful or she is creative. If Lucretia is forceful,
then she will be a good executive. It is not possible that Lucretia is
both efficient and creative. If she is not efficient, then either she is
forceful or she will be a good executive.
    Can we conclude that Lucretia will be a good executive?

In attacking these and other problems, it is not always necessary
to make formal use of the rules of logic. But it is essential to have a
clear understanding of the statement of the problem and to be able to
determine what conclusions can be deduced from what is stated. It is in
developing these abilities that a study of the rules of symbolic logic
can be helpful.
    We begin our study by explaining what is meant by a statement.

STATEMENTS

In the standard theory of logic, a **statement** or **proposition** is a
declarative sentence that has one "truth value"—it is either true or
false but not both. If it is true, we say that it has truth value T (true);
if it is false, it has truth value F (false).
    For example, the following are statements:

    The moon is farther from the earth than it is from the sun.

$2 + 3 = 6.$
$2 + 3 = 5.$
George Washington was the first president of the United States.
John Smith will be president of the United States in 1984.
It is now raining in Philadelphia.

Each of the above sentences is a statement, as each has a truth value. The truth value of each of the first two sentences is F; the next two have truth value T. Although we cannot (at the time of this writing) determine the truth value of the fifth sentence, it does have a truth value—it is either true or false. The truth or falsity of the sixth sentence will depend on the weather conditions in Philadelphia at the time that the sentence is uttered. Nevertheless, at any given time, it is either true or false.

The following sentences are not statements:

Will it rain tomorrow?
Get out of here.
This statement is false.⎫
$2 + x = 5.$ ⎬(see Problem 3 below.)

PRACTICE PROBLEMS 2.A

1. [A] Which of the following are statements?
    (a) George Washington was not the first president of the United States.
    (b) Is there life on Mars?
    (c) Read this book.
    (d) Mr. Dunstan's wife is Inez.
    (e) The trash is collected on Mondays and Fridays.

2. Which of the following are statements?
    (a) Half past five.
    (b) The elevators are not running.
    (c) How do I get to City Hall?
    (d) The rain in Spain stays mainly in the plain.
    (e) I think she's got it.

3. [A] (a) Why is the sentence "$2 + x = 5$" not a statement?
    (b) Why is the sentence "This statement is false" not a statement?

Let us now consider Problem 2.1 (page 38). If you were unable to solve it before, try it again. Think about what A actually said.

In order to find out whether or not C is telling the truth, we must first determine what A actually said. One might expect that what he said must depend on whether he was a Truthteller or a Liar. But let us see. Consider each of the possibilities.

If A were a Truthteller, then he would tell the truth and say, "I am a Truthteller."

On the other hand, if A were a Liar, then he would lie and say, "I am a Truthteller."

In either case, A must have said, "I am a Truthteller." Hence, B told the truth and C lied. Thus, C was a Liar.

Solution of
Problem 2.1

VARIABLES AND CONNECTIVES

We now return to our discussion of symbolic logic, aided by an analogy with algebra.

Just as we use the variables x, y, z, . . . , in algebra, to represent general but unknown numbers, we use the **propositional variables** p, q, r, . . . , to represent general but unknown statements. In algebra, we may assign a specific value to x, y, and so on; for example, "let $x = 1$." Similarly, p may represent a specific statement. For example, if we wish p to represent the statement "It is raining," we write

p: It is raining.

Continuing the analogy, numbers or variables may be combined in algebra by using the familiar operations ($+$, $\times$, $-$, $\div$). Similarly, we may combine statements or propositional variables by using **connectives**. The most commonly used connectives are: "and" ($\wedge$), "or" ($\vee$), "if _____ , then _____" ($\rightarrow$), "if and only if" ($\leftrightarrow$), and "not" ($\sim$). For example, if p represents the statement "It is raining" and q represents "I am carrying an umbrella," then we can form a new statement "It is raining and I am carrying an umbrella," which we represent by $p \wedge q$.

The five connectives mentioned above enable us to combine statements to form new ones. Let us examine the logical meaning of these new, compound statements.

NEGATION

Actually, "not" is not a connective in the same sense as the other four. Whereas the others are **binary** connectives—that is, they actually connect two statements or propositional variables, "not" is **unary**—it relates only to one statement or propositional variable, rather than two. If p is any statement, then "not p" (written symbolically as $\sim p$) is just

p	$\sim p$
T	F
F	T

FIGURE 2.1

the statement "It is not the case that p." The statement $\sim p$ is called the **negation** of the statement p.

For example, the negation of the statement "It is raining" is the statement "It is not the case that it is raining," or, more simply, "It is not raining."

Obviously, if a statement p is true, then its negation $\sim p$ is false, and vice versa. This can be exhibited by means of a visual aid called a **truth table** (Figure 2.1).

This table in a sense defines the symbol "$\sim$". It tells exactly under what circumstances $\sim p$ is true, and under what circumstances it is false. Observe that the truth value of $\sim p$ depends only on the truth value of the statement p and not on the specific statement that p represents. The table has two lines, as p can have two possible truth values.

In defining the other four connectives, we deal with two statements, p and q. Since each can have truth value T or F, there are four possible cases that must be considered. Figure 2.2 shows these in two ways.

			p	q
T	T	case 1	T	T
	F	case 2	T	F
F	T	case 3	F	T
	F	case 4	F	F

FIGURE 2.2

Note that, by the Multiplication Principle (Chapter 1), there are eight cases for three statements, p, q, r; sixteen cases for four statements; and so on.

"AND"—CONJUNCTION

The meaning of the word "and" is well known. The statement "It is raining and I am carrying an umbrella," when true, gives us two pieces of information: one, that it *is* raining; two, I *am* carrying an umbrella.

IT IS RAINING AND I AM CARRYING AN UMBRELLA.

In general, the statement "p and q," represented symbolically by $p \wedge q$, is true when p and q are both true and only then. This information may be exhibited visually in a truth table (Figure 2.3).

The statement "p and q" is called the **conjunction** of the statement p and the statement q. It is sometimes expressed in English by replacing "and" with a word such as "but," "however," "moreover," "furthermore," or "nevertheless." For example, "Roses are red, but violets are blue" has exactly the same logical meaning as "Roses are red, and violets are blue."

p	q	$p \wedge q$
T	T	T
T	F	F
F	T	F
F	F	F

FIGURE 2.3

"OR"—DISJUNCTION

The meaning of the word "or" presents more of a problem as it is used in two different ways in the English language. If Grover says "Either I lost my wallet or I left it home," he is really asserting that there are two possibilities:

He lost his wallet.

He left his wallet home.

Grover's intention is to assert that one or the other of the two possibilities occurred, but not both; that is, the possibility that Grover left his wallet home and lost it is excluded, and "or" is being used in the **exclusive** sense.

On the other hand, the statement "Either the accountant made a mistake or the teller is a thief" allows three possibilities:

The accountant made a mistake and the teller is honest.

The accountant did not make a mistake and the teller is a thief.

The accountant made a mistake but (and) the teller is a thief anyway.

In this case, "or" is being used in the **inclusive** sense; that is, the speaker is including the possibility that both component parts occurred. When "or" is used in the inclusive sense, it means that *at least one* of the component parts occurs; when it is used in the exclusive sense, it means that *exactly one* of the component parts must hold.

This lack of precision in the use of the word "or" can create problems of communication and understanding. The speaker may intend the word to have the exclusive meaning, and the listener may interpret it inclusively; or the other way around.

This ambiguity is not tolerable in the realm of deductive reasoning. Propositions must be precisely stated in order for us to be able to draw conclusions from them. For this reason, logicians and mathematicians have decided to adopt the following convention.

The word "or" is to be used in the inclusive sense, unless otherwise indicated. Thus, for any particular statements, p, q, if we say "Either p occurred or q occurred," we include the possibility that both occurred. If we wish to exclude this possibility, we must say "Either p occurred or q occurred but not both." We call the statement

"p or q" used in the inclusive sense the **disjunction** of the statement p and the statement q. We denote it symbolically by $p \vee q$. Thus, the truth table of $\vee$ is as shown in Figure 2.4.

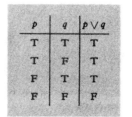

p	q	$p \vee q$
T	T	T
T	F	T
F	T	T
F	F	F

FIGURE 2.4

IT IS RAINING OR I AM CARRYING AN UMBRELLA.

CONDITIONAL AND BICONDITIONAL STATEMENTS

A statement of the form "If p, then q" is called a **conditional** statement. The "if" part is called the **antecedent** of the statement and the "then" part is called the **consequent**. Such a statement is denoted by $p \rightarrow q$. Many mistakes in reasoning are made because this type of statement is often misunderstood.

Imagine that a teacher told his class, "If you pass the final exam, then you will certainly pass the course." Roberta passed the final exam, so she can be confident that she will pass the course. But Ruth failed the final exam. What expectations can she have? The answer is that, based on the teacher's statement alone, she can have no expectations. The teacher said what would happen if she passed the final, but not what would happen if she failed it. If he decides to fail Ruth, there can be no claim that he acted contrary to his previous statement; but even if he decides to pass Ruth anyway, he still cannot be accused of telling a lie.

This is the situation with conditional statements in general. The statement "If p, then q" is easily accepted as true if p and q both turn out to be true; and it is clearly false if p turns out to be true and q turns out to be false. However, if p turns out to be false, then the statement makes no claims whatsoever about q and hence cannot be said to be false; for this reason a conditional statement in which the antecedent turns out to be false is considered to be true regardless of the truth or falsity of the consequent.

Contrast this situation with the following: The teacher said to Ruth, "You will pass the course if and only if you pass the final exam." Now, if Ruth fails the final exam, then she knows that she will fail the course; and of course, if she passes the final exam, then she has every right to expect that she will pass the course.

Notice the difference between the conditional "if, then" and the **biconditional** "if and only if" statements. The latter, denoted symbolically by $p \leftrightarrow q$, makes a claim regardless of the truth value of p; whereas the former makes a claim only in the case that the antecedent, p, is true.

The truth tables for the conditional and the biconditional are shown in Figures 2.5 and 2.6.

IF IT IS RAINING, THEN I AM CARRYING AN UMBRELLA.

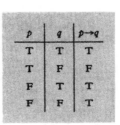

p	q	$p \rightarrow q$
T	T	T
T	F	F
F	T	T
F	F	T

FIGURE 2.5

IT IS RAINING IF AND ONLY IF I AM CARRYING AN UMBRELLA.

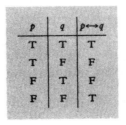

p	q	$p \leftrightarrow q$
T	T	T
T	F	F
F	T	F
F	F	T

FIGURE 2.6

There are many different ways in which a conditional statement can appear in English. For example, each of the following statements has the same logical meaning as "if p, then q":

q, if p

p only if q

p is a sufficient condition for q

q is a necessary condition for p

q, whenever p

q, provided that p

not p, unless q★

q, unless not p

★ It will be shown (see observation 9 on page 48) that an "If _____, then _____" statement can also be translated as an "or" statement. Thus "not p, unless q" can be translated as "q or not p." For this reason, the word "unless," like the word "or," is sometimes used in more than one sense. For example, the statement "I won't go unless you go" might be interpreted in the biconditional sense "I will go if and only if you go." However, in this text the word "unless" is to be used in the conditional sense: "If you don't go, then I won't," which, as we'll see below (page 54), is equivalent to "If I go, then you go."

Similarly, the biconditional statement "p if and only if q" sometimes appears in the form "p is a necessary and sufficient condition for q."

PRACTICE PROBLEMS 2.B

1. **A** Suppose that

p: Jarvis received a B grade on the final examination
q: Jarvis passed the course

are both true statements. Find the truth values of each of the following statements:
(a) If Jarvis received a B on the final, then he passed the course.
(b) If Jarvis received an F on the final, then he failed the course.
(c) If Jarvis received an F on the final, then he passed the course.
(d) If Jarvis passed the course, then he received an A on the final.
(e) If Jarvis failed the course, then he received an A on the final.

2. Suppose that

p: Chico earns \$4.50 an hour
q: Chico makes more money per week than Terry does

are both true statements. Which of the following are true?
(a) If Chico earns \$3.00 an hour, then he does not make more money per week than Terry does.
(b) If Chico earns \$4.50 an hour, then Terry makes more money per week than Chico does.
(c) If Chico makes more money per week than Terry does, then Chico earns \$4.50 an hour.
(d) If Chico makes more money per week than Terry does, then Chico earns \$5.00 an hour.
(e) If Terry makes more money per week than Chico does, then Chico earns \$5.00 an hour.

3. **A** Write each of the following statements symbolically and then express each in English in an "if _____, then _____" form:
(a) The paraffin test will be positive (t) provided that you recently fired a gun (f).
(b) I will go to the beach tomorrow (g), unless it rains (r).
(c) Hearing you say so (h) is sufficient for me to believe it (b).
(d) I'll leave (l), if you ask me nicely (a).
(e) Tex cries (c) whenever he sees a cowboy movie (s).

4. Write each of the following statements symbolically and then express each in English in an "if _____, then _____" form:
(a) You must eat (e) if you want to grow (g).

(b) It is necessary for you to acquire a driver's license (*a*) in order
for you to be permitted to drive a car (*d*).
(c) A rectangle is a square (*r*) only if all four sides are the same
length (*s*).
(d) Being a multiple of four (*m*) is a sufficient condition for a
number to be even (*e*).

5. $\boxed{A}$ If *p* and *q* are the statements
p: Alice passed math
q: Bob passed math,
translate each of the following into symbols and indicate in which
cases each is true:
(a) If Alice passed math, then so did Bob.
(b) If Bob passed math, then Alice did.
(c) Alice and Bob both passed math.

6. If *r* and *s* are the statements
r: Alice passed history
s: Carol passed history,
translate each of the following into symbols and indicate in which
cases each is true:
(a) Either Alice passed history or Carol did not pass it.
(b) Alice did not pass history.
(c) Alice did not pass history unless Carol passed it.

7. $\boxed{A}$ If *p* and *q* are the statements
p: Jack is a good golfer
q: Jack is not a good tennis player,
translate each of the following into English:
(a) $p \leftrightarrow q$ (b) $p \wedge q$ (c) $\sim q$ (d) $\sim p$ (e) $q \rightarrow p$.

8. If *r* and *s* are the statements
r: Macbeth is the Thane of Cawdor
s: Ruth is the Sultan of Swat,
translate each of the following into English:
(a) $r \rightarrow s$ (b) $s \rightarrow r$ (c) $\sim r$ (d) $r \vee s$ (e) $r \leftrightarrow s$.

DRAWING CONCLUSIONS

We are now ready to discuss ways in which we can draw conclusions
from statements of the type above. These conclusions or observations
follow directly from the truth tables defining the given statements.

From the definition of $\sim$ we make the following observations:

Observation 1. Either p is true or $\sim p$ is true (not both).

Observation 2. It is not possible for both p and $\sim p$ to be true simultaneously. (Thus, if an assumption leads to a situation where p and $\sim p$ are both found to be true, then we say that the assumption has led to a contradiction and the assumption must be false.)

The definition of $\wedge$ leads to

Observation 3. If $p \wedge q$ is true, then p must be true and q must be true.

Observation 4. If $p \wedge q$ is false, then at least one of p, q is false. Hence, in particular, if $p \wedge q$ is false and p is true, then q must be false; and, similarly, if $p \wedge q$ is false and q is true, then p must be false.

The definition of $\vee$ leads to

Observation 5. If $p \vee q$ is false, then p must be false and so must q.

Observation 6. If $p \vee q$ is true, then at least one of p, q is true. Hence, in particular, if $p \vee q$ is true and p is false, then q must be true; and, similarly, if $p \vee q$ is true and q is false, then p must be true. (Note, however, that if $p \vee q$ is true and if p is true, we *cannot* conclude that q is true or that it is false.)

The definition of $\rightarrow$ leads to

Observation 7. If $p \rightarrow q$ is true and p is true, then q must be true. (Note, however, if $p \rightarrow q$ is true and q is true, p could be true or false.)

Observation 8. If $p \rightarrow q$ is true and q is false, then p must be false.

Observation 9. If $p \rightarrow q$ is true, then either p is false or q is true, or both. (Note that we cannot in this case conclude explicitly that p is false and q is true.)

Observation 10. If $p \rightarrow q$ is false, then p must be true and q must be false.

Finally, from the definition of $\leftrightarrow$ we can observe that

Observation 11. If $p \leftrightarrow q$ is true, p and q must have the same truth value.

Observation 12. If $p \leftrightarrow q$ is false, then p and q have opposite truth values.

PRACTICE
PROBLEMS
 2.C

1. [A] Given that "If Archie was early, then no one was home" is a true statement, what can we conclude if we find out that
   (a) Archie was early.
   (b) Someone was home.

2. Given that "Bonnie is tired and Carrie is sleeping" is a false statement, what can we conclude if we find out that
(a) Bonnie is tired.
(b) Carrie is sleeping.
(c) Bonnie is tired if and only if Carrie is sleeping.

COMPOUND STATEMENTS

As we mentioned earlier, statements and connectives can be used as building blocks to form new statements. For example, from the simple statements

e: Lucretia is efficient
f: She is forceful
g: She will be a good executive,

we can form the compound statement, "If Lucretia is not efficient, then either she is forceful or she will be a good executive."

This statement may be represented symbolically as

$$(\sim e) \to (f \vee g).$$

Note the importance of the parentheses to indicate exactly which statements are connected by each connective.

The types of conclusions that we were able to draw in the preceding section can be extended to more complicated statements such as the one above. For example,

If the above statement is known to be true, and if we find out that Lucretia is not efficient, then we can conclude that either she is forceful or else she will be a good executive.

If the given statement is known to be true, and we find out that Lucretia is not forceful and that she will not be a good executive, then we can conclude that she must be efficient.

---

1. [A] If p, q, r, and s are the statements
p: George Washington slept here
q: George Washington is the father of our country
r: Paul Revere woke George Washington
s: The Declaration of Independence was signed on July 4, 1776,
translate each of the following into symbols:
(a) If George Washington is the father of our country, then he didn't sleep here.
(b) If the Declaration of Independence was signed on July 4, 1776, then either George Washington did not sleep here or else Paul Revere woke him.

PRACTICE PROBLEMS 2.D

(c) Paul Revere did not wake George Washington unless George Washington slept here.

(d) If Paul Revere woke George Washington if and only if George Washington slept here, then the Declaration of Independence was signed on July 4, 1776 and George Washington is not the father of our country.

2. If p, q, r, and s are the statements

p: Roses are red

q: Violets are blue

r: Sugar is sweet

s: You are sweet,

translate each of the following into symbols:

(a) If roses are red and violets are blue, then sugar is sweet and so are you.

(b) You are sweet and sugar is too, unless roses are not red or violets are not blue.

(c) Roses are red only if violets are blue, but if sugar is sweet then you are not sweet too.

3. [A] If p, q, r, and s are the statements

p: The poet is dreaming

q: The poet is jesting

r: Titania is the queen of the fairies

s: Oberon is the king of the fairies,

translate each of the following into English:

(a) $p \to (\sim r)$ (b) $(\sim p) \vee (\sim r)$ (c) $\sim (p \wedge r)$

(d) $[\sim (p \vee q)] \to (s \to r)$

4. If p, q, and r are the statements

p: Jack and Jill went up the hill

q: Little Bo Peep has lost her sheep

r: Little Tommy Tittlemouse lived in a little house,

translate each of the following into English:

(a) $(\sim p) \leftrightarrow (\sim q)$ (b) $\sim (p \vee q)$ (c) $(q \to p) \to r$

5. [A] Given that "If either Archie is early or Janice is absent, then no one is home" is a true statement, what can we conclude if we find out that

(a) Archie is early.

(b) Someone is home.

(c) Either Archie is early or no one is home.

6. Given that "If today is Thursday and the sun is shining, then either the robins are singing or it is not summer" is a true statement,

what can we conclude if we find out that

(a) It is summer and the robins are not singing.

(b) Today is Thursday and the robins are singing.

(c) Today is Thursday and the robins are not singing.

7. **A** If it's true that three krimmls are worth one glunk and that if Xcag Zemph is the ruler of Mars then the Martian canals are empty, then it's also true that I am not a Martian. Given that the above statement is true, what can we conclude if we find out that

(a) Xcag Zemph is the ruler of Mars and I am a Martian.

(b) The Martian canals are empty and I am a Martian.

(c) Three krimmls are worth one glunk and the Martian canals are empty.

(d) Three krimmls are worth one glunk and Xcag Zemph is the ruler of Mars.

(e) Three krimmls are worth one glunk and I am not a Martian.

**We are ready to discuss the solution of Problem 2.2
(page 39). If you were unable to solve it before, try it again
before reading on.**

For reference purposes, we label the statements as follows:

1. Alice: If I passed math, then so did Bob.

2. Alice: I passed English if and only if Carol did.

3. Bob: If I passed math, then so did Alice.

4. Bob: Alice did not pass history.

5. Carol: Either Alice passed history or I did not pass it.

6. Carol: If Bob did not pass English, then neither did Alice.

7. Each of the three siblings passed at least one subject.

8. Each subject was passed by at least one of the three siblings.

9. Carol did not pass the same number of subjects as either of her siblings.

Solution of Problem 2.2

What conclusions can we draw from these statements about each of the three subjects?

Statements 1 and 3 deal with math: Since statement 1 is true, it follows, by observation 7 (page 48), that if "Alice passed math" is true, then "Bob passed math" is also true. Similarly, by statement 3 and observation 8, if "Alice passed math" is false, then "Bob passed math" is also false. Therefore, since statements 1 and 3 must be true simultaneously, there are two possibilities:

Alice and Bob both passed math.

Neither Alice nor Bob passed math.

This information can be recorded in a chart (Figure 2.7).

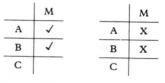

	M
A	✓
B	✓
C	

	M
A	X
B	X
C	

FIGURE 2.7

Now consider statements 2 and 6, which pertain to English. By statement 2, together with observation 11, exactly two cases are possible:

Alice and Carol both passed English.

Neither Alice nor Carol passed English.

These conclusions may be combined with the previous case to give four possibilities (Figure 2.8):

I

	M	E
A	✓	✓
B	✓	
C		✓

II

	M	E
A	✓	X
B	✓	
C		X

III

	M	E
A	X	✓
B	X	
C		✓

IV

	M	E
A	X	X
B	X	
C		X

FIGURE 2.8

Statement 6 together with observation 9, implies that either:

Bob passed English (and Alice did or did not pass it)

or

Alice did not pass English (and Bob did not either).

Of these, only the case that both Bob and Alice passed English is consistent with charts I and III of Figure 2.8 (since, in these charts, Alice passed English). Statement 6 does not lead to any conclusion about charts II and IV. However, statement 8 does: Since at least one of the three siblings passed English, Bob must pass English (Figure 2.9).

I

	M	E
A	✓	✓
B	✓	✓
C		✓

II

	M	E
A	✓	X
B	✓	✓
C		X

III

	M	E
A	X	✓
B	X	✓
C		✓

IV

	M	E
A	X	X
B	X	✓
C		X

FIGURE 2.9

Now, for history: By statement 4,
    Alice did not pass history.
Since statement 5 is true, it follows from observation 6 that
    Carol did not pass history.
This information, together with statement 8, implies that
    Bob passed history.
We now have Figure 2.10.

I	M	E	H
A	✓	✓	X
B	✓	✓	✓
C		✓	X

II	M	E	H
A	✓	X	X
B	✓	✓	✓
C		X	X

III	M	E	H
A	X	✓	X
B	X	✓	✓
C		✓	X

IV	M	E	H
A	X	X	X
B	X	✓	✓
C		X	X

FIGURE 2.10

In this figure, chart IV violates statement 7 and so cannot hold. By applying statements 7 and 8 to charts II and III, we obtain Figure 2.11.

I	M	E	H
A	✓	✓	X
B	✓	✓	✓
C		✓	X

II	M	E	H
A	✓	X	X
B	✓	✓	✓
C	✓	X	X

III	M	E	H
A	X	✓	X
B	X	✓	✓
C	✓	✓	X

FIGURE 2.11

In charts II and III of this new figure, statement 9 is violated. Only chart I remains, and, by statement 9, Carol must pass only one subject (Figure 2.12).

	M	E	H
A	✓	✓	X
B	✓	✓	✓
C	X	✓	X

FIGURE 2.12

Thus, Bob passed all three subjects, Alice passed math and English, and Carol passed only English. You should now check that this solution is consistent with all nine statements.

LOGICAL IMPLICATION AND EQUIVALENCE

So far we have discussed the logical meaning of certain terms and how we can draw logical conclusions from statements involving them.

If we can logically deduce S from knowing R, then it must follow that S is true whenever R is true; that is, all conceivable circumstances that would make R true must also make S true. We say that R **logically implies** S, and write this symbolically as $R \Rightarrow S$.

For example, in our previous discussion of drawing conclusions, we saw that if $p \to q$ is true and if p is true, then we can conclude that q is true; that is,

$$[(p \to q) \wedge p] \Rightarrow q.$$

Similarly we can express each of our other observations above as a logical implication. For example, observation 6 (page 48) becomes

$$[(p \vee q) \wedge (\sim p)] \Rightarrow q.$$

Sometimes, we have two statements R and S such that not only does R logically imply S but also S logically implies R; that is, whenever R is true, then S is true and whenever S is true, then R is true. In this case we say that R and S are **logically equivalent** and we express this symbolically as $R \Leftrightarrow S$.

For example, the statements $p \to q$ and $(\sim q) \to (\sim p)$ are logically equivalent. One way to see this is to consider truth tables (Figure 2.13).

Observe that $p \to q$ and $(\sim q) \to (\sim p)$ are true under exactly the same circumstances.

p	q	$p \to q$
T	T	T
T	F	F
F	T	T
F	F	T

p	q	$(\sim q) \to (\sim p)$
T	T	T
T	F	F
F	T	T
F	F	T

FIGURE 2.13

We can also see the logical equivalence of $p \to q$ and $(\sim q) \to (\sim p)$ without actually drawing truth tables: The statement $p \to q$ is true except in the case that p is true and q is false; the statement $(\sim q) \to (\sim p)$ is true except in the case that $(\sim q)$ is true and $(\sim p)$ is false—that is, p is true and q is false. Thus these statements are true in exactly the same circumstances, and hence are logically equivalent.

The statement $(\sim q) \to (\sim p)$ is called the **contrapositive** of the statement $p \to q$. Since they are logically equivalent, these two statements are identical in meaning and may be used interchangeably.

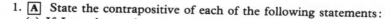

PRACTICE PROBLEMS 2.E

1. [A] State the contrapositive of each of the following statements:
   (a) If Jonas has at least two nickels in his wallet, then he has at least ten cents.
   (b) If the door is locked, no one can enter.

(c) If I go to the concert, I'll hear some music.

(d) If the three angles of a triangle are equal, then the triangle is equilateral.

2. State the contrapositive of each of the following statements:

(a) If you sit in the sun, then you get a tan.

(b) If Juanita reads the newspaper, she will know the news of the day.

(c) If the measure of an angle is 30°, then the angle is acute.

(d) If my wife is not home, then no one answers the phone.

3. [A] Which of the following statements are equivalent to each other?

(a) If Sondra attended the lecture, she fell asleep.

(b) If Sondra didn't attend the lecture, she didn't fall asleep.

(c) If Sondra fell asleep, then she attended the lecture.

(d) If Sondra didn't attend the lecture, then she fell asleep.

(e) If Sondra didn't fall asleep, then she didn't attend the lecture.

4. Which of the following statements are equivalent to each other?

(a) If it will rain, then the roof will leak.

(b) If it won't rain, then the roof won't leak.

(c) If the roof won't leak, then it won't rain.

(d) If the roof will leak, then it will rain.

(e) If it won't rain, then the roof will leak.

ARGUMENTS AND VALIDITY

In most situations that require deductive reasoning, we are given certain clues or information and are required to determine what conclusions can logically be drawn from this information: That is, we are given certain statements $P_1, P_2, \ldots, P_k$ and are required to find another statement, C, which follows from the given statements.

In general, a set of statements $P_1, P_2, \ldots, P_k, C$, one of which (C) is alleged to follow from the others, is called an **argument**. The statement, C, which is alleged to follow from the others is called the **conclusion** of the argument; the other statements ($P_1, P_2, \ldots, P_k$) are called the **premises**. We sometimes indicate an argument as follows

$$P_1$$
$$P_2$$
$$\vdots$$
$$P_k$$
$$\therefore \overline{C} \;(\therefore \text{ means ``therefore.''})$$

For example,

> If Lucretia is forceful, then she will be a good executive
> Lucretia is forceful
> _____
> ∴ Lucretia will be a good executive.

is an argument. Its premises are "If Lucretia is forceful, then she will be a good executive" and "Lucretia is forceful"; and its conclusion is "Lucretia will be a good executive."

If the conclusion of an argument does logically follow from the premises, then we say that the argument is **valid**. A valid argument is one in which the conclusion must be true whenever all the premises are true. This may be expressed in terms of logical implication:

$$(P_1 \wedge P_2 \wedge \cdots \wedge P_k) \Rightarrow C.$$

In general, to show that an argument is valid, we must show that the conclusion is true in all cases in which the premises are all true. Thus, we assume that all premises are true and we must show that it then follows that the conclusion is true.

To show that an argument is not valid, it is enough to find one case in which all the premises are true but the conclusion is false.

For example, the argument

"If Lucretia is forceful, then she will be a good executive. Lucretia is forceful. Therefore, she will be a good executive."

is valid. To see this, let p and q represent the statements

p: Lucretia is forceful

q: She will be a good executive.

Then the argument is of the form

$$\frac{\begin{array}{c} p \rightarrow q \\ p \end{array}}{\therefore q}$$

As we have already seen that $[(p \rightarrow q) \wedge p] \Rightarrow q$, this argument is valid.

On the other hand, the argument

"If Lucretia is forceful, then she will be a good executive. Lucretia will be a good executive. Therefore, Lucretia is forceful."

is not valid. It has the form

$$\frac{\begin{array}{c} p \rightarrow q \\ q \end{array}}{\therefore p}$$

But here it is possible for both premises to be true while the conclusion is false. Specifically, if p is false (Lucretia is not forceful) and q is true (she will be a good executive), then both premises are true but the conclusion is false. Hence the argument is not valid.

Note that the validity of an argument depends only on the form of the argument and not on the specific statements involved. Thus, for example, any argument that can be represented in the form

$$\frac{\begin{array}{c} p \to q \\ p \end{array}}{\therefore q}$$

is valid, and arguments of the form

$$\frac{\begin{array}{c} p \to q \\ q \end{array}}{\therefore p}$$

are not valid.

As an example of a form of a valid argument involving three variables, consider the following:

$$\frac{\begin{array}{c} p \to q \\ q \to r \end{array}}{\therefore p \to r}$$

This form of argument is important because it not only appears often in everyday reasoning but also is used in most mathematical proofs. We leave it as an exercise for you to show that arguments of this form are valid.

---

1. [A] What are the premises and what is the conclusion of each of the following arguments? Represent each argument symbolically.
   (a) Either Rachelle is brilliant or she has a wonderful personality. If she has a wonderful personality, then she has an active social life. Therefore, either Rachelle is brilliant or she has an active social life.
   (b) If Bessie is a cow, then she moos. Bessie is not a cow. Therefore, Bessie doesn't moo.
   (c) Either Lucretia is forceful or she is creative. She is forceful. Therefore, she is not creative.
   (d) If Lee Wong is a student, then he is bright. If Lee Wong is bright, then he studies logic. Therefore, if Lee Wong is a student, then he studies logic.
2. What are the premises and what is the conclusion of each of the following arguments? Represent each argument symbolically.
   (a) Either Lucretia is forceful or she is creative. She is not forceful. Therefore she is creative.
   (b) If Lucretia is not efficient, then either she is forceful or she will

PRACTICE PROBLEMS 2.F

be a good executive. Lucretia is not forceful. Lucretia will not be a good executive. Therefore, Lucretia is efficient.

(c) If Bessie is a cow, then she moos. Bessie moos. Therefore, Bessie is a cow.

(d) If Bessie is a cow, then she moos. Bessie doesn't moo. Therefore, Bessie is not a cow.

(e) If Bessie is a cow, then she moos. Bessie is not a cow. Therefore, she doesn't moo.

3. $\boxed{A}$ Determine whether each of the following is the form of a valid argument:

(a) $\dfrac{\begin{array}{c} p \vee q \\ \sim p \end{array}}{\therefore\ q}$ (b) $\dfrac{\begin{array}{c} p \rightarrow q \\ q \rightarrow r \end{array}}{\therefore\ p \rightarrow r}$ (c) $\dfrac{\begin{array}{c} p \vee q \\ p \end{array}}{\therefore\ \sim q}$ (d) $\dfrac{\begin{array}{c} p \rightarrow q \\ \sim p \end{array}}{\therefore\ q}$

4. Determine whether each of the following is the form of a valid argument:

(a) $\dfrac{\begin{array}{c} p \rightarrow q \\ \sim q \end{array}}{\therefore\ p}$ (b) $\dfrac{\begin{array}{c} \sim(p \wedge q) \\ p \end{array}}{\therefore\ \sim q}$ (c) $\dfrac{\begin{array}{c} p \rightarrow q \\ q \rightarrow p \end{array}}{\therefore\ p \wedge q}$ (d) $\dfrac{\begin{array}{c} p \rightarrow q \\ q \rightarrow p \end{array}}{\therefore\ p \leftrightarrow q}$

5. $\boxed{A}$ For each of the arguments in Problem 1, determine whether or not it is valid.

6. For each of the arguments in Problem 2, determine whether or not it is valid.

7. $\boxed{A}$ For each of the indicated conclusions, determine whether the argument is valid:

> If Alice passed math, then so did Bob
> <u>If Bob passed math, then so did Alice</u>

Therefore

(a) Alice and Bob both passed math.

(b) Alice passed math if and only if Bob passed it.

8. For each of the indicated conclusions, determine whether the argument is valid:

> Alice passed English if and only if Carol did
> <u>If Bob did not pass English, then neither did Alice</u>

Therefore

(a) If Bob passed English, then Carol did not.

(b) If Bob passed English, then Carol did.

(c) If Bob did not pass English, then neither did Carol.

9. Show that arguments of the form

$$p \to q$$
$$q \to r$$
$$\overline{\therefore \ p \to r}$$

are valid. (Hint: Under what circumstances will the conclusion be false? Is it possible for both premises to be true in this case?)

We are now ready to solve Problem 2.3 (page 39). If you haven't already solved it, try it again now.

In this problem we are given certain information and asked whether or not a particular conclusion follows from this information; that is, we are being asked if a particular argument is valid.

We present two methods of attacking this problem: on the left, an informal, verbal approach; on the right, a symbolic approach. In the latter approach, we begin by representing the simple statements of the problem by symbols.

Solution of Problem 2.3

Verbal approach

Symbolic approach

f : Lucretia is forceful
g : Lucretia will be a good executive
e : Lucretia is efficient
c : Lucretia is creative

The premises of the argument are:

P_1: Either Lucretia is forceful or she is creative.

P_2: If she is forceful, then she will be a good executive.

P_3: Lucretia is not both efficient and creative.

P_4: If she is not efficient, then either she is forceful or she will be a good executive.

The desired conclusion is:

C: Lucretia will be a good executive

We may assume that all the premises are true. Therefore, by P_1, there are two cases to consider:

The premises may be expressed as:

$P_1: f \vee c$

$P_2: f \to g$

$P_3: \sim(e \wedge c)$

$P_4: (\sim e) \to (f \vee g)$

and the conclusion to be tested:

$$C : g$$

We may assume that all the premises are true. Therefore, by P_1, there are two cases to consider:

Case 1. Lucretia is forceful.

Case 2. Lucretia is not forceful.

Suppose Lucretia is forceful. Then, by the second premise, she will be a good executive (and so the desired conclusion follows).

If Lucretia is not forceful, then, from the first premise, she must be creative.

Since she cannot be both efficient and creative (by P_3), she cannot be efficient.

Therefore, by P_4, either she is forceful or she will be a good executive.

Since we have assumed that Lucretia is not forceful, we can conclude that she will be a good executive.

Case 1. f is true.

Case 2. f is false.

In case 1, it follows from P_2 (by observation 7, page 48) that g must be true, and so the desired conclusion follows.

In case 2, it follows from P_1 (by observation 6) that c must be true.

By P_3 and observation 4, e must be false.

But, then, by P_4 and observation 7, $f \vee g$ is true.

But, in case 2, f is false and so, by observation 6, g is true.

In both cases the conclusion is true and so the argument is valid.

In this problem we did not seem to gain very much by representing the problem symbolically. However, for problems in which the wording is very involved, the use of letters to represent statements can help us eliminate the complicated language and get down to the basics of the problem. It is for this reason that we have introduced the symbolic approach above.

THE CHAPTER IN RETROSPECT

This chapter has been a brief and informal introduction to symbolic logic. You have been introduced to the logical meanings of the connectives and some of the basic rules of inference, so that you will be able to determine whether or not particular conclusions follow from a given set of statements. Obviously, this knowledge is important to the reasoning process.

In applying the principles of symbolic logic, you seldom have to point them out. As long as you are satisfied that each conclusion you draw logically follows from what you knew previously (and that this could be justified formally if it were necessary to do so), you can usually suppress formalism. In attempting the exercises at the end of the chapter, just be sure that you are completely convinced by each step of your arguments. You may find that representing statements symbolically may prove helpful; if this is the case, don't hesitate to do so. And, of course, to present your solutions to others, you must explain your reasoning, and symbols are often helpful in doing so.

Notice that no attempt has been made in this chapter to list a set of steps that must be followed to reason through every problem with which you may be faced. Every problem must be attacked in its own way.

Nevertheless, to attack any problem no matter what the setting (in business, social science, economics, etc.), you must clearly understand the problem and must be able to recognize when one statement follows from another. It is in carrying out these tasks that this chapter may prove helpful.

Exercises

Truthtellers and Liars

2.1. One summer solstice, while searching for the Southern Shortcut, Silas Seeker, that superman of seafarers, was shipwrecked during a sudden severe squall. Auspiciously, Silas sighted a small sandy isle in the distance and succeeded in swimming ashore. Exhausted, he slipped into a sound sleep.

While he slept, Silas dreamt that he was discovered by two trolls, both of whom, fortunately, were able to converse in English. The two were identical in all aspects of their appearance; however, one belonged to a clan whose members always tell the truth, the other to a clan whose members always lie.

When Silas awoke, he discovered two trolls standing over him, exactly as in his dream.

"Where am I?" he inquired.

"The Isle of Hamlock," replied the first troll.

"The Isle of Grindle," replied the second.

"And what are your names?" asked Silas.

"I am Glog and he is Glum," responded the first.

"No, I am Glog and he is Glum," answered the second.

Just then a third troll appeared. Hoping to shed some light on the matter, Silas pointed to the two and asked, "Which of them can I believe?"

"He and I belong to the same clan," replied the third troll, pointing to the first.

"That's true; they do belong to the same clan," said the second troll.

Assuming that Silas' dream was accurate, who is telling the truth and what was the name of the island?

2.2. [A] Silas lived on the island for six years (see Exercise 2.1), but during all that time he never learned how to visually tell the difference between members of the two clans.

One day Silas met two trolls. The first claimed that they belonged to different clans, whereas the second proclaimed that the first was a liar.

To which clan did each of the two belong?

2.3. [H] During his sojourn on the island, Silas gave names to the two clans (see Exercises 2.1 and 2.2). The truthtellers he called Truthfuls and the liars he called Liars.

One day, Silas met three trolls. He inquired to which clans they belonged.

"All of us are Liars," said the first.

"No, only two of us are Liars," amended the second.

"That's not true either," corrected the third. "Only one of us is a Liar."

To which clan does each of the three belong?

★ **2.4.** [S] [A] Shortly after Silas completed building a raft to take him off the island (see Exercises 2.1 to 2.3), five trolls arrived with food and water for his voyage. As usual, Silas inquired about their clans.

"Three of us are Truthfuls," said the first.

"That's right, three of us are Truthfuls," the second agreed.

"No, only two of us are Truthfuls," said the third.

"The first three are all lying," said the fourth.

The answer of the fifth man left no doubt in Silas' mind as to the clan affiliation of each troll. To which clan did they each belong?

2.5. [A] One hundred and fifty years after Silas Seeker left the island (see Exercises 2.1 to 2.4), his great-great-granddaughter, Heidi Seeker, found Silas' diary and decided to revisit the site of her ancestor's salvation.

When she arrived on the island, she was surprised to discover that there were now three clans instead of two. During the century and a half that had elapsed, some intermarriage between the two groups had occurred and the progeny of these mixed marriages comprised a third group of people who always alternated between telling the truth and lying. Thus, when a member of this third group (whom Heidi named Alternators) makes a series of statements, the first statement might be truthful or a lie, but the veracity of each subsequent statement would alternate. In other words, if the first statement were truthful, then the second would be a lie, the third would be true, and so on.

When Heidi first arrived on the island, she encountered a group of three trolls. One was very tall, one was very short, and the third was of medium height. Heidi introduced herself.

"Welcome," said the shortest of the three. "I am Gladden. He (indicating the tallest of the three) is Lowax. And he (indicating the remaining troll) is Grout."

"He is, as he says, Gladden. But I am Kaut," corrected the troll of medium height. "And the tall troll is Tildeau."

The tall troll objected: "I am Lowax. But the short one is Waldar. And he (pointing to the troll of medium height) is Gaut."

If at least one of them was a Truthful, what were the names of the three trolls?

2.6. [H] Later that day, Heidi (see Exercise 2.5) met another group of three trolls—one from each clan. Heidi asked to which clans they belonged. The spokestroll of the group replied:

"I am a Liar; the lady is a Truthful; the young gentletroll is a mixed breed."

To which clans did each belong?

2.7. [A] After Heidi (see Exercise 2.5) had been on the island for a few days, she realized that each individual's name was woven into his or her clothing and easy to see.

When Heidi was introduced to Winken, Blinken, and Finken, she asked her usual question. The replies she received were:

Winken: I am an Alternator.

Blinken: Winken is a Liar.

Finken: Winken is a Truthful.

If no two of the three belonged to the same clan, to which did each belong?

2.8. [H] When Heidi (see Exercise 2.5) asked her usual question to Hocus, Pocus, Crocus, and Cloy, only Hocus and Pocus replied.

"I am a Truthful; Pocus is a Liar; Crocus is a Liar; Cloy is a Truthful," was Hocus' response.

"No, Hocus is a Liar; I am a Truthful; Crocus is an Alternator; Cloy is a Truthful," corrected Pocus.

To which clan did Hocus, Pocus, and Crocus belong?

2.9. Each year, the leaders of the three clans (see Exercise 2.5) participate in an athletic competition to determine which of the three should govern the island for the new year. After this year's competition, the following statements were made by the competitors:

Ange: I came in second; Benge came in last.

Benge: Ange came in last; Conge won.

Conge: I came in last; Ange won.

What was the order of finish in the competition?

2.10. [A] The inhabitants of the planet Logos invented robots that respond to any question, provided that the proper coin is dropped in the slot in the robot's chest. Actually, there are two different model robots. One variety, Mendibles, respond truthfully when a coin made of flacus is inserted, but lie if any other kind of coin is used. The second model, Lawbakes, respond truthfully to coins made of flacus or grenjo, but lie whenever a telic coin is inserted.

On the first visit of Jake Cooke (Captain of the star cruiser *Ganymede*) to Logos, he was confronted by two robots. The Captain found a supply of coins that were all obviously made of the same material, although he had no idea what the material in question was.

Inserting a coin in the first robot, Captain Cooke asked, "What model robot are you?"

"A Mendible," was the reply.

"And what kind of coin is this?" asked the Captain, inserting another coin.

"Grenjo," answered the robot.

Turning his attention to the second robot, Captain Cooke inserted a coin and asked what model the robot was.

"A Lawbake," was the response he received.

Which model was each of the robots, and what kind of coin was the Captain using?

★ ★ **2.11.** H On his way to the capital city of Logos, Captain Cooke (see Exercise 2.10) came to a fork in the road. Not knowing which road to take, he decided to ask a passing robot. However, the Captain only had one coin, and he was not sure of what material the coin was made.

How could he determine with absolute certainty which road to take?

2.12. The Turner Triplets have an annoying habit—whenever a question is asked of the three of them, two tell the truth and the third lies. When I asked them which of them was born first, they replied as follows:

Werner: Virna was born first.

Virna: I am not the oldest.

Myrna: Werner is the oldest.

Which of the Turner Triplets was born first?

2.13. A When Mr. Cotter entered the early morning detention room at Benedict Arnold High School, he found the school's three biggest troublemakers—Phil Fagin, Ben Tolliver, and Spike Sykes—sitting like angels in their seats, with wide grins on their faces. On the blackboard was a very uncomplementary picture of Mr. C.

"Okay, who's the artist?" the teacher bellowed.

"I didn't do it," Fagin answered. "I was out of the room when it happened. Ben did it."

"I'm innocent," protested Tolliver. "The picture was already on the board when I arrived. Fagin lied when he said I did it."

"One of us did it," Sykes admitted. "But I had nothing to do with it. Phil is innocent, too."

If each of the three students made two true statements and lied once, who is the culprit, and who was present when the picture was drawn?

2.14. Each year the Chamber of Commerce in Las Vegas, Nevada sponsors a contest to see who can build the tallest structure using only a deck of old playing cards. This year's contest was very exciting. Four contestants, Ace Arlington, Jack Johnstone, Emery Queen, and King Collins had surpassed the 3-foot high mark. Suddenly, an underground nuclear test in Los Alamos shook the building and caused all of the card houses to collapse. It was decided to award the prize to the person who was in the lead at the time that the shockwaves struck, but there was a lack of consensus as to who that person was.

Ace said: I beat King, but Queen's house was taller than mine.

Jack said: My building was taller than Ace's, but Queen won.

Queen said: Jack won. I came in second.

King said: Ace won. I came in second.

Fortunately, they were able to determine the winner by viewing the videotape replay. It turned out that each of the four made one correct and one incorrect statement.

What was the order of finish of the four contestants?

★ **2.15.** [S] [A] On Wednesday, May 5, 1976, the wicked Simon Legrew was murdered at his home in a Boston suburb. The police were able to place the time of death at between 11:10 and 11:30 PM. They arrested four suspects—Jeeves, the butler; Fifi, the French maid; Julia, the cook; and Jessica, Mr. Legrew's private secretary. Under questioning, the four suspects made the following statements:

Jeeves: I didn't do it. Jessica did it. Mr. Legrew was blackmailing Jessica. Fifi and I were watching television together from 10:10 PM until 12:30 AM.

Fifi: I'm innocent. Jeeves and I were watching television together at the time of the murder. Jessica was being blackmailed. I saw Jessica speaking to Mr. Legrew at 9:30 PM on the night of the murder.

Julia: I'm innocent. Jessica was being blackmailed. Jeeves murdered Mr. Legrew. I saw Jessica leave the house at 10:00 PM.

Jessica: I did not kill Mr. Legrew. I was not being blackmailed. I was in Chicago during the entire night of the murder. Fifi is the murderess.

Each of the four suspects made two true statements and told two lies.

Whodunnit?

2.16. The hop-off at the Toad County Frog Jump featured four participants—Harold Hawkins' frog Hopalong, Leopold Lohmann's leaper Longjump, Gerald Glauberman's entry Gribbit, and Carlotta Cantrell's bull Croaker. After the contest, the following statements were made:

Harold: Croaker won. Hopalong came in second. Hopalong beat Gribbit by five inches. Gribbit beat Longjump.

Leopold: Longjump won. Croaker beat Gribbit. Gribbit did better than Hopalong. Hopalong beat Croaker.

Gerald: Harold took home the prize money for first place. Gribbit came in second. Longjump came in third. Croaker was a distant last.

Carlotta: Gribbit won. Croaker came in second. Hopalong beat Longjump. Longjump finished last.

If each of the four participants made two true statements and told two lies, what was the order of finish in the race?

2.17. [A] Probably the first real evidence of the existence of the Abominable Snowman is a photograph taken by the renowned mountaineer Sir Hilary Edmund and his party.

Before the photo was developed, Sir Hilary was asked to describe the beast. "It was over 7 feet tall, with long white fur, and 6 toes on each foot," replied Edmund.

"Him a liar," objected famed Sherpa guide, Nenzing Torkay, grinning. "The Snowman has no fur at all, is under 5 feet tall, and has hooves."

The other three members of the expedition also disagreed in their descriptions.

Monte Everesto said that the Snowman was 6 feet tall, had long white fur, and hooves.

Matty Horne claimed the Snowman was over 7 feet tall, had brown fur, and had 5 toes on each foot.

And Snowsov "Killer" Manjaro estimated

the beast's height at under 5 feet and attributed brown fur and 6-toed feet to the creature.

These discrepancies in the descriptions were probably due to the fact that the beast was sighted during a severe snowstorm. When the picture was developed, it proved that each of the five mountain climbers was correct about exactly one aspect of the Snowman's appearance.

What does an Abominable Snowman look like? (Adapted from [10], Problem 16)

Matching Problems

2.18. [A] Marvin Mandlebaum shares a medical office with three women—Claudine Coates, Melissa Mavis, and Rosalinde Rowe. One of the four doctors is a cardiologist; another is a gastroenterologist; a third is an endocrinologist; and the fourth is a hematologist.

According to the rumors that I heard when I was a patient in the hospital,

Dr. Coates, the cardiologist, and the gastroenterologist were all sorority sisters in college.

The cardiologist and the gastroenterologist were once legally married to each other.

Dr. Mavis and the hematologist are engaged to be married to each other.

Marvin and his girlfriend play bridge each week with the gastroenterologist and her husband.

If exactly three of these statements are true, what is each person's medical specialty?

2.19. [A] Five playmates, Jason, Danny, Ethan, Mark, and Mindy, had birthdays within two weeks of each other, so their parents decided to give them a joint party. In order to cut down on presents, it was decided that each child should bring one gift. The first to arrive would give his or her present to the second; the second would give his or her

present to the third; and so on; the fifth giving his or her present to the first.

After the party, the children were asked what they had received.

Jason said, "I received the present Ethan brought."

Ethan said, "I received Mark's present."

Mindy said, "Jason received my present."

Mark said, "I received Ethan's present."

Danny said, "I gave my present to Ethan."

As is to be expected with children, not all of these statements are accurate. In fact, exactly three are correct.

From whom did each child receive his or her present?

★ 2.20. [S] [A] Four men were asked about their yearly incomes. Their names are Earl, Moe, Luis, and Randy and their professions are architect, carpenter, plumber, and mason (not necessarily in that order). Each made two statements; but the only statements whose correctness can be depended on are statements in which the speaker specifically names his *own* profession.

Other statements may or may not be true.

Earl: The plumber makes three times as much as the carpenter. The architect makes more money than I do.

Moe: The carpenter makes more money than the plumber. Luis is either the mason or the architect.

Luis: I make more than the architect. The carpenter makes less than each of the others.

Randy: The plumber makes twice as much as the carpenter. I make more than the mason.

Match each person with his profession.

★ 2.21. Three women (Abby, Janice, Linda) and two men (Martin, Roberto) are a singer, a dancer, a comic, a television writer, and a theatrical agent, although not necessarily in that order.

Abby said: I'm not the comic. The writer and the dancer are happily married. The singer and the agent are engaged to be married.

Janice said: The singer is my cousin. The writer and the dancer are siblings. The comic and the agent share an apartment.

Linda said: I am not the writer. The singer and the agent hate each other. The dancer and the comic frequently work together.

Martin said: The singer owes me $10. The writer and the dancer are not related and have never met. The comic and the agent are next-door neighbors in an apartment house.

Roberto said: The singer saved my life once. The agent lives alone in an old mansion. The dancer and the comic have never met.

If the only certainty is that every statement in which an individual alludes to his or her own profession is true, who is who?

Connectives and Arguments

2.22. [A] Amy, Betty, Carmen, and Dee spent all their money at the corner store. One girl spent her dime on candy; another spent her quarter to buy a ball; a third bought a coloring book for thirty-five cents; and the fourth spent forty cents on two comic books. Each paid for her purchase with the exact change. Upon leaving the store, the girls made the following true statements.

Amy said: "If I had a quarter, then so did Betty. Betty had a nickel if and only if Carmen did. If I had a dime, then so did Dee."

Betty said: "If I had a quarter, then so did Amy. If I had a nickel, then so did Dee."

Carmen said: "If I had a dime, then so did Amy. If Amy did not have a quarter, then neither did I."

Dee said: "If I had a quarter, then so did Carmen. Betty had a dime. None of us had any pennies."

What did each girl buy?

★ ★ 2.23. [H] Prove that if there are more red cards in the top half of an ordinary deck of playing cards than there are black cards in the bottom half, then somewhere in the deck there are seven consecutive cards of the same color. ([41], Vol. 26, p. 167)

★ ★ 2.24. [S] [A] When the first astronaut to visit the planet Mars returned to Earth, he was asked to describe the inhabitants of the "red planet." Still suffering from the effects of interplanetary travel, he answered in the following correct, but confusing, manner.

"It is not true that if Martians are green then they either have three heads or else they cannot fly, unless it is also true that they are green if and only if they can fly and that they do not have three heads."

Assuming that all Martians look alike and that they have at least one of the three characteristics referred to:

Do Martians have three heads?

Are they green?

Can they fly?

★ 2.25. [A] Melvin Muddle was having trouble with his car, so he brought it to his mechanic. After a careful examination of the engine, the mechanic said, "It's hard to be certain, but either it's true that if the spark plugs and the points are O.K. then you need a new distributor cap, or else it's true that if the points are O.K. but you need new spark plugs then your distributor cap is fine, but not both."

Assuming that at least one of the three items mentioned by the mechanic needs replacing, what should Melvin replace?

2.26. Stanley Plumb, Bing Cherry, and Walter Mellin decided to form a recording group—The Three Fruits. Not only do they all have wonderful singing voices, but each also plays either the guitar or the banjo.

If Stanley and Walter can both play the guitar, then so can Bing. If Bing cannot play the guitar, then Walter can; but if Walter plays the banjo, then Stanley does not. Either

Stanley or Bing, but not both, can play the guitar.

Only one of the three can play both the banjo and the guitar. Which one?

2.27. [A] Nat E. Newcomb owns two suits, one blue and one brown. Whenever he wears his blue suit and a blue shirt, he also wears a blue tie. He always wears either a blue suit or white socks. He never wears the blue suit unless he is also wearing either a blue shirt or white socks. Whenever he wears white socks, he also wears a blue shirt.

Today, Nat is wearing a gold tie. What else is he wearing?

2.28. [H] Archie, Brian, and Joaquim own a Ford, a Chevrolet, and a Chrysler, but not necessarily in that order. One car is blue, one is green, and the third is brown.

Archie does not own the Ford, and his car is not blue.

If the Chevy does not belong to Archie, then it is green.

If the blue car either is the Ford or belongs to Brian, then the Chrysler is green.

If the Chevy is either green or brown, then Brian does not own the Ford.

Identify each person's car in terms of make and color.

★ 2.29. [A] When Luigi opened the box from his new toy airplane, a piece of paper fell out. It read as follows:

If this plane is able to fly more than 25 feet high and if this is model #25, then four penlight batteries are required. But if this is model #25, then either the plane is not able to fly more than 25 feet high or else four penlight batteries are not required. On the other hand, if this plane is able to fly more than 25 feet high, then, if four batteries are required, this must be model #25.

Assuming that either it was model #25 or that four penlight batteries were required, could Luigi's plane fly more than 25 feet high?

★ 2.30. [H] [A] Pamela Potter's pease porridge is putrid provided that Pablo Picasso painted potted palms. Either Pablo Picasso painted potted palms, or Peter Piper did not pick a peck of pickled peppers. There are two possibilities: Either Peter Piper picked a peck of pickled peppers or else it is impossible that both Pablo Picasso did not paint potted palms and that Pamela Potter's pease porridge is not putrid.

Is Pamela's porridge putrid?

★ 2.31. [H] Unless Manfred Mensa is not telepathic, it is inconceivable that he is not a marvelous magician. Either electronic devices are not used in Manfred's act, or else, if mind reading is not a form of magic, then Manfred is not a marvelous magician. If Manfred is not wealthy, then either he is not a marvelous magician or else he uses electronic devices in his act. Either mind reading is not a form of magic, or else Manfred is not telepathic.

If Manfred is telepathic, is he wealthy?

★ ★ 2.32 [H] [A] If life begins at eighty and Attila the Hun died at the age of seventy-nine, then Attila the Hun never lived and he was not reincarnated as a snake. However, if life does not begin at eighty and the emotional development of primates parallels that of reptiles, then either Attila the Hun never lived or the female viper is not deadlier than the male. It is well known that unless the emotional development of primates does not parallel that of reptiles, either Attila the Hun never lived or else he died at seventy-nine; it is also true that the emotional development of primates parallels that of reptiles provided that the female viper is deadlier than the male. On the other hand, if the female viper is not deadlier than the male, then it is not possible that either life begins at eighty or that Attila the Hun died at seventy-nine.

Attila the Hun lived.

Did he die at the age of seventy-nine? Does the emotional development of primates parallel that of reptiles? Was Attila the Hun reincarnated as a snake?

3

From Words to Equations:
Algebraic Recreations

Interest in recreational mathematics goes back many centuries. For example, Problem 3.2 below is taken from the *Greek Anthology*, a collection of number problems in epigrammatic form, that was assembled by Metrodorus in about the year 500. According to Eves [14], there is reason to believe that many of these problems are even older in origin. At the time that they were originally proposed, most such problems were solved by trial and error. In fact, there are probably many people today who would still use the hit and miss approach. However, these problems are essentially algebra problems, which can be set up and solved quite simply, as we shall see.

The word algebra is derived from the title of a book on the subject (as it existed at the time), *Hisâb al-jabr w'al-muqâbalah,* which was written by the Persian mathematician Al-Khowarizmi (c. 825). The term al-jabr apparently meant "restoring," as in restoring the balance between the two sides of an equation by subtracting from one side what has been subtracted from another.

The beginnings of the development of symbolic algebraic notation can probably be traced back to Diophantus of Alexandria (c. 275), who introduced abbreviations for some of the quantities and operations that occur most frequently. However, it was not until the sixteenth and seventeenth centuries that the symbolism of today began to come into its own.

Algebra, as it applies to recreational problem solving, has two aspects—the translation of the information of the problem into one or more equations (or inequalities), and the algebraic solution of these equations.

The translation begins when one or more variables—x, y, a, r, or any other letters we deem suitable—are introduced to represent the unknown quantities of the problem.

After appropriate variables have been chosen, they can be used in equations to express the given relationships between the unknown quantities of the problem. These equations may be of any degree and may contain any number of variables. In most of the recreational problems considered in this chapter, the equations obtained are solvable by the techniques of elementary algebra. These techniques, when applicable, enable us to do more than just find a solution. We will be able to determine whether or not solutions exist; and, if there are solutions, we will be able to find all of them. In particular, if there is a **unique solution** (that is, only one solution), then the method will indicate that this is the case.

You are probably familiar with the basic algebraic techniques. If this is not the case, you might want to look at Appendix A for a review of some of these techniques before reading on.

The main difficulties, if there are any, in solving a verbal algebraic problem, usually lie in setting up the equations, that is, in translating from English into symbols. It is this aspect of algebraic problems that is briefly considered in this chapter.

The following Sample Problems are used to illustrate our discussion. Try them first, then read on.

Problem 3.1

The magician said to a volunteer from the audience:

Pick a number, but don't tell me what it is.

Add 15 to it.

Multiply your answer by 3.

Subtract 9.

Divide by 3.

Subtract 8.

Now tell me your answer.

"Thirty-two," replied the volunteer.

At this point, the magician immediately guessed the number that the volunteer had originally chosen.

What was this number, and how did the magician know so quickly?

SAMPLE PROBLEMS

Problem 3.2

Demochares has lived one-fourth of his life as a boy, one-fifth as a youth, one-third as a man, and has spent thirteen years in his dotage.
   How old is the gentleman?

Problem 3.3

Bob, Mary, and Sue all have birthdays on the same day. Bob's present age is two years less than the sum of Sue's and Mary's present ages. In five years, Bob will be twice as old as Mary will be then. Two years ago, Mary was one-half as old as Sue was.
   How old is each of them now?

Problem 3.4

In a Grand Prix automobile race, Speedy Ryder averaged 110 miles per hour for the first half of the course and 130 miles per hour for the second half. Cannonball Carter maintained a constant speed of 120 miles per hour throughout the race.
   Who won the race?

Problem 3.5

Alphonse and Gaston, working together, can complete a job in thirty-six minutes. Working alone, Alphonse would require half an hour more than Gaston would to complete the same job.
   How long would it take Alphonse, working alone, to complete the job?

INTRODUCING VARIABLES

Before we consider the solutions of these problems, let us examine the steps followed in translating a verbal problem into one or more equations. As with any problem presented in words, you must first read it carefully to make sure you understand exactly what information is given and what you are required to find.

   The next step is to introduce letters, such as x, y, z, to represent unknown quantities. These letters are called **variables.** It is frequently a good idea to choose variables that are reminiscent of the quantities they represent. For example, we may use n to represent a *n*umber, b to represent *B*ob's age, a to represent a number of *a*pples, and so on.

   Many algebra problems involve more than one unknown. You may introduce a variable for each unknown quantity. Be careful, though, not to use the same letter to represent two different quantities, unless you know that these quantities are equal to each other. Sometimes you may be able to reduce the number of variables by making use of known relationships to represent some of the unknowns in terms of others. For example, if b represents Bob's current age, then Bob's age five years from now may be represented by $b + 5$, his age ten years ago may be represented by $b - 10$, and his age when he is twice as old as he is now may be represented by $2b$.

1. $\boxed{A}$ Each of the following refers to two quantities. Introduce a variable to represent the first of these quantities, and then represent the second quantity in terms of this same variable.
   (a) A number; the number obtained by adding 15 to the original number.
   (b) A number; three less than twice the number.
   (c) A number; the product of the number and two less than the number.
   (d) Demochares' age; one-fifth of his age.
   (e) The number of minutes it takes Alphonse to do a job; how much—what portion—of the job Alphonse can do in one minute.

2. Each of the following refers to three quantities. Introduce variables to represent the first two of these quantities, and then represent the third quantity in terms of the variables used for the first two.
   (a) Sue's age; Mary's age; the sum of Sue's and Mary's ages.
   (b) Sue's age; Mary's age; the difference between their ages five years from now. (Sue is older than Mary.)
   (c) Sue's age; Mary's age; half of Sue's age two years ago.
   (d) The number of minutes it takes Alphonse to do the job; the number of minutes it takes Gaston to do the job; how much— what portion—of the job they can do in one minute if they work together.

PRACTICE PROBLEMS 3.A

We are now ready to present the solution of Problem 3.1 (page 71). The trick lies in using algebra to keep track of the volunteer's number at each step in the procedure.

If you were not able to solve this problem previously, try it again now.

Solution of Problem 3.1

Since we don't know what number the volunteer chose, we represent it by some variable, say n. Then we can keep track of the volunteer's number at each stage by making a chart.

Magician says	Volunteer obtains
Pick a number	n
Add 15	$n + 15$
Multiply by 3	$3(n + 15) = 3n + 45$
Subtract 9	$3n + 36$
Divide by 3	$n + 12$
Subtract 8	$n + 4$

The final answer obtained by the volunteer is $n + 4$, four more than the number originally chosen. Therefore, all the magician has to do is subtract four from the answer announced by the volunteer. That is,

$$n + 4 = 32$$
$$n = 28.$$

After each of the unknown quantities of a problem has been represented in terms of variables, the next step is to use the clues of the problem to symbolically express the relationship(s) between these quantities by means of equations or inequalities. Note that some of these relationships may have already been used in representing the quantities symbolically.

As an illustration consider Problem 3.2 (page 72).

Try it again now if you haven't already solved it.

Solution of Problem 3.2

The key relationship between the quantities in this problem is:

Years as a boy + Years as a youth + Years as a man

+ Years in his dotage = Demochares' current age.

We can proceed in two ways:

Let d = Demochares' current age
b = the number of years he spent as a boy

Let d = Demochares' age. Then, from the statement of the problem, we may represent the number of years he spent as a

y = the number of years he spent as a youth

m = the number of years he spent as a man.

Then, the given relationships of the problem yield the following equations

$$b = \tfrac{1}{4}d$$

$$y = \tfrac{1}{5}d$$

$$m = \tfrac{1}{3}d$$

$$b + y + m + 13 = d.$$

This is a system of four equations in four unknowns. We may solve it by substituting for b, y, and m in the fourth equation to obtain

$$\tfrac{1}{4}d + \tfrac{1}{5}d + \tfrac{1}{3}d + 13 = d.$$

Solving this equation,

$$13 = d - \tfrac{1}{4}d - \tfrac{1}{5}d - \tfrac{1}{3}d = \tfrac{13}{60}d.$$

Hence, $d = 60$.

boy by $\tfrac{1}{4}d$, the number of years he spent as a youth by $\tfrac{1}{5}d$, and the number of years he spent as a man by $\tfrac{1}{3}d$.

The key relationship of the problem may be expressed as the equation

$$\tfrac{1}{4}d + \tfrac{1}{5}d + \tfrac{1}{3}d + 13 = d.$$

Solving this equation,

$$13 = d - \tfrac{1}{4}d - \tfrac{1}{5}d - \tfrac{1}{3}d = \tfrac{13}{60}d.$$

Hence, $d = 60$.

Note that both approaches to the problem eventually come down to the solution of the same equation: $\tfrac{1}{4}d + \tfrac{1}{5}d + \tfrac{1}{3}d + 13 = d$. Since this equation has a unique solution, we know that there is a unique solution to the problem. That is, Demochares was 60, he spent 15 years as a boy, 12 years as a youth, 20 years as a man, and 13 years in his dotage.

Observe that although the second approach to Problem 3.2 is slightly shorter, both approaches to the problem are equally correct. The difference lies in the work saved by representing all of the unknowns in terms of one variable (using three of the relationships of the problem in the process). But, in any problem, if you do not see how to cut down on the number of variables you must introduce, there is still nothing to worry about. You cannot go wrong by introducing too many variables; the worst that can happen is that the resulting system of equations will be more complicated than it might otherwise have been.

Sometimes, it is not completely clear how to express some of the relationships between the variables in terms of equations. One technique that is often helpful is to write each sentence in words and, underneath, to translate each part into its symbolic form. For example, suppose B and M represent Bob's and Mary's current ages respectively; then the sentence "In five years, Bob will be twice as old as Mary will be" may be translated as follows:

In five years, Bob will be twice as old as Mary will be

$$B + 5 \qquad = \qquad 2 \qquad (M + 5)$$

Observe that the words *will be, was, are, is,* and so on are often represented by " = ."

If you are unsure whether you need to add, subtract, or perform some other arithmetical operation, try substituting specific values for the unknowns to see what effect each operation would have. For example, in translating "Bob's age is two less than the sum of Sue's and Mary's ages,"

Bob's age is two less than the sum of Sue's and Mary's ages

$$B \qquad = \qquad ? \qquad S + M$$

on which side of the equation does the " -2 " for "two less than" belong? Try an example: If Sue is five and Mary is three, then Bob clearly would be six.

$$6 = (5 + 3) - 2 \qquad\qquad (\text{not } 6 - 2 = 5 + 3)$$

Thus $\quad B = S + M - 2 \qquad$ (rather than $B - 2 = S + M$).

PRACTICE PROBLEMS 3.B

1. $\boxed{\text{A}}$ Write the equation or inequality obtained by translating each of the following sentences into symbols:
   (a) A number is equal to three less than its square.
   (b) The square of a number is twelve more than the number.
   (c) One more than a number is less than or equal to one-fourth of the square of the number.
   (d) Felippe can seal more envelopes in an hour than can Sean and Misha together.

2. Write the equation or inequality obtained by translating each of the following sentences into symbols:
   (a) The difference between the cube of a number and the number is six less than ten times the number.
   (b) Two years ago, Mary was half as old as Sue was.
   (c) Six apples and five pears weigh more than seven pears and five apples.
   (d) One half of Pedro's sheep added to one-third of his horses is equal in number to half of his horses minus half his sheep.

We are now ready to solve Problem 3.3 (page 72). Try it again yourself, before reading on.

Let B, M, and S represent the present ages of Bob, Mary, and Sue respectively. We have already expressed the first two clues of the problem as equations:

$$B = S + M - 2$$

$$B + 5 = 2(M + 5)$$

The third clue gives:

$$\underbrace{\text{Two years ago, Mary}}_{M - 2} \underbrace{\text{was}}_{=} \underbrace{\text{one-half}}_{\frac{1}{2}} \text{ as old as } \underbrace{\text{Sue was}}_{S - 2}$$

$$M - 2 = \tfrac{1}{2}(S - 2)$$

Solution of Problem 3.3

We have obtained a system of three equations in three unknowns. This system can be solved either by the method of substitution or by the method of elimination. We present both methods of solution below (also see Appendix A).

Method of substitution

(1) $B = S + M - 2$
(2) $B + 5 = 2(M + 5)$
(3) $M - 2 = \tfrac{1}{2}(S - 2)$

Using (1) to substitute for B in (2), we obtain

$$(S + M - 2) + 5 = 2(M + 5).$$

Simplifying,

$$S + M + 3 = 2M + 10$$

(4) $S = M + 7.$

Using (4) to substitute for S, (3) becomes

$$M - 2 = \tfrac{1}{2}((M + 7) - 2).$$

Simplifying

$$M - 2 = \tfrac{1}{2}(M + 5)$$

$$2M - 4 = M + 5$$

$$M = 9.$$

Now, using (4),

$$S = M + 7 = 9 + 7 = 16$$

Method of elimination

We rewrite the equation in the form

(1′) $B - S - M = -2$
(2′) $B \quad\ - 2M = 5$
(3′) $\quad - S + 2M = 2$

Subtract (1′) from (2′). The system becomes

(1′) $B - S - M = -2$
(2″) $\quad\quad S - M = 7$
(3′) $\quad - S + 2M = 2$

Now add (2″) to (3′):

(1′) $B - S - M = -2$
(2″) $\quad\quad S - M = 7$
(3″) $\quad\quad\quad\quad M = 9$

From (3″), $M = 9$. Then from (2″),

$$S = M + 7 = 9 + 7 = 16.$$

and using (1),   Finally, from (1′),

$$B = S + M - 2$$ $$B = S + M - 2$$

$$= 16 + 9 - 2 = 23.$$ $$= 16 + 9 - 2 = 23.$$

Thus, Bob's age is 23, Sue's is 16, and Mary's is 9.

Note again that the algebraic method allows us to conclude that the solution is unique.

Both Problems 3.2 and 3.3 are examples of age problems, one of the many types of verbal problems. Most types of verbal problems primarily require the ability to set up and solve equations. Some types, however, require a bit of outside knowledge such as elementary formulas from geometry or the laws of motion. For example, uniform motion problems (often called rate-time-distance problems) require knowledge of the fact that rate = distance divided by time or distance = rate multiplied by time.

Problem 3.4 (page 72) is such a problem. If you haven't already solved it, try it again now.

Solution of Problem 3.4

Chances are that if you have never solved a problem like this before, then your first impression was that the race was a tie.

But before you jump to conclusions, consider a slight variation of the problem. Suppose Cannonball maintained a constant speed of 120 miles per hour throughout the race but that Speedy averaged 60 miles per hour for the first half of the course and 180 miles per hour for the second half. Would it still seem that the race was a tie? Suppose further that the course was 120 miles long. Then Cannonball finished after one hour; but, at the end of that first hour, Speedy was just finishing the first half of the course. Clearly the race was not a tie. But why not? Certainly 120 is the average of 60 and 180. The answer is that Speedy spent more time at the lower speed than he did at the higher speed. To finish the first half of the course at 60 miles per hour took Speedy 1 hour; but to finish the second half at 180 miles per hour required only $\frac{1}{3}$ hour = 20 minutes. Thus Speedy spent more time at the lower speed than at the higher speed, and so his average speed was less than 120 miles per hour.

The same type of reasoning applies in general. The amount of time each driver spent at each speed is important. This becomes clear if we ask ourselves what it means for one racer to win the race. It means that he took less time to complete the course than the other racer did. This observation enables us to attack the problem algebraically, in terms of inequalities.

In the Speedy versus Cannonball race, suppose the total distance for the course is d miles.

Let t_c represent the number of *hours* it takes Cannonball to complete the course,

and

let t_s represent the number of hours it takes Speedy. Then, using the formula

$$\text{rate} \cdot \text{time} = \text{distance}$$

or, equivalently

$$\text{time} = \frac{\text{distance}}{\text{rate}}$$

we find that

$$t_c = \frac{d}{120}.$$

To find t_s, we consider the two parts of the course during which Speedy's speed was constant. That is,

$$t_s = t_1 + t_2,$$

where t_1 is the time required for Speedy to complete the first half of the course and t_2 is the time required for him to complete the second half. Since each half is $d/2$ miles,

$$t_1 = \frac{d/2}{110} = \frac{d}{220} \qquad t_2 = \frac{d/2}{130} = \frac{d}{260}$$

and

$$t_s = t_1 + t_2 = \frac{d}{220} + \frac{d}{260} = \frac{d}{20}\left(\frac{1}{11} + \frac{1}{13}\right) = \frac{d}{20}\left(\frac{24}{143}\right) = \frac{6d}{715}.$$

The question now is, who completed the course first? Is

$$t_s < t_c \qquad \text{or is} \qquad t_s > t_c \qquad \text{or does} \qquad t_s = t_c\,?$$

Since

$$\frac{1}{120} = \frac{6}{720} < \frac{6}{715}, \qquad \frac{d}{120} < \frac{6d}{715}.$$

That is, $t_c < t_s$, and so Cannonball won the race.

There are several important aspects of the solution above. The first is that we used t_c, t_s, t_1 and t_2 to represent four different times. It wasn't necessary to do this: We could have chosen c, s, b, and e (or any four other letters) to represent Cannonball's time, Speedy's total time, Speedy's beginning time (for the first half of the course) and Speedy's ending time (for the second half); but we decided to use the mnemonic t to help indicate that we were dealing with periods of time. As we could not represent four different times by the exact same symbol t, we introduced the **subscripts** c, s, 1, 2 so that we could differentiate between the different times. The symbol t_c is read "t sub c" or simply "t c"; the symbol t_1 is read "t sub 1" or "t one"; and so on. In general, the decision about whether to use subscripts or different letters to represent the unknowns of a problem is up to you; we recommend that subscripts be used when several of the unknowns in a problem represent the same kinds of quantities (several time periods, several distances, etc.).

The second important observation in the solution of Problem 3.4 relates to the notion of **dimensional consistency**. When we introduced the variable t_c to represent Cannonball's time, why did we let t_c represent a number of hours rather than a number of minutes or seconds? The answer is that since distance was being measured in miles and rate was being measured in miles per hour, then time should be measured in hours; that is,

$$\text{time} = \frac{\text{distance}}{\text{rate}} = \frac{\text{miles}}{\text{miles/hour}} = \text{hours}.$$

In general, in any equation involving physical measurement, you must make sure to maintain consistency. If you measure time in hours in one part of a problem, then you must measure it in hours throughout. This may require conversion from one unit to another (for example, writing $\frac{1}{3}$ hour rather than 20 minutes, or $\frac{1}{2}$ mile rather than 2640 feet). Similarly, as in the solution above, the units used for one quantity may determine the units to be used for another. For example, if rate is measured in feet per second, then distance should be measured in feet and time should be measured in seconds.

Dimensional consistency can sometimes be used to help check the reasonableness of an equation. Thus, if you write an equation in which hours are added to miles, then something is wrong. Only quantities expressed in units of the same type can be added to each other. Similarly, the units on both sides of an equation must balance. For example, in the formula rate · time = distance

$$\frac{\text{miles}}{\text{hour}} \cdot \text{hours} = \text{miles},$$

so that both sides of the equation are expressed in terms of miles. Thus, if the units on one side of an equation result in an expression in terms of miles, then the other side of the equation must also be a number of miles and not, for example, a number of miles per hour.

The solution of Problem 3.4 also leads to a third consideration. What was the key to being able to attack the problem algebraically? In this problem, it was the realization that what had to be compared were the times of the two racers. In general, you must ask yourself the question, "What kind of quantities must I consider in this problem in order to get a meaningful equation or inequality?"

For example, in attacking Problem 3.5, there is a tendency to proceed as follows:

Let A = the amount of time (in minutes) that it would take Alphonse, working alone, to complete the job

and

let G = the amount of time it would take Gaston alone.

Then

$$A + G = 36 \text{ minutes.}$$

But is this really meaningful? Is the sum of Alphonse's time and Gaston's time equal to the time it takes the two of them together? Certainly not. Together, it should take them less time than it takes either of them working alone. In this problem, then, it is not meaningful to add times. But in that case, how can we approach the problem? What kinds of quantities should we compare? How can we represent the fact that they are working together? How about adding the amount of work they each do in the same time period?

Think about this suggestion. Then, if you haven't already solved Problem 3.5 (page 72), try it again.

Upon rereading the problem, we observe that the information in one of the clues is stated in terms of minutes and that the information in the other clue is stated in terms of hours. To maintain dimensional consistency, we must choose one of these units and stick to it. Which we choose is a matter of personal preference. The problem can be solved either way, and each approach has its own advantages:

Solution of Problem 3.5

Let A = the number of minutes it would take Alphonse, working alone, to complete the job; and let G = the number of minutes it would take Gaston working alone.	Let a = the number of hours it would take Alphonse, working alone, to complete the job; and let g = the number of hours it would take Gaston working alone.

How much of the job could Alphonse do in one minute?

Since it takes him A minutes to do the whole job, he does $\frac{1}{A}$ of the job in one minute.

Similarly, Gaston does $\frac{1}{G}$ job in one minute.

Together, they do $\frac{1}{A} + \frac{1}{G}$ job in a minute. (Note, here it is meaningful to add.)

Therefore, in 36 minutes they do $36\left(\frac{1}{A} + \frac{1}{G}\right)$ job.

But we are told that they do the whole job in 36 minutes. Thus

$$36\left(\frac{1}{A} + \frac{1}{G}\right) = 1$$

or, equivalently,

(1) $\frac{1}{A} + \frac{1}{G} = \frac{1}{36}$.

How much of the job could Alphonse do in one hour?

Since it takes him a hours to do the whole job, he does $\frac{1}{a}$ of the job in one hour.

Similarly, Gaston does $\frac{1}{g}$ job in one hour.

Together, they do $\frac{1}{a} + \frac{1}{g}$ job in an hour. (Note, here it is meaningful to add.)

Observe that 36 minutes $= \frac{3}{5}$ hour.

Therefore, in 36 minutes, they do three-fifths of what they can do in an hour—that is,

$$\frac{3}{5}\left(\frac{1}{a} + \frac{1}{g}\right) \text{ job.}$$

Since we are told that they do the whole job in 36 minutes,

(1') $\frac{3}{5}\left(\frac{1}{a} + \frac{1}{g}\right) = 1$.

We are also told that Alphonse, working alone, requires one-half hour (30 minutes) more than Gaston does. Hence,

(2) $A = G + 30$.

Using (2) to substitute for A in (1), we get

$$\frac{1}{G + 30} + \frac{1}{G} = \frac{1}{36}.$$

Combining over a common denominator,

$$\frac{G}{G(G + 30)} + \frac{G + 30}{G(G + 30)} = \frac{1}{36}$$

$$\frac{2G + 30}{G(G + 30)} = \frac{1}{36}.$$

(2') $a = g + \frac{1}{2}$.

Using (2') to substitute for A in (1'), we get

$$\frac{3}{5}\left(\frac{1}{g + \frac{1}{2}} + \frac{1}{g}\right) = 1.$$

Simplifying,

$$\frac{3}{5}\left(\frac{2}{2g + 1} + \frac{1}{g}\right) = 1$$

$$\frac{3}{5}\left(\frac{2g}{g(2g + 1)} + \frac{2g + 1}{g(2g + 1)}\right) = 1$$

Cross-multiplying,

$$36(2G + 30) = G(G + 30)$$

$$72G + 1080 = G^2 + 30G$$

$$G^2 - 42G - 1080 = 0.$$

This may be solved by factoring or by using the quadratic formula:

$$(G - 60)(G + 18) = 0$$

$$G = 60, G = -18$$

(Or, using the formula,

$$G = \frac{42 \pm \sqrt{(42)^2 - 4 \cdot 1 \cdot (-1080)}}{2}$$

$$= \frac{42 \pm \sqrt{1764 + 4320}}{2}$$

$$= \frac{42 \pm \sqrt{6084}}{2}$$

$$= \frac{42 \pm 78}{2}$$

$$= \frac{42 + 78}{2} \text{ or } \frac{42 - 78}{2}$$

$$= 60 \text{ or } -18.)$$

$$\frac{3}{5}\left(\frac{4g + 1}{g(2g + 1)}\right) = 1$$

$$3(4g + 1) = 5g(2g + 1)$$

$$12g + 3 = 10g^2 + 5g$$

$$10g^2 - 7g - 3 = 0.$$

This may be solved by factoring or by using the quadratic formula:

$$(g - 1)(10g + 3) = 0$$

$$g = 1, g = -\tfrac{3}{10}$$

(Or, using the formula,

$$g = \frac{7 \pm \sqrt{7^2 - 4 \cdot 10 \cdot (-3)}}{2 \cdot 10}$$

$$= \frac{7 \pm \sqrt{49 + 120}}{20}$$

$$= \frac{7 \pm \sqrt{169}}{20}$$

$$= \frac{7 \pm 13}{20}$$

$$= \frac{7 + 13}{20} \text{ or } \frac{7 - 13}{20}$$

$$= 1 \text{ or } -\tfrac{3}{10}.)$$

The negative answer is **extraneous** (it satisfies the equations but does not satisfy the conditions of the problem) since a negative time does not make sense. It may therefore be discarded.

Thus $G = 60$, and so

$$A = G + 30 = 90$$

That is, Alphonse, working alone, could do the job in 90 minutes.

Thus $g = 1$, and so

$$a = g + \tfrac{1}{2} = 1\tfrac{1}{2}$$

That is, Alphonse, working alone, could do the job in an hour and a half.

Note that both methods lead to the same answer (90 minutes = $1\tfrac{1}{2}$ hours).

Each method has some minor disadvantages. The first results in an equation with relatively large numbers, so that factoring or taking the necessary square root is more difficult; the second requires more work with fractions, but, overall, seems to result in easier computations. But whichever units you would choose, the mathematical steps in the solution of the problem are basically the same.

THE CHAPTER IN RETROSPECT

In this chapter, you have gained experience in translating from words to equations. Even though your experience was won by solving problems of recreational mathematics, the techniques of this chapter are useful in many fields. Algebraic problems arise in many settings— business, economics, social science, and others (see, for example, Exercise 3.6). Once you recognize that a problem is algebraic in nature, the basic steps for solving it are the same, regardless of the setting:

1. Represent the unknowns by variables.

2. Set up the equations relating these variables. Be consistent with the dimensions (in fact, the dimensions can sometimes give a clue as to how to set up the equation).

3. Solve the equations or show that there is no solution.

In this chapter, we have considered only problems that can be solved by elementary methods. Sometimes a problem can be set up algebraically, but the method of solution is not obvious or not known. Then more mathematics might be needed (some examples of this type will be considered in Chapter 4). In general, though, the algebraic approach is very useful.

Exercises

Problems Involving One Variable

3.1. The union election was quite close. Two-fifths of the votes went to Millie, five-twelfths of the votes went to Jacob, and the remaining 33 votes were cast for Harriet. How many votes were cast in all?

3.2. [S] [A] To celebrate their wedding anniversary, Derek and Lucy decided to spend the night out on the town. After dinner, which incidentally cost them one-third of the money they had brought with them, they purchased two theater tickets at $9.80 each. After the show, they spent one-fourth of their remaining bankroll for a taxi ride around the city. They then went to a night club, where their total tab came to $23.10. The cab fare home cost them one-half of their remaining money. After they had tipped the driver a dollar, they had $4.10 left.

How much did their celebration cost them?

3.3. A local charity had set a goal of $70,000 in its fund drive. This morning, when I was solicited for a donation, I asked how the drive was doing. I was told that one-third of the amount that has already been collected is equal to three-fifths of the amount still needed.

How much money is still needed?

★ **3.4.** ⬚H⬚ ⬚A⬚ Mildred, Ethel, Samuel, and Leonard purchased a bag of candy which they divided as follows:

First, Samuel took one candy plus one-third of the candy remaining. Next, Mildred took one candy plus one-third of the candy then remaining. Then Ethel took one candy plus one-third of the candy then remaining. Leonard received the remainder of the candy.

If Mildred and Ethel together received seven pieces of candy more than Samuel, how many pieces of candy did Leonard receive?

3.5. ⬚S⬚ ⬚A⬚ Two mendicants decided to spend their day's proceeds to buy some Muscatelle. Tom was able to buy seven bottles, while Don could afford only five. Just as they were about to begin, Mack showed up and asked to join them. He offered to pay $8.40. If all three men shared the wine equally, how much of the $8.40 should Tom receive?

3.6. Ann Mann has decided to join her brothers Stan and Dan in a business venture.

They plan to form a trucking company, with twelve trucks in all. Stan is contributing seven vans and Dan five. Ann has agreed to pay her brothers $120,000 in compensation. How should Stan and Dan divide this money?

Quadratic Equations in One Variable

3.7. ⬚H⬚ Farmer Gray bought a number of sacks of wheat seed, three times as many sacks of corn seed, and six times as many sacks of soybean seed (as of wheat). Coincidentally, the price per sack of each kind of seed was the number of sacks of that kind of seed that Farmer Gray bought.

If Farmer Gray spent a total of $184, how many sacks of seed did he buy?

3.8. ⬚A⬚ Farmer Gray owns two rectangular fields of the same area. One is 700 yards longer than it is wide; the second is 450 yards shorter than the first, and 400 yards wide.

What are the dimensions of each field?

3.9. The floor of a square room is to be tiled according to the pattern in Figure 3.1.

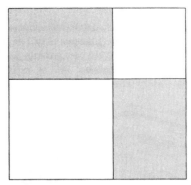

FIGURE 3.1

Both white sections are themselves square and the larger white square has exactly eight tiles more on each side than the smaller one.

If 1000 white tiles are needed in all, how many gray tiles are required?

★ **3.10.** S A Assuming that there is a unique number x that may be expressed in the following form:

$$x = \sqrt{1 + \sqrt{1 + \sqrt{1 + \cdots}}},$$

where all roots are positive square roots and where ... means "and so on, ad infinitum," find x.

★ **3.11.** H A Assuming that there exists a unique number x that may be expressed in the following form:

$$x = 2 + \cfrac{15}{2 + \cfrac{15}{2 + \cfrac{15}{2 + \cdots}}},$$

find the value of x.

Systems of Equations

3.12. A If a towel and a washcloth together cost $2.20, and the towel costs $1.20 more than the washcloth, how much does the washcloth cost?

3.13. A Whenever a cattle drive crosses Solly Shepherd's land, he charges a toll of 10¢ per riderless animal and 25¢ per cowboy and horse combination. Yesterday, as a drive was passing, Solly counted a total of 4248 legs

(including riders, horses, and cattle) and 1078 heads. How much money did Solly collect?

3.14. H A Calvin Carpenter needed screws of three different sizes. He bought the same number of screws of each size at $.03, $.04, and $.05 per screw respectively.

Wendy Woodall spent the same amount of money as Calvin, but divided her money equally among screws of the three sizes. In all, Wendy purchased six more screws than Calvin did. How much money did they each spend?

3.15. S A At the weekly meeting of Plumps Anonymous, only six members—Ada, Brendan, Corinne, Darryl, Eva, and Floyd—flunked the weigh-in; they had not lost the prescribed number of pounds.

Ada and Brendan together tipped the scales at 647 lb.

Brendan and Corinne together weighed 675 lb.

The sum of Corinne's and Darryl's weights was 599 lb.

Darryl's and Eva's weights totaled 583 lb.

If Floyd weighed 370 lb and the sum of the weights of all six was 1927 lb, how much did each weigh?

3.16. Because each of the people in Exercise 3.15 had not lost enough weight, each had to pay a fine. Ada and Brendan together paid $1.80; Brendan and Corinne paid $4.30; Corinne and Darryl paid $4.00; Donald and Eva paid $3.80; Eva and Floyd paid $3.10; Ada and Eva paid $2.80.

How much was each fined?

3.17. A Minski, Pinski, Dubinski, and their wives went shopping. Each of the three couples spent a total of $51. The sum of the amounts spent by Minski, Pinski, and the wife of Dubinski was $43. Minski, Dubinski, and the wife of Pinski together spent $93. Pinski, Dubinski, and the wife of Minski spent $81.

How much did each spend?

3.18. [A] The paper currency of the Kingdom of Bonoria bears the pictures of the country's monarchs. The one-bonor note carries the picture of Queen Griselda the Good. Other notes not exceeding 100 bonors bear the pictures of King Randolph the Rotten, Queen Carrie the Charming, Queen Bonita the Beautiful, King Gerald the Gross, King Waldo the Wicked, and King Hilary the Hairy.

Together the notes on which Bonita's and Carrie's pictures appear are worth 102 bonors. One Gerald, Waldo, and Randolph together give 73 bonors. Hilary and Bonita give 22 bonors. Hilary and Carrie give 120 bonors. Hilary, Gerald, and Randolph give 43 bonors. And Carrie, Waldo, and Randolph give 168 bonors.

On what size note does Waldo's picture appear?

★ **3.19.** [A] The silver currency of the Kingdom of Bonoria (see Exercise 3.18) consists of glomeks, nindars, and morms. Four glomeks are equal in value to seven nindars; and one glomek and one nindar together are worth thirty-three morms.

On my last visit to Bonoria, I entered a bank, handed the teller some glomeks and nindars, and asked him to change them into morms.

"Do you think that I am a magician?" he replied. (Bonorians are noted for their warped sense of humor.) "Well, let's see," he continued. "If you had twice as many glomeks, I could give you 120 morms; and if you had twice as many nindars I could give you 114 morms."

How many morms did he give me?

3.20. A merchant borrowed 120 dinars to go into business selling baubles, bangles, and beads. Each item she purchased cost her 1 dinar. She stocked some of each item, but kept some of the borrowed money in reserve for emergencies (such as bribing the palace guard, etc.). Had she spent all of her reserve money on baubles, then she would have had as many baubles as she had bangles and beads together; had she spent all of her reserve money on bangles, then she would have had twice as many bangles as she had baubles and beads; had she spent all of her reserve money on beads, then she would have had four times as many beads as baubles and bangles together.

How many of each item did she purchase?

3.21. [A] The Fastbinder Family Circle decided to have a catered get-together this year. The cost was to be divided equally among all of the individuals attending. If everyone that was expected had shown up, the cost per person would have been $8. As it was, eight people canceled out at the last minute, so each person's share had to be raised to $9.

How many people attended the get-together?

3.22. [A] When the wagon train left Fort McConnell and headed West, each of the wagons contained the same number of people. By the time the caravan had crossed the Mississippi River, eight wagons and four people had unfortunately been lost, so that there were now exactly two more people per wagon than there were when they started.

Crossing the Plains, they lost three more wagons and eight more people during an Indian attack. But one baby was born on the way, so that, when the wagon train arrived in California, each wagon carried three more people than it did at first.

How many of the people who left Fort McConnell with the train survived the trip?

Age Problems

3.23. My age is the difference between twice my age four years hence and twice my age four years ago. How old am I?

3.24. Two years ago, I was three times as old

as I was eight years before I was half as old as I will be six years from now. How old am I?

3.25. [S] [A]   Sue Ling is three times as old as Chin Lee was when Sue was as old as Chin is now. When Chin is as old as Sue is now, Sue will be 56. How old are Sue and Chin now?

★ **3.26.** [H] [A]   Mutt is half as old as Jeff will be when Mutt is twice as old as Jeff was when Mutt was half as old as Jeff is now. In five years, the sum of Mutt's and Jeff's ages will be 100.

How old are Mutt and Jeff now?

★ **3.27.**   When Erica was two years old, Leroy was four times as old as Miriam. When Miriam was twice as old as Erica, Leroy was three times as old as Miriam. How old was Erica when Leroy was twice as old as Miriam?

★ **3.28.** [H] [A]   The butcher is twice as old as he was when the baker was half as old as was the candlestick maker. The candlestick maker is twice as old as the butcher was when the baker was half as old as the butcher is now. Twelve years ago the baker was half as old as the butcher will be twelve years from now.

How old is each of them now?

★ **3.29.**   My nephew Seymour has four pet hamsters—Norma, Izzy, Susie, and John—whose ages he measures in months. Norma is three times as old as John. In three months Izzy will be as old as the other three combined.

In five months Izzy will be three times as old as Susie will be. When Izzy is twice as old as Norma, Susie will be one and one half times as old as John.

How old are the hamsters now?

3.30.   Lee is two years older than Brenda. The product of their ages is 575. How old are they?

Uniform Motion Problems

3.31. [A] When the tortoise raced the hare, the former maintained a constant pace of one mile per hour throughout the race, while the latter, being overconfident, wasted much time and averaged only a half mile per hour for the first half of the course.

How fast must the hare run over the second half of the course in order to win?

3.32. [S] [A]   Two trains are traveling toward each other on adjacent tracks. The first train is 270 feet long and is moving at 45 miles per hour. The second train is 192 feet long and is moving at 60 miles per hour.

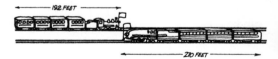

How much time elapses from the moment the trains first meet until they completely pass each other? (Note that 1 mile per hour = 88/60 feet per second.)

3.33. [H] [A] At 8:00 AM a train leaves Topeka for Santa Fe and another train leaves Santa Fe for Topeka. The trains maintain constant speeds with no stops. The first train requires five hours to complete the trip, and the second train requires seven hours.

At what time do the trains pass each other?

★ 3.34. [A] A stagecoach leaves Deadwood headed for Tombstone at the same instant that another coach leaves Tombstone for Deadwood. They meet along the way at a point which is 24 miles closer to Deadwood than it is to Tombstone. Since each driver wants to return to the town from which he left, the drivers trade coaches. That is, the driver from Deadwood takes the coach from Tombstone on to Deadwood, completing the trip in nine more hours; and the driver from Tombstone takes the coach from Deadwood on to Tombstone, arriving 16 hours after the changeover.

Assuming that each coach travels at a constant speed throughout the entire trip (ignore the few minutes required for the changeover of the drivers), what is the distance from Deadwood to Tombstone?

3.35. [H] At 8:00 AM, a train leaves Washington, D.C., bound for Miami. At the same moment, another train leaves Miami headed for Washington.

Each train maintains a uniform speed throughout its trip and makes no stops until it arrives at its destination.

If the trains pass each other at 5:00 PM, and

the train from Miami arrives in Washington at 11:00 PM, at what time does the train from Washington arrive in Miami?

3.36. [A] A ferry is crossing the English Channel from Dover to Calais at a speed of 6 knots (nautical miles per hour). A second boat is crossing from Calais to Dover at a speed of 10 knots.

How far apart are the two boats one hour before they pass each other?

3.37. [H] [A] Two driverless trains race toward each other on the same track; one is traveling at a constant speed of 60 miles per hour and the other at 70 miles per hour. A fly starts at the front of one train and flies toward the second train. When he reaches it, he turns around and flies back to the first train, whence he turns around, and so on. The fly maintains a constant speed of 110 miles per hour, losing no time in changing direction.

If the trains are now 65 miles apart, how far will the fly fly before he is crushed when the trains collide?

3.38. [S] [A] Two cyclists are holding a race on a straight track. They begin at opposite ends of the track. Each races to his opponent's end, turns around, races back, turns around, and so on. The winner is the cyclist who first overtakes his opponent.

The race begins. The cyclists first pass each other (in opposite directions) at a point 150 yd from the northern end of the track. They next pass each other (in opposite directions) at a point 100 yd from the southern end.

If they maintain constant speeds and lose no time in turning

(a) How long is the track?

(b) Who will win the race?

(c) At what point will the winner overtake the loser?

★ ★ **3.39.** H A One of the ski lifts at Snow-flake Mountain climbs the mountain in a line that parallels one of the ski runs. Both the run and the lift are two miles long, and a chair passes the starting point of the lift every 10 seconds.

A skier starts her run just as a chair arrives at the top of the lift and another chair starts back down. She arrives at the bottom just as a chair is starting up the mountain and another chair is completing its descent. Counting these two chairs and the two at the start of her run, she sees 97 chairs on their way up the mountain and 61 chairs on their way down.

If the chair that started down the slope at the same time that the skier did is still on its way down when she reaches the bottom of the slope, what is her average speed?

3.40. Jeremiah Jordan wanted to test the power of his new outboard motor. He took his boat out on the Mississippi River and noted that, with the throttle wide open, he could go

5 miles upstream in 9 minutes and he could go 5 miles downstream in 7 minutes.

How many minutes would it take him to go 5 miles if there were no current?

3.41. H A As Rosalind Rowe rowed upstream on the Raritan River, she noticed a piece of driftwood floating by. She continued rowing upstream for half an hour, when she suddenly decided that her husband would find the driftwood quite decorative. Rosalind quickly turned around, headed downstream after the ornament, and caught up to it 2 miles downstream of the spot at which she originally saw it.

Assuming that Rosalind maintained a constant rate of rowing throughout, how fast was the current in the river?

★ **3.42.** S A Just as Liz and Irv stepped off the express train, they could hear the local pulling in to the upper platform of the station. They quickly ran up the escalator, Irv taking three steps for each two that Liz took. Unfortunately, the doors of the local closed just as Irv reached the top step of the escalator.

Since they had to wait for another local to arrive, Liz and Irv had time to reflect about the escalator. Liz remarked that it had taken her 24 steps to reach the top; Irv noted that it had taken him 30. Assuming that Liz and Irv each climbed the escalator at a constant rate, how many steps would be visible if the escalator stopped running?

★ **3.43.** H A The absentminded professor walked up the escalator in the department store at a constant rate of one step per second. Just as he reached the top, he realized that he had left his briefcase on the counter where he had just purchased a pocket calculator. Fortunately, no one else was on the up escalator, so the professor decided to run down. He managed three steps per second on his descent. After he retrieved his briefcase, the professor decided to repeat the procedure. He noted that he took 18 steps on the way up and 90 steps on the way down.

How many steps of the escalator are visible at one time?

★ ★ **3.44.** [S] [A] At 4:15 PM, things looked bleak for supercilious Pilious Flog and his three traveling companions. According to the terms of their bet, they must all arrive at Winchester Cathedral by 6:00 PM; but they were still 19.3 kilometers from their destination and could only sustain a jog of 4 kilometers per hour.

Just then, Flog flagged down a passing motorcycle, which had room enough for one passenger. Unfortunately, none of the four men could drive, but Pilious was able to enlist the aid of the motorcyclist. Three of the men would continue jogging while the motorcyclist would drive one of the men part of the way to the cathedral. That man would continue on foot, and the driver would return for another of the voyagers, to whom he would then give a ride until they caught up with the first passenger, and so on.

By the time these arrangements were completed, it was 4:20 PM.

Assuming that the motorcycle travels at the rate of 56 kilometers per hour and that the pedestrians continue to jog at 4 kilometers per

hour, did Pilious and his friends have enough time to win their bet? Explain.

★ **3.45.** [H] [A] Harriet Harrier was in a hurry. She had to reach 34th St. and 6th Ave. as soon as possible. When she entered the Continental Avenue station both an EE Local and an F express train were waiting. This left Harriet several options:

1. She could take the F all the way to 34th St.; but the F train spends two minutes in the Queens Plaza station waiting for a local to arrive.

2. She could take the EE all the way; but then she would still have one station to go at the time that the F train would be arriving at 34th St.

3. She could take the EE to Queens Plaza and change for the F. Actually the F would reach Queens Plaza by the time the EE had only gone 10 stations; but, owing to the F's stopover at Queens Plaza, the EE would arrive just in time for Harriet to change for the F.

4. She could take the F to Queens Plaza and arrive just in time to catch an earlier EE. This last route is the fastest, saving one full minute over the time required to travel the entire route by F train.

Assuming that the trains in question maintain uniform speeds (in terms of stations per minute) except for the two minute period when the F train is stationary, how many stations are there between Continental Avenue and 34th St? And how long will Harriet's trip take if she travels by the fastest route?

★ **3.46.** [A] Comm. Dale E. Muter travels to and from work by train. His train arrives at his hometown station at 6 PM each evening, and his wife always arrives promptly at this time to pick him up. One day, Comm. Muter left work early and arrived at the station at 5 PM. Not wishing to disturb his wife, he started to walk home along the route that she always drove. When he was one-quarter of the way home, he met his wife. They proceeded home at their usual driving speed and arrived home 12 minutes earlier than usual.

If the Muters live 12 miles from the station, how fast was Comm. Muter walking?

Work Problems

3.47.   The representative from Leonardo and Winslow Paint While You Wait Co. came to give me an estimate on the cost of painting my house.

"If we work with our regular crew," she said, "we can have the job done in eight hours. But if we hire three extra people, it will be done in six hours."

Assuming that all the painters work at the same rate, how long would it take one of them working alone to paint the house?

3.48.   Arnold, Hank, and Wilhelm are the three mailmen in the town of Littlewood. On an average day, if Arnold is not working, Hank and Wilhelm, working together, can sort the mail in 36 minutes; if Hank is not working, Arnold and Wilhelm can sort the mail in an hour; if Wilhelm is not working, Arnold and Hank can sort the mail in half an hour.

(a) [A] How long will it take all three of the men, working together, to sort the mail on an average day?

(b) If Arnold is on vacation and Wilhelm calls in sick, how long will it take Hank,

working alone, to sort the mail on an average day?

3.49. [A] Bertram Bovine has two drinking troughs for his cows. Both troughs have vertical sides a foot deep, but one is rectangular in shape and the other is circular.

Half of Bertram's herd of cows can drain the rectangular trough in 24 minutes, and the other half of the herd can drain the circular trough in 28 minutes.

If both troughs are filled and half the herd drinks from each, in how many minutes will the water in the circular trough be twice as deep as the water in the rectangular trough?

3.50. [A] If, on the average, a chicken and a half can lay an egg and a half in a day and a half, how many average eggs are laid by seven average chickens in seven average days?

★ **3.51.** [A] Peter Piper and his wife Pepper have a vegetable garden and a fruit orchard. Working together they can collect the harvest from the garden in 3 hours, whereas Pepper, working alone, requires 12 hours. Furthermore, together they can harvest the orchard in 2 hours, whereas Peter, alone, takes 10 hours.

It would seem that to harvest both the garden and the orchard, they should first spend 3 hours in the garden and then 2 hours in the orchard—a total of 5 hours in all. However, Peter is much more skillful at picking vegetables and Pepper is better at picking fruit, so that they can save time by having Peter work in the garden and Pepper work in the orchard until one of them finishes. That person would then help the other.

How much time can they save in this manner?

★ ★ **3.52.** [H] Gilbert Greenfield owns a pasture on which the grass has been cut to a uniform height of two inches. The grass in the pasture grows uniformly at a constant rate.

In addition to owning the pasture, Mr. Greenfield also owns a cow, a horse, and a sheep.

The grass existing and growing in the pasture is sufficient to feed all three animals grazing together for 20 days. It would feed the cow and the horse alone for 25 days; the cow and the sheep alone for $33\frac{1}{3}$ days; and the horse and sheep alone for 50 days.

How long would it sustain
(a) [A] The cow alone?
(b) The horse alone?
(c) [A] The sheep alone?

Miscellaneous Problems

3.53. [A]
At a college class reunion from dear old N.Y.U.
I met fifteen former classmates—counting men and women too.
More than half of them were doctors, and the rest all practiced law.
Of the former, more were females; of that, I'm pretty sure.
More abundant still were women who received a law degree.
And these statements are still truthful, even if you include me.
If my friend, a noted lawyer, left his wife and children home,
Can you draw any conclusion about me from this short poem?

3.54. [H] Leroy Jefferson went shopping. He bought some shirts and some pairs of slacks, paying a whole number of dollars for each. Actually, the price per shirt, in dollars, was equal to the number of shirts that Leroy bought; and the price per pair of slacks was equal to the number of pairs of slacks he purchased. If Leroy spent $39 more on slacks than he did on shirts, and if three times the number of shirts that Leroy bought is equal to seven more than the number of pairs of slacks he purchased, how many items of each type did Leroy buy?

3.55. [S] [A] Chef Jeff had just finished pouring one pint of vinegar into one large cruet and one pint of oil into another when he was called

to the telephone. During his absence, his child, Julia, decided to take over. She removed one level tablespoon of oil from the cruet of oil and carefully mixed it into the cruet of vinegar. Then, before any separation could take place, she took one tablespoon from the mixture and mixed it into the cruet of oil. When Jeff returned, did he find more oil in the vinegar or more vinegar in the oil? (A pint is equivalent to 32 tablespoons.)

3.56. Malcolm Milquetoast owns a dairy farm. The milk that he gets from his cows contains a cream content of 4 percent. In order to produce skimmed milk, he must reduce the amount of cream to 1 percent.

(a) [A] How much cream must be removed from 100 gallons of natural milk to produce skimmed milk?

(b) How many gallons of natural milk must farmer Milquetoast collect in order to produce 100 gallons of skimmed milk?

3.57. A grocer has 3 lb of almonds which she purchased for $1.05 per pound and 5 lb of cashew nuts for which she paid $6.25. If she can purchase peanuts at $.80 per pound, how many pounds of peanuts should she buy in order to be able to produce a mixture which she can sell for $1.30 per pound to give her a total profit of $7.00?

3.58. [A] The Blubber boys were slightly overweight. The four together tipped the scales at 1160 lb. Whelan, the oldest and the heaviest of the lot, weighed 20 lb more than

Wally, who in turn weighed 20 lb more than Bubba, who in turn weighed 20 lb more than Bobby, the baby of the family.

Three of the boys were married. Beulah weighs as much as her husband. Barbra's husband weighs twice as much as Barbra does; Belinda's husband weighs one and one half times as much as Belinda.

Altogether, the seven Blubbers weigh 1780 lb. Who is married to whom? And how much does each Blubber weigh?

3.59. [A] Using an honest balance scale, you find that a tweezer and a bar of soap weigh as much as three combs.

A bar of soap weighs the same as three toothbrushes and a comb.

Three tweezers weigh the same as a comb and a toothbrush.

How many toothbrushes are needed to balance a bar of soap?

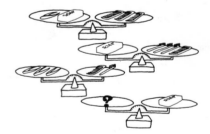

3.60. [H] Jeff, Cookie, Chuck, Pearl, and Bruce went shopping together. Pearl spent the most money, and Bruce spent the least, Jeff spent more than Chuck, who in turn spent more than Cookie. The total amounts spent by each pair of individuals were $7, $10, $11, $14, $15, $16, $17, $18, $20, and $24 (for example, Jeff and Pearl together spent $24).

How much did each person spend?

★ **3.61.** [H] [A] A dishonest grocer owns a balance scale for which one arm is slightly longer than the other. Whenever she purchases produce from the local farmers, she places the produce in the pan on the shorter arm of the scale, so that the produce seems lighter than it actually is. When the grocer sells produce, she naturally places it in the pan which is further from the fulcrum (the point of balance.)

Figure 3.2 shows how the grocer weighs six tomatoes and a sack of green beans when she buys them, and then how she weighs the tomatoes and green beans separately when she sells them.

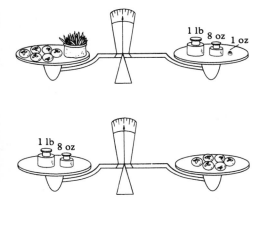

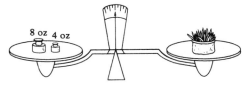

FIGURE 3.2

The principle is that the force exerted by an object on a balance scale is the product of the weight of that object and the length of the arm on which the pan is located. Thus, for example, if an object balances with a 1 lb weight on an honest scale, then that same object will balance with a 2 lb weight if the pan on which the object is located is twice as

far from the fulcrum as the other pan which contains the known weight.

(a) What is the ratio of the arm lengths of the scale?

(b) How much do the six tomatoes weigh?

(c) How much does the bag of beans weigh?

Ancient (and Not Quite as Ancient) Problems

The problems below can be found in the following sources, all of which predate the eighteenth century.

1. The *Rhind Papyrus* (c. 1550 BCE).
2. *Greek Anthology* (c. 500) assembled by Metrodorus.
3. *Ganita-Sāra Sangraha* (c. 850) by Mahāvīra.
4. *Triparty en la science des nombres* (1484) by Chuquet.
5. *Algebra Christophori Clavii Bambergensis* (1608) by Clavius.

3.62. [A] *(Rhind Papyrus)* Mass, its 2/3, its 1/2, its 1/7, its whole, it makes 33.

3.63. *(Greek Anthology)* I am a brazen lion, a fountain; my spouts are my two eyes, my mouth, and the flat of my right foot. My right eye fills a jar in two days [1 day = 12 hours], my left eye in three, and my foot in four; my mouth is capable of filling it in six hours. Tell me how long all four together will take to fill it.

3.64. *(Greek Anthology)* This tomb holds Diophantus. Ah, how great a marvel! The tomb tells scientifically the measure of his life. God granted him to be a boy for the sixth part of his life, and adding a twelfth part to this, He clothed his cheeks with down; He lit him the light of wedlock after a seventh part, and five years after his marriage He granted him a son. Alas! late-born wretched child; after attaining the measure of half his father's [full] life, chill Fate took him. After consoling his grief by this science of number for four years, Diophantus ended his life.

How old was Diophantus when he died?

3.65. (Mahāvīra) The mixed price of 9 citrons and 7 fragrant wood-apples is 107; again, the mixed price of 7 citrons and 9 fragrant wood-apples is 101. O you arithmetician, tell me quickly the price of a citron and of a wood-apple here, having distinctly separated those prices well.

3.66. [A] (Mahāvīra) One-fourth of a herd of camels was seen in the forest; twice the square root of that herd had gone to the mountain slopes; three times five camels remained on the river bank. What is the numerical measure of that herd of camels?

3.67. (Chuquet) A merchant visited three fairs. At the first he doubled his money and spent $30, at the second he tripled his money and spent $54, at the third he quadrupled his money and spent $72, and then had $48 left. How much money had he at the start?

3.68. [A] (Clavius) If I were to give 7 cents to each of the beggars at my door I would have 24 cents left. I lack 32 cents of being able to give them 9 cents apiece. How many beggars are there, and how much money have I?

Tricks

Each of the following exercises describes a card trick or a number guessing trick that is based on the use of algebra. Your task is to describe

what the magician must do to find the chosen card or number, and to explain how and why the trick works.

3.69. [A] Magician: "Think of a number. Double it. Add 8. Multiply by 5. Subtract 3. Double the result. Subtract 31. Multiply by 5. Tell me your result."

Subject: "915."

Magician: "Then the number you started with is _____."

How can the magician quickly find the number that the subject originally selected? Explain the trick.

3.70. Magician: "Select a card from this ordinary deck. Counting the Ace as one, Jack as 11, Queen as 12 and King as 13, double the value of the card. Add 3. Multiply by 5. If the selected card is a club, add 1; if it is a diamond, add 2; if a heart, add 3; and if it is a spade, add 4. Now tell me your result."

Subject: "79."

Magician: "The card you selected is the _____."

What is the selected card? And how does the trick work?

3.71. [H] [A] Magician: "Write an odd number greater than one and less than ten. Now write an even number between one and ten. Subtract the smaller of the two chosen numbers from the larger to give a third number. Add the two chosen numbers to give a fourth number. Multiply the third and fourth numbers. Tell me your result."

Subject: "39."

Magician: "The numbers you chose were _____ and _____."

Find the selected numbers and explain how the trick works.

★ **3.72.** [H] [A] Magician: "Here is an ordinary deck of playing cards. Place it on the table, face down. Select a number from 10 to 26. Remove the selected number of cards from the deck, turn them face up, and replace them on top of the deck. Shuffle the deck to distribute the face up cards throughout the deck. Again remove the selected number of cards from the top of the deck and place them in my hands, behind my back. Although I have no way of knowing how many face up cards there now are in my hand and how many remain in the rest of the deck, I shall try to turn some of the cards in my hands over, so that the number of face up cards in my hands is exactly equal to the number of face up cards in the rest of the deck."

After manipulating the cards behind his back for a few moments, the magician produced them and, sure enough, the number of face up cards in his packet was equal to the number of face up cards in the rest of the deck.

How was the magician able to produce this effect?

★ **3.73.** [S] [A] Magician: "Here is an ordinary 52 card deck of playing cards. You may shuffle them if you like. I will turn my back. While my back is turned, you will make a number of piles as follows: Turn over the top card of the deck. If it is a picture card, replace it in the middle of the deck and turn over the new top card. Continue doing this until you obtain a card numbered between 1 (Ace) and 10. When you have such a card, place it face up on the table. Beginning with the number of the card on the table, start counting to yourself until you reach 12. With each count, take one card from the deck and place it face up on the top of the pile that you are creating. When you reach 12, turn the pile over (so that the cards are face down with the card you originally selected on top), and begin a new pile. Thus, for example, if you turn up an 8, place a card on top of the 8 and silently count '9'; then place another card on top of the pile and count '10'; then place another card and count '11'; then place a final card and count '12.' Then turn the pile over and begin a new pile. Continue doing this until you do not

have enough cards remaining to complete a new pile. Give these remaining cards to me."

Subject: "Here they are. Now what?"

Magician: "I will now turn around. I see that you have six piles and five cards left over. Remove the top card from each of your piles and add the values of these cards together. You will obtain a total of _____ ."

How was the magician able to determine the total, and how does the trick work in general?

★ **3.74.** [H] [A] Magician: "I have here an ordinary 52 card deck of playing cards. You may shuffle them if you like. Now divide the deck into two equal halves and give me one half. I'm fanning the cards in my half of the deck. Select one; look at it; and now put it back here." (The magician has cut his pile into two parts; the spectator places his card on top of the indicated part and then the magician places the remaining part on top of the selected card.)

"Now, turn up the top three cards of your half on the deck. On top of each of these cards, place as many cards as are necessary to complete the count to 10. Thus, if you turn up a 7, you need three more cards to complete the count—8, 9, and 10. If you turn up a 10 or a picture card, no cards are needed to complete the count. (If you do not have enough cards to complete the count, you may use the top few cards from my half of the deck.) After you have completed the count with all three cards, place the remaining cards from your half of the deck on top of my half, which I have placed on the table.

"Next, add up the three cards you have turned up from your deck. (Count picture cards as 10.) Remove from the top of the stack of cards on the table as many cards as the sum you have obtained. The next card will be the card you originally selected."

How did the magician know?

★ **3.75.** [A] Magician: "This is another variation of the previous trick. Again I have an ordinary 52 card deck, which I am shuffling. I now take the top eleven cards and place them face down on the table. Choose four of these cards while I write a prediction on this piece of paper. Turn the four chosen cards face up, and I will return the other seven cards to the bottom of the deck. With each of the four face up cards, complete the count to 10 (see Exercise 3.74) by taking cards from the top of the deck. Now add up the four numbers (counting picture cards as 10) and count that many cards from the top of the deck. Turn up the next card. You will note it is exactly the card I predicted."

How did the magician know which card to predict?

★ **3.76.** [S] The magician handed an ordinary 52 card pack of cards to his first subject and asked her to remove a packet containing between one and five cards. He then asked the subject to pass the remaining cards to a second subject, who was instructed to remove a packet of between 10 and 20 cards. The remaining cards were then returned to the magician.

Magician: "I will now form a pile face up on the table by removing one card at a time from my packet. With each card I place on the table I will count. When I call out the number equal to the number of cards you originally selected, say nothing but note the card that I place on the table at that time. That will be your card for this trick. To make sure that you both have adequate opportunities to observe your cards, I will continue the count until I reach 26. I will then pick up the pile of cards from the table and place it, face down, on the bottom of my packet."

After the magician had completed this procedure, he divided his packet in half (leaving the extra card, if there was one, with the bottom half). The bottom card of the bottom half was then discarded, and the two halves were placed side by side on the table—the top half face down and the bottom half face up.

Magician: "I will now simultaneously remove the top card from each pile, and will continue doing so until you (the second subject) tell me that the next face up card is your card. At that point the top face down card will be yours (the first subject)."

As usual, the magician was correct. How did he know this would happen?

★ **3.77.** [H] [A] Magician: "I have here an ordinary 52 card deck of playing cards which you may shuffle if you like. Now remove from the top of the deck a packet containing between one and 20 cards. Count the number of cards in the packet, put the packet in your pocket, and return the remainder of the deck to me.

PICK A NUMBER BETWEEN ONE AND TEN.

"Now I will slowly deal a pile of 20 cards, one at a time, face up on the table. At the moment that my pile contains the same number of cards as your packet, say nothing, but note the card that is momentarily showing on the top of my pile.

"Now that I have completed my pile, pick a number between one and 10."

Subject: "Nine."

Magician: "I shall deal a second pile containing nine cards, face down on the table. I now place my original pile face down upon the second pile and then I place the remaining cards in my hand face down upon this combined pile. This places your selected card somewhere in the middle of the deck, so that I can have no idea where it is. I will attempt to find your selected card by handling one card at a time, without seeing the face, until the tingle in my fingers tells me that I have located the correct one.

"Aha! This is your card."

The magician was correct, as usual. How did he know which card to choose?

3.78. [H] The magician handed an ordinary 52 card deck of playing cards to Jack, the volunteer from the audience, and asked him to shuffle the deck, to remove a packet of no more than 10 cards from the top of the deck, to count the number of cards in the packet and, using that number, to locate and memorize the card that is in that position in the remaining part of the deck. (If, for example, Jack's packet contains six cards, then he should note the sixth card in the remaining portion of the deck.) The entire deck—with the exception of Jack's packet—was then returned to the magician.

Magician: "Now pick a number between 10 and 20."

Jack: "14."

Magician: "I want you to deal a pile of 14 cards, one card at a time, face down on the table, as follows."

The magician thereupon placed the top card of his packet face down on the table, counting "1." He placed the second card face down on top of the first, counting "2." And he continued in this manner until he had a pile of 14 cards. He then picked up this pile, placed it back on top of the deck and handed the deck to Jack.

Magician: "Oh, yes, before proceeding, please return your packet to the top of the deck."

Jack did as he was told, and, after he had completed a pile of 14 cards, returned the remainder of the deck to the magician. The magician tapped the deck three times, placed the deck on the table, and asked Jack to name his card. The subject said, "Jack of hearts."

Magician: "Would you please turn over the top card of the deck and hold it up to the audience."

Naturally, the card was the Jack of hearts. How did the magician know?

3.79. [H] [A] Magician: "I will let you be the magician for this last trick. Here is an ordinary 52 card deck of playing cards. Hand it to me and tell me to shuffle it."

Alice: "Here, please shuffle these cards."

Magician: "Now take the cards back, fan them and tell me to pick a card, look at it, and replace it in the deck."

Alice: "Pick a card, look at it, and replace it in the deck."

Magician: "Now ask someone in the audience for a number between 1 and 10."

Voice from the audience: "6."

Magician: "Count 6 cards off the top of the deck, placing each one face down on the table on top of the previous one. Then ask me my card."

Alice: "What was your card?"

Magician: "The four of diamonds. Now turn over the next card and it should be the four of diamonds."

Alice: "Nope. It's the six of hearts."

Magician: "I know what's wrong. You forgot to blow on the cards. Pick up that packet of six cards from the table and put it back on top of the deck. Then ask the audience for a number between 10 and 20."

Alice (to audience): "Would someone please pick a number between 10 and 20.

Audience: "14."

Magician: "Count off 14 cards, face down, one at a time, as before. Then blow on the next card, look at it. It should be the diamond four."

Alice: "Nope, wrong again."

Magician: "Can't you do anything right? Put the packet of 14 cards back on top of the deck, and I'll show you how it's done. Let me pick 8. I'll now count off 8 cards, blow on the deck, and, lo and behold, the next card is the four of diamonds."

The magician was correct. How was the trick done?

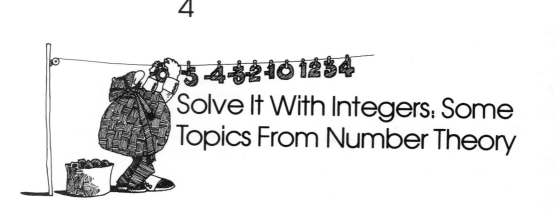

4

Solve It With Integers: Some Topics From Number Theory

The theory of numbers is a field of study that has excited the interest of mathematicians for centuries. It is primarily concerned with properties of the positive integers (1, 2, 3, . . .).

In early western and oriental cultures, whole numbers were originally of interest for ritual and computational purposes. The study of their properties, however, seems to have begun with the Pythagoreans, a school of scholars led by Pythagoras (c. 572 BCE), of Pythagorean Theorem fame. They felt that certain whole numbers had mystical significance.

Over the following two centuries, mathematicians of the time considered various types of whole numbers: for example, primes (which we will discuss below), perfect numbers (numbers such as 6 that are equal to the sum of their proper divisors: $6 = 1 + 2 + 3$), and deficient numbers (numbers such as 8 that exceed the sum of their proper divisors: $8 > 1 + 2 + 4$). Euclid (300 BCE) was able to characterize all even perfect numbers, but to this day it is not known whether or not any odd perfect numbers exist.

It was not until the 1600s that Fermat, a lawyer and mathematician, gave importance to the theory of numbers as a field of mathematics in its own right. His insight gave us many results, such as "Every odd prime can be expressed as a difference of two squares in exactly one way." He also left us the famous conjecture that for n, a positive integer greater than two, $x^n + y^n = z^n$ has no positive integer solutions for x, y, and z.

Many of the problems in the field of number theory are simple to state but difficult to solve. Some, such as the question of odd perfect numbers and Fermat's famous conjecture, have in fact withstood all attempts at solution to this time. However, many of these attempts have resulted in significant contributions to our mathematical knowledge.

In this chapter, we are primarily interested in some elementary aspects of number theory that will be useful in solving recreational problems. And, of course, we are interested in the types of problems to which this theory can be applied. Some of these are illustrated in the following sample problems. (Since these problems relate to mathematical techniques with which you may as yet be unfamiliar, you may find some of them quite difficult. You should try them, anyway. Even if you fail to solve any of them, you will become aware of what difficulties exist and will better appreciate the techniques when they are presented.)

SAMPLE PROBLEMS

Problem 4.1
Heloise Homemaker prepared 84 canapés for a dinner party she was making. Although she never ate at her own parties, she always prepared precisely enough hors d'oeuvres for each of her guests to consume the exact same quantity.

What possibilities are there for the number of guests she has invited?

Problem 4.2
Henry, Eli, and Cornelius and their wives Gertrude, Katherine, and Anna (not necessarily in that order) each purchased at least one animal at a farm auction. Coincidentally, each of the six spent an exact number of dollars for each animal equal to the number of animals he or she purchased. Henry purchased 23 animals more than Katherine did, and Eli spent $11 per animal more than Gertrude. Also, each husband spent a total of $63 more than his wife.

Who was married to whom?

Problem 4.3
(a) Given a 3 gallon jug and a 5 gallon jug (without any markings), is it possible to get exactly 1 gallon of water from a well? If so, how? If not, why not?

(b) What if we have only a 4 gallon and a 6 gallon jug? Is it possible to get 1 gallon of water? If so, how? If not, why not?

Problem 4.4
It was a strange lapse on the part of the bank teller. Evidently, he misread the check, for he handed out the amount of dollars in cents and

the amount of cents in dollars. When the error was pointed out to him, he became flustered, made an absurd mathematical mistake, and handed out a dollar, a dime, and a cent more. But the depositor declared that he was still short of his due. The teller pulled himself together, doubled the amount he had already given to the depositor (that is, he handed the depositor an additional amount equal to the total amount he had given him previously), and so settled the transaction to everyone's satisfaction.

What was the amount called for by the check? ([45], Problem 134)

How are these problems different from the ones we considered in Chapter 3?

Let us examine Problems 4.1 and 4.2.

Setting Up Problem 4.1

On the surface, this looks like an algebra problem, and, algebraically, it may be set up as follows:

Let x = the number of people Heloise could have invited, and

let y = the number of canapés each person would eat. Then

$$xy = 84. \qquad (4.1)$$

This equation has many solutions; in fact, infinitely many. For each nonzero value we choose for x, we can just let $y = 84/x$ and get a solution. For example, $x = 5$, $y = \frac{84}{5}$ is a solution to equation 4.1; so is $x = -2$, $y = -42$. But neither of these gives a solution to Problem 4.1. Each guest cannot eat a fractional number ($\frac{84}{5}$) of canapés; and there cannot be a negative number of guests. That is, in solving Problem 4.1, we are interested only in solutions of the equation for which x and y are positive whole numbers. An obvious solution is $x = 1$, $y = 84$. Another is $x = 2$, $y = 42$. Are there more solutions and, if so, how can we be sure of finding all of them?

Before we try to answer this question, let us observe that Problem 4.2 (page 101) leads to a similar situation.

Setting Up Problem 4.2

This problem is a slightly modified version of one which, according to Dudeney in his *Amusements in Mathematics* ([13], pp. 26–27), first appeared in the 1700s in *Ladies Diary*.

Again, we may begin to attack the problem algebraically. Let the number of animals each person bought be represented by the first letter of his or her name. By the conditions of the problem, this initial will also represent the number of dollars he or she spent per animal. Hence, the total amount spent by each person is represented by the square of that person's initial. For example, Henry spent H^2 dollars, (H animals at H dollars each gives H^2 dollars).

Since each man spent $63 more than his wife, we have, for each couple, the following equation:

$$x^2 - y^2 = 63, \tag{4.2}$$

where x represents the number of animals the husband purchased and y represents the number of animals purchased by his wife. (Note that equation 4.2 applies to all three couples. As we do not as yet know who is married to whom, we use x and y rather than specific initials.)

Since each person purchased a whole number of animals, x and y must be positive integers (whole numbers), and so we are again interested only in integer solutions of this equation.

This, then, is the basic difference between the problems of Chapter 3 and many of the problems of this chapter. Algebraically, the setup is similar; but here we are interested only in whole number solutions, and we need other techniques to find them.

DIOPHANTINE EQUATIONS

We will return shortly to the solutions of Problems 4.1 and 4.2. In the meantime, let us consider the general type of equation for which we want integer solutions. Such an equation is called a Diophantine equation, named after the Greek mathematician, Diophantus of Alexandria (c. 275).

Definition 4.1 A **Diophantine equation** is an equation in which all coefficients are integers and in which we are interested only in integer solutions.

Any equation with integer coefficients may or may not be considered as a Diophantine equation, depending on our point of view.

For example, the (linear) equation $3x + 5y = 1$ has an infinite number of solutions in general ($x = \frac{1}{3}$, $y = 0$; or $x = 1$, $y = -\frac{2}{5}$; and so on). It also has integer solutions (such as $x = 2, y = -1$; or $x = -3$, $y = 2$); therefore, as a Diophantine equation, it has solutions. On the other hand, the (linear) equation $4x + 6y = 1$ also has an infinite number of solutions in general (for example, $x = \frac{1}{4}$, $y = 0$; $x = 1$, $y = -\frac{1}{2}$). However, we will soon prove that it has no integer solutions; so, as a Diophantine equation, it has no solutions.

As another example, consider the equation $x^3 + y^3 = z^3$.* This also has many solutions in general, such as $x = y = 1$, $z = \sqrt[3]{2}$; however, if we consider this as a Diophantine equation—that is, if we are only interested in integral solutions—then it is known that there are no solutions other than $x = 0$, $y = z$, or $y = 0$, $x = z$, or $z = 0$, $y = -x$. Similarly, equations 4.1 and 4.2 have infinitely many solutions in general; however, to solve the problems from which these equations arise, we must consider them as Diophantine equations.

Diophantine equations frequently arise in problems such as 4.1 and 4.2 in which noninteger values for the unknowns would be meaningless. (For example, x might represent a number of animals or a number of people.)

There are no general techniques that may be used to solve all Diophantine equations; however, there are methods that may be applied to certain types of Diophantine equations. Later in the chapter, we will investigate a method for attacking linear Diophantine equations. But, for now, we are interested in the quadratic equations 4.1 and 4.2.

In order to consider the solutions of these equations, we need some information about whole numbers and their properties.

DIVISIBILITY

Let N refer to the set of positive integers; that is, $N = \{1, 2, 3, 4, \ldots\}$.

Definition 4.2 If m and n represent integers, $m \neq 0$, we say that **m divides n** (written symbolically as $m|n$) if there exists an integer q such that $n = mq$. (If m does not divide n, we write $m \nmid n$.)†

For example, $2|10$ because $10 = 2 \cdot 5$; but $3 \nmid 10$ because there is no integer q such that $3 \cdot q = 10$.

The following statements are equivalent:

m is a divisor of n

---

* The claim that the Diophantine equation $x^n + y^n = z^n$ has no positive integral solutions for any integer $n > 2$ has become known as Fermat's Last Theorem. Although it is known to be true for many values of n, including all values of n between 3 and 125,000, in general it is still an open question as to whether or not it is true for all $n > 2$. (While reading a book on Diophantus' work, Fermat noted in the margin that he had a proof of this result, but that the proof was too lengthy to write in the margin. Unfortunately, no trace of his proof has been found.) For an interesting discussion of the history of this problem, see "Fermat's Last Theorem" by H. Edwards, *Scientific American*, October 1978, pp. 108–122.

† Caution: $m|n$ is not the same as m/n. The latter is sometimes used to denote $\dfrac{m}{n}$.

m is a factor of n

n is a multiple of m

and are sometimes used instead of "m divides n."

Note that "m divides n" refers to the operation of division but is defined in terms of multiplication.

When we write $n = mq$, we are factoring n into two factors, each one being a divisor of n.

Definition 4.2 applies to negative as well as positive integers. However, for the remainder of this chapter when we speak of the divisors of a number n, we will mean the positive divisors of n.

The following theorem, which follows directly from the definition of divisibility, expresses some important properties of divisibility:

Theorem 4.1 Let a, b, c, and d be integers.
(i) If $a|b$ and $b|c$, then $a|c$ (transitivity of divisibility).
(ii) If $a|b$, then $a|bc$.
(iii) If $a + b = c$ and if d divides two of a, b and c, then d also divides the third.

We prove (i) below and leave (ii) and (iii) as an exercise (Practice Problems 4.A). The remainder of this chapter presents a number of theorems, usually without proof. Proofs and other related results may be found in [30] or [47].

Proof of (i):
Since $a|b$, there is an integer d such that $b = ad$.
Since $b|c$, there is an integer e such that $c = be$.
Therefore, $c = be = (ad)e = a(de)$.
Since de is an integer, $a|c$.

As examples of Theorem 4.1:
$3|12$, and $12|84$, therefore, by (i), $3|84$.
$3|6$, therefore, by (ii), $3|6x$ for any integer x.
$3|6x$ and $3|15y$ by (ii), therefore if $k + 6x = 15y$, then $3|k$ by (iii).

---

1. Suppose $k = 6x + 12y$ for some integers x and y.
    (a) $\boxed{A}$ Explain why we can conclude that $2|k$, $3|k$, and $6|k$, even though we do not know the specific values of x and y.
    (b) Show by a counterexample that 12 need not divide k.

2. Prove part (ii) of Theorem 4.1.

3. $\boxed{A}$ Prove part (iii) of Theorem 4.1.

PRACTICE
PROBLEMS
4.A

PRIME NUMBERS

Consider the set N of positive integers. Every integer n in N, other than 1, has at least two divisors in N—namely, n itself and 1. (Note that $n = n \cdot 1$.)

Definition 4.3 A positive integer p that has exactly two divisors in N is called a **prime number**. (The two divisors are 1 and p.)

Definition 4.4 A positive integer other than 1 that is not a prime number is called **composite**.

For example, 2 is a prime; it has two divisors, 1 and 2. Similarly, 3, 5, 7, 11, 13, and 17 are primes.

On the other hand, 4, 6, 8, and 9 are composite.

    4 has three divisors: 1, 2, 4
    6 has four divisors: 1, 2, 3, 6
    8 has four divisors: 1, 2, 4, 8
    9 has three divisors: 1, 3, 9

Note that 1 is neither prime nor composite.

Note also that, since a composite number n is not prime, it has at least one divisor m other than 1 and itself; that is, $n = mq$ where neither m nor q is 1. If m is not prime, then it is composite and hence has a divisor r other than 1 and itself. By Theorem 4.1 (i), r is also a divisor of m. If r is not prime, it has a divisor other than 1 and itself. Etc. Each new divisor is smaller than the previous one, so the process cannot continue indefinitely. Eventually we must reach a divisor of n that is a prime number. We thus have the following theorem.

Theorem 4.2 Every positive integer n other than 1 is divisible by a prime.

✩ THE INFINITUDE OF PRIMES

How can we find all primes? As a matter of fact, we can't, because there are infinitely many primes. This fact was already known to Euclid, the famed geometer (c. 300 BCE). His proof is interesting to us because it uses an idea that we discussed in Chapter 1—if an assumption leads to a contradiction, then the fact assumed must be false. The proof also makes use of Theorems 4.1 and 4.2, and proceeds as follows:

There are two possibilities—either there are a finite number of primes or there are an infinite number. We assume that there are a finite number, k, of them and see what happens. Call the k primes $p_1, p_2, \ldots, p_k$. Let

$$n = p_1 \cdot p_2 \cdot \ldots \cdot p_k + 1.$$

By Theorem 4.2, n must be divisible by one of the primes p_i. By the definition of divisibility, the product $p_1 \cdot p_2 \cdot \ldots \cdot p_k$ is also divisible by p_i. That is,

$$p_k | (p_1 \cdot p_2 \cdot \ldots \cdot p_k + 1) \text{ and } p_i | (p_1 \cdot p_2 \cdot \ldots \cdot p_k),$$

so that, by Theorem 4.1 (iii),

$$p_i | 1.$$

This is impossible. Hence, our assumption leads to a contradiction and the only alternative remaining is that there are infinitely many primes.

☆ THE SIEVE OF ERATOSTHENES

Although it is not possible to find all primes, it is possible to find all primes less than any given number. One way of doing this is to use the sieve method, so called because the composite numbers drop out, leaving only primes in the "sieve." This method, which was devised by Eratosthenes of Alexandria (c. 200 BCE), proceeds as follows (we illustrate using $n = 50$):

Since 1 is neither prime nor composite, we list all integers from 2 to n:

2　3　4　5　6　7　8　9　10　11　12　13　14　15　16　17　18
19　20　21　22　23　24　25　26　27　28　29　30　31　32　33　34　35
36　37　38　39　40　41　42　43　44　45　46　47　48　49　50

The first number on the list is 2, a prime. We circle it. The other multiples of 2 are composite (since they are divisible by 1, 2, and themselves), so we cross them out.

The first uncircled number remaining is 3. Since it is not a multiple of 2, it must be prime. We therefore circle 3 and cross out its multiples, which must be composite. (Note that every third integer after 3 is a multiple of 3; note also that some multiples of 3, such as 6 and 12, that are also multiples of 2 have already been crossed out.) This leaves:

②　③　4̶　5　6̶　7　8̶　9̶10̶　11　12̶　13　14̶　15̶　16̶　17　18̶
19　20̶　21̶　22̶　23　24̶　25　26̶　27̶　28̶　29　30̶　31　32̶　33̶　34̶　35
36̶　37　38̶　39̶　40̶　41　42̶　43　44̶　45̶　46̶　47　48̶　49　50̶

None of the uncircled numbers remaining can be multiples of 2 or 3 (or they would have been crossed out), and so, by part (ii) of Theorem 4.1, they cannot be multiples of any numbers that are multiples of 2 or 3. Therefore, the first uncircled number remaining, 5, cannot be a multiple of any smaller number (except for 1) and, hence, must be prime. We circle it and cross out all of its multiples (every fifth number after 5). (Again note that many of these have already been crossed out. In fact, as $2 \cdot 5$ and $4 \cdot 5$ have been crossed out as multiples of 2, and $3 \cdot 5$ has been eliminated as a multiple of 3, $5 \cdot 5$ is the first multiple of 5 that has not been crossed out previously. In general, when using the sieve method, the first multiple of a new prime p that will not have been crossed out previously is $p \cdot p = p^2$.)

The first uncircled number now remaining, 7, is a prime. We circle it and cross out its multiples.

We continue in this manner until all numbers on the list have either been circled or crossed out. (Actually, in our example, once we finish crossing out the multiples of 7, no further numbers need be crossed out. Each time we circle a new prime $p > 7$, its smallest multiple that has not already been crossed out, p^2, is larger than 50 and hence not on the list. Thus, all numbers remaining at this point are prime. In general, once we have eliminated all multiples of primes less than or equal to $\sqrt{n}$, all numbers remaining in the sieve are prime. We will return to this point shortly.)

In our example, we are left with

② ③ 4̸ ⑤ 6̸ ⑦ 8̸ 9̸ 1̸0̸ ⑪ 1̸2̸ ⑬ 1̸4̸ 1̸5̸ 1̸6̸ ⑰ 1̸8̸
⑲ 2̸0̸ 2̸1̸ 2̸2̸ ㉓ 2̸4̸ 2̸5̸ 2̸6̸ 2̸7̸ 2̸8̸ ㉙ 3̸0̸ ㉛ 3̸2̸ 3̸3̸ 3̸4̸ 3̸5̸
3̸6̸ ㊲ 3̸8̸ 3̸9̸ 4̸0̸ ㊶ 4̸2̸ ㊸ 4̸4̸ 4̸5̸ 4̸6̸ ㊼ 4̸8̸ 4̸9̸ 5̸0̸

So the primes less than 50 are: 2, 3, 5, 7, 11, 13, 17, 19, 23, 29, 31, 37, 41, 43, and 47.

PRACTICE PROBLEMS 4.B

1. Use the sieve method to find all primes up to 200.

MORE ABOUT PRIMES

Why are we interested in primes? The main reason is that, in a multiplicative sense, primes are the building blocks of the integers. This may be seen from the following theorems:

Theorem 4.3 If a prime divides the product of two positive integers, then it divides at least one of them. That is, if p is a prime and $p \mid mn$, then either $p \mid m$ or $p \mid n$ or both.

For example, if p is a prime and $p \mid 60$, then—since $60 = 6 \cdot 10$—either $p \mid 6$ ($p = 2$ or 3) or $p \mid 10$ ($p = 2$ or 5). Thus $p = 2, 3,$ or 5.

Theorem 4.4 (Unique Factorization Theorem) Every positive integer other than 1 can be factored into prime factors in exactly one way (except possibly for the order of the factors).

For example,

$6 = 2 \cdot 3 \, (= 3 \cdot 2)$

$3 = 3$

$24 = 2 \cdot 2 \cdot 2 \cdot 3 \, (= 2 \cdot 2 \cdot 3 \cdot 2 = 2 \cdot 3 \cdot 2 \cdot 2 = 3 \cdot 2 \cdot 2 \cdot 2).$

The uniqueness of the theorem may be expressed more precisely if we use exponents and write the factorization in the form

$$n = p_1^{r_1} p_2^{r_2} \cdots p_k^{r_k},$$

where $p_1 < p_2 < \cdots < p_k$ are distinct primes. In other words, if we write the primes in increasing order, using exponents to collect powers of the same prime, then the factorization is unique. For example, instead of writing $2 \cdot 2 \cdot 2 \cdot 3 \cdot 3$, we write $2^3 \cdot 3^2$. Thus,

$$24 = 2^3 \cdot 3^1 \qquad \text{or simply } 2^3 \cdot 3,$$

$$300 = 2 \cdot 2 \cdot 3 \cdot 5 \cdot 5 = 2^2 \cdot 3 \cdot 5^2,$$

and so on.

How do we find the prime factorization of a number? We answer this question on the left below, while, on the right, we apply the technique to finding the prime factorization of 44296:

To find the prime factorization of a number n, we could first try to divide by 2. If $2 | n$, then $n = 2 \cdot m$, and we then try to divide m by 2. If $2 | m$, then $m = 2 \cdot k$, and $n = 2^2 \cdot k$, etc. We continue until we obtain $n = 2^r \cdot t$, where r is the largest power of 2 that divides n; that is, $2 \nmid t$. (Note, if $2 \nmid n$, then $r = 0$ and $t = n$.)

We next find the largest power of 3 that divides t. We obtain $t = 3^s \cdot u$, and $n = 2^r \cdot 3^s \cdot u$, where $2 \nmid u$ and $3 \nmid u$.

In this manner, we continue to factor out the primes 5, 7, 11, etc. (Note: It is not necessary to try 4, 6, or any other composite

$n = 44296$

$44296 = 2 \cdot 22148$
$ = 2 \cdot 2 \cdot 11074$
$ = 2 \cdot 2 \cdot 2 \cdot 5537$
$ = 2^3 \cdot 5537$

$2 \nmid 5537$

$3 \nmid 5537$
$44296 = 2^3 \cdot 3^0 \cdot 5537 = 2^3 \cdot 5537$
(Note $3^0 = 1$, and hence is omitted.)

$5 \nmid 5537$
$5537 = 7 \cdot 791$
$ = 7 \cdot 7 \cdot 113$
$ = 7^2 \cdot 113$

number because, if for example, 4|u, then 2 would certainly divide it but does not.) | Thus, $44296 = 2^3 \cdot 7^2 \cdot 113$.

In general, we would now try the next prime, 11, to see if it divides 113, etc. In this example, it is not necessary to do so. This follows from the general rule that, *to determine whether a number is a prime, it is necessary to test only those primes up to and including the square root of the number as possible factors of the number*. Thus, suppose that 113 could be factored into $r \cdot s$. Then either r or s would have to be less than or equal to $\sqrt{113}$. (If both were greater than $\sqrt{113}$, then rs would have to be greater than $\sqrt{113} \cdot \sqrt{113} = 113$.) But we have already considered all possible prime divisors less than $\sqrt{113}$ and have seen that none of these divides 113; hence, 113 is a prime.

Thus, $44296 = 2^3 \cdot 7^2 \cdot 113$ is the prime factorization of 44296.

Since, by Theorem 4.4, the factorization of 44296 is unique, we would obtain the same result if we first tried to divide by 7:

$$44296 = 7 \cdot 6328$$

and then proceeded to factor 6328; or, if we somehow knew that 14 divides 44296, we could have written

$$44296 = 14 \cdot 3164$$

and then have factored 14 and 3164 to obtain

$$44296 = 2 \cdot 7 \cdot 2^2 \cdot 7 \cdot 113$$
$$= 2^3 \cdot 7^2 \cdot 113.$$

Although we were able to find the prime factorization of 44296 in a few steps, the fact is that the process of finding the prime factorization of a number is sometimes long and tedious, particularly if the number has a large prime factor.

PRACTICE
PROBLEMS
 4.C

1. [A] Find the prime factorization of each of the following:
   (a) 15120   (b) 2183   (c) 409   (d) 72814
2. Find the prime factorization of each of the following:
   (a) 3816   (b) 353   (c) 64680   (d) 6006   (e) 44304
   (f) your social security number.

Theorem 4.4 is so important in our number system that it is often called the **Fundamental Theorem of Arithmetic**. Why? How does the prime factored form of a number help us? One way is that it gives us a method of telling quickly whether a number is a perfect square, perfect cube, etc.

For example, is 42250000 a perfect square? Note that $42250000 = 2^4 \cdot 5^6 \cdot 13^2$, which is seen by the laws of exponents to be $(2^2 \cdot 5^3 \cdot 13)^2$.

In general, a number is a perfect square if and only if each exponent is even when the number is expressed in prime factored form. Similarly, a number is a perfect cube if and only if each exponent in the prime factored form of the number is divisible by 3.

---

1. [A] Indicate whether each of the following numbers is a perfect square and/or a perfect cube.
   (a) 15120   (b) 46656   (c) 1728.

2. Indicate whether each of the following numbers is a perfect square and/or a perfect cube.
   (a) 784   (b) 353   (c) 8000.

3. [A] (a) For each of the perfect squares in Problem 1, find the square root.
   (b) For each of the perfect cubes in Problem 1, find the cube root.

4. (a) For each of the perfect squares in Problem 2, find the square root.
   (b) For each of the perfect cubes in Problem 2, find the cube root.

5. [A] (a) If $11x$ is a square, what can be said about x?
   (b) If $9y$ is a square, what can be said about y?

6. Find a number that is a perfect fifth power.

7. [A] Give a rule that indicates when a number is a perfect nth power.

PRACTICE PROBLEMS 4.D

Another important consequence of Theorem 4.4 is the following theorem, which provides us with a method for finding all divisors of a given number.

Theorem 4.5 If $n = p_1^{r_1} p_2^{r_2} \ldots p_k^{r_k}$ is the prime factorization of a positive integer n ($p_1 < p_2 < \cdots < p_k$ are prime numbers), then
(i) every divisor m of n is of the form

$$m = p_1^{s_1} p_2^{s_2} \ldots p_k^{s_k},$$

where $0 \le s_1 \le r_1, 0 \le s_2 \le r_2, \ldots, 0 \le s_k \le r_k$.
(Note the possibility that some of the s_i's may be 0; that is, some of the p_i's may not appear in the factorization of m.)

(ii) The number of divisors of n is $(r_1 + 1)(r_2 + 1) \cdots (r_k + 1)$.

For example, does 84 divide 392? Writing the prime factorization of these numbers, we find that

$$84 = 2^2 \cdot 3 \cdot 7 \quad \text{and} \quad 392 = 2^3 \cdot 7^2.$$

From Theorem 4.5, it is now clear that 84 does not divide 392, because 3 appears in the prime factorization of 84 but does not appear in the prime factorization of 392. (Note: $\dfrac{392}{84} = \dfrac{2^3 \cdot 7^2}{2^2 \cdot 3 \cdot 7} = \dfrac{2 \cdot 7}{3}$, but there is no 3 in the numerator to cancel the 3 in the denominator.)

Similarly, $112 = 2^4 \cdot 7$ does not divide $392 = 2^3 \cdot 7^2$. The exponent of 2 in the factorization of 112 is too big—it is bigger than 3, the exponent of 2 in 392 ($\dfrac{392}{112} = \dfrac{2^3 \cdot 7^2}{2^4 \cdot 7} = \dfrac{7}{2}$, but there are not enough 2's in the numerator to cancel all the 2's in the denominator). On the other hand, $28 = 2^2 \cdot 7$ does divide $392 = 2^3 \cdot 7^2$. Each exponent in the factorization of 28 is less than the corresponding exponent in the factorization of 392.

Theorem 4.5 even enables us to find the quotient $392 \div 28$:

$$392 \div 28 = \frac{392}{28} = \frac{2^3 \cdot 7^2}{2^2 \cdot 7} = \frac{2 \cdot 7}{1} = 14.$$

Theorem 4.5 can also be used to find all divisors of a number. For example, to find the divisors of 24, we could proceed by trial and error (test each positive integer less than or equal to 24 to see if it divides 24), or we could use Theorem 4.5. Since $24 = 2^3 \cdot 3$, every divisor must be expressible in the form $2^a 3^b$, where $0 \le a \le 3$, and $0 \le b \le 1$.

The divisors may be listed systematically by use of a tree diagram (see Chapter 1 for a discussion of tree diagrams). This is illustrated in Figure 4.1.

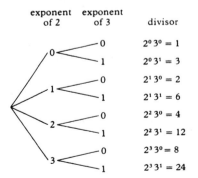

FIGURE 4.1

Note that part (ii) of Theorem 4.5 tells us that there should be $(3 + 1)(1 + 1) = 8$ divisors in all. This may also be seen from the tree diagram and the use of the Multiplication Principle (see page 19).

PRACTICE PROBLEMS 4.E

1. $\boxed{A}$ How many divisors does 6480 have?
2. How many divisors does 44296 have?
3. $\boxed{A}$ Find the divisors of each of the following:
   (a) 392   (b) 353   (c) 112
4. Find the divisors of each of the following:
   (a) 96   (b) 315   (c) 1032   (d) 199
5. $\boxed{A}$ (a) Given the square number $2^2 \cdot 13^4 \cdot 17^6$, how many divisors does it have?
   (b) What can be said about the parity (oddness or evenness) of the number of divisors of a perfect square?
   (c) What can be said about the parity of the number of divisors of a number that is not a perfect square?

We are now ready to solve the Diophantine equations 4.1 and 4.2 (pages 102 and 103). Try them again if you have not already solved them.

Solution of Problem 4.1 (Continued)

Since $xy = 84$ and x and y are positive integers, x and y must be divisors of 84.

The prime factorization of 84 is

$$84 = 2^2 \cdot 3 \cdot 7.$$

Hence, by Theorem 4.5, the divisors of 84 are

$$2^0 \cdot 3^0 \cdot 7^0 = 1, \quad 2^1 \cdot 3^0 \cdot 7^0 = 2, \quad 2^2 \cdot 3^0 \cdot 7^0 = 4,$$
$$2^0 \cdot 3^1 \cdot 7^0 = 3, \quad 2^1 \cdot 3^1 \cdot 7^0 = 6, \quad 2^2 \cdot 3^1 \cdot 7^0 = 12,$$
$$2^0 \cdot 3^0 \cdot 7^1 = 7, \quad 2^1 \cdot 3^0 \cdot 7^1 = 14, \quad 2^2 \cdot 3^0 \cdot 7^1 = 28,$$
$$2^0 \cdot 3^1 \cdot 7^1 = 21, \quad 2^1 \cdot 3^1 \cdot 7^1 = 42, \quad 2^2 \cdot 3^1 \cdot 7^1 = 84.$$

Thus, we can have

$x = 1, y = 84$	$x = 84, y = 1$
$x = 2, y = 42$	$x = 42, y = 2$
$x = 3, y = 28$	$x = 28, y = 3$
$x = 4, y = 21$	$x = 21, y = 4$
$x = 6, y = 14$	$x = 14, y = 6$
$x = 7, y = 12$	$x = 12, y = 7.$

Therefore, Heloise could have invited 1, 2, 3, 4, 6, 7, 12, 14, 21, 28, 42, or 84 guests.

Solution of Problem 4.2 (Continued)

How is the discussion in this chapter relevant to this problem? Observe that $x^2 - y^2$ may be factored into $x^2 - y^2 = (x - y)(x + y)$. Therefore, equation 4.2 becomes

$$(x - y)(x + y) = 63.$$

Since x and y are to be integers, $x - y$ and $x + y$ must be a pair of divisors of 63 whose product is 63. But $63 = 3^2 \cdot 7$, so, by Theorem 4.5, the only divisors of 63 are

$$3^0 \cdot 7^0 = 1 \qquad 3^1 \cdot 7^0 = 3 \qquad 3^2 \cdot 7^0 = 9$$
$$3^0 \cdot 7^1 = 7 \qquad 3^1 \cdot 7^1 = 21 \qquad 3^2 \cdot 7^1 = 63.$$

Arranging in pairs those numbers whose product is 63, we get

$$1 \text{ and } 63 \qquad 3 \text{ and } 21 \qquad 7 \text{ and } 9.$$

Since x and y are positive, $x - y < x + y$. Therefore, we have three possible cases:

$$\text{(a) } x - y = 1 \quad \text{(b) } x - y = 3 \quad \text{(c) } x - y = 7$$
$$x + y = 63 \qquad x + y = 21 \qquad x + y = 9$$

In each case, we have two equations in two unknowns, which we can solve simultaneously to obtain:

$$\text{(a) } x = 32, y = 31; \quad \text{(b) } x = 12, y = 9; \quad \text{(c) } x = 8, y = 1.$$

Thus, equation 4.2 has three sets of solutions. The number of purchases made by each couple in Problem 4.2 must satisfy equation 4.2, where x represents the number of animals purchased by the husband and y represents the number of animals purchased by his wife. We can therefore say that the three men (in some order) purchased 32, 12, and 8 animals and that their wives (in the same order) purchased 31, 9, and 1 animals.

Since we are told that Henry purchased 23 animals more than did Katherine and that Eli spent $11 per animal more than Gertrude,

$$H = K + 23$$

$$E = G + 11.$$

The only possibility that is consistent with what we have discovered is

$$H = 32, \quad K = 9, \quad E = 12, \quad G = 1,$$

leaving

$$C = 8, \quad A = 31.$$

Thus, Henry (32) is married to Anna (31); Eli (12) is married to Katherine (9); and Cornelius (8) is married to Gertrude (1). This completes the solution of Problem 4.2.

LINEAR DIOPHANTINE EQUATIONS

In the solution above, we were fortunate that $x^2 - y^2$ factored easily into $(x - y)(x + y)$, so that we were able to use Theorem 4.5. Higher order Diophantine equations do not behave this nicely in general, and may not prove to be as easy to solve. However, there is one kind of Diophantine equation that can always be handled—a linear Diophantine equation.

A linear Diophantine equation in two unknowns is an equation of the form

$$ax + by = c, \tag{4.3}$$

where a, b, and c are known integers, and where we are interested only in integer solutions—integers x and y which satisfy the equation. Similarly, a linear Diophantine equation in three unknowns is of the form

$$ax + by + cz = d,$$

etc.

In this chapter we restrict our attention to linear Diophantine equations in two unknowns.

It follows from Theorem 4.1 that if x and y are integers and if d is any integer that divides both a and b, then $d|ax$, $d|by$, and so $d|(ax + by)$. Thus, if equation 4.3 has a solution, then any divisor of a and b must also divide c. For example, the equation $8x + 12y = 17$ has no solution since the left side is divisible by 4 for any values of x and y, whereas the right is not divisible by 4.

---

1. $\boxed{A}$ Are there integers x and y such that $3x + 6y = 5$? Explain.

2. Are there integers x and y such that $4x + 6y = 7$? Explain.

3. $\boxed{A}$ If there are integers x and y such that $3x + 6y = k$, what can be said about k?

PRACTICE PROBLEMS 4.F

Continuing our discussion, we make the following definitions:

Definition 4.5 A **common divisor** of a and b is a positive integer that divides both a and b.

For example, 1, 2, 3, 4, 6, and 12 are common divisors of 24 and 36.

Definition 4.6 The **greatest common divisor** of a and b, denoted here by $\gcd(a, b)$, is the greatest of all the common divisors of a and b.

For example, $\gcd(24, 36) = 12$,

$$\gcd(9, 17) = 1,$$
$$\gcd(4, 16) = 4.$$

Definition 4.7 If $\gcd(a, b) = 1$, then a and b are said to be **relatively prime**.

PRACTICE PROBLEMS 4.G

1. **A** (a) Find the greatest common divisors of each of the following pairs of numbers:

   (i) 16 and 28   (ii) 18 and 26   (iii) 37 and 54   (iv) 15 and 39

   (b) Which of the pairs of numbers above are relatively prime?

2. (a) Find the greatest common divisors of each of the following pairs of numbers:

   (i) 34 and 51 (ii) 12 and 35 (iii) 8 and 74 (iv) 113 and 197

   (b) Which of these pairs of numbers are relatively prime?

3. **A** (a) If $a = 3^4$ and $b = 3^7$, find $\gcd(a, b)$.
   (b) If $a = p^r$ and $b = p^s$, find $\gcd(a, b)$.
   (c) If $a = 3^4 \cdot 5^6$ and $b = 3^2 \cdot 5^7$, find $\gcd(a, b)$.
   (d) If the prime factorizations of a and b respectively are

   $$a = p_1^{r_1} p_2^{r_2} \ldots p_k^{r_k} \quad \text{and} \quad b = p_1^{s_1} p_2^{s_2} \ldots p_k^{s_k}$$

   and if, for each i, $m_i = min\{r_i, s_i\}$—that is, the smaller of r_i and s_i—find an expression for $\gcd(a, b)$ as a product of powers of primes.
   (e) Use the result in (d) to find $\gcd(12600, 15120)$.

4. (a) If $a = 3^4$ and $b = 5^7$, find $\gcd(a, b)$.
   (b) If $a = 3^4 \cdot 5^2$ and $b = 2^3 \cdot 5^7$, find $\gcd(a, b)$.
   (c) Find $\gcd(3816, 15120)$.

As we observed previously, a necessary condition for equation 4.3 (page 115) to be solvable is that c is divisible by every common divisor of a and b. In particular, $\gcd(a, b)$ must divide c.

Thus, for example, the equation $4x + 6y = 1$ has no integral solutions, since $2 = \gcd(4, 6)$ does not divide 1.

It turns out that $\gcd(a, b)|c$ is also a sufficient condition for equation 4.3 to be solvable. That is, we have the following theorem:

Theorem 4.6 Given integers a, b, and c, the Diophantine equation $ax + by = c$ is solvable if and only if $\gcd(a, b)|c$.

We will shortly return to the question of how to find the solutions of equation 4.3 (when they exist). First let us consider Problem 4.3 (page 101).

If you haven't solved it already, try it again now.

Problem 4.3 is one of a famous category of "decanting" problems. It is easily found, by trial and error, that one method of obtaining 1 gallon in part (a) of the problem is as follows (see Figure 4.2).

Fill the 3 gallon jug with water and empty it into the 5 gallon jug.

Refill the 3 gallon jug from the well and use part of its contents to fill up the 5 gallon jug (which already has 3 gallons in it). This leaves 1 gallon in the 3 gallon jug (as well as 5 gallons in the 5 gallon jug, which we may spill out). We will see later that this is not the only solution of the problem.

Part (b) cannot be solved so easily. After many trials, you begin to feel that 1 gallon of water cannot be obtained using only a 4 gallon jug and a 6 gallon jug. In every case, there seem to be an even number of gallons left in each jug. Is this because 4 and 6 are both even numbers?

To answer this question, let us reconsider Problem 4.3 in a different light.

Since the jugs have no markings, if we ever partially fill a jug from the well, we will have no idea of exactly how much water it contains. Similarly, if we pour from one jug to the other, leaving them both partially filled, we will not know exactly how much water is in either. Finally, as we can never have two partially filled jugs whose exact contents are known, pouring water from a partially filled jug back into the well is senseless, because then either both jugs are empty (our starting position) or one is empty and one is full (a situation that can be reached directly in one step).

Therefore, the only reasonable moves we could make at any step are:

1. Completely fill an empty jug from the well. (Filling a partially filled jug is the same as first emptying a partially filled jug and then filling it completely—a move that we have seen makes no sense.)

Solution of Problem 4.3

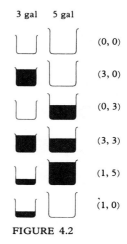

3 gal   5 gal

(0, 0)

(3, 0)

(0, 3)

(3, 3)

(1, 5)

(1, 0)

FIGURE 4.2

2. Completely empty a full jug into the well.

3. Pour from one jug into the other, completely filling or completely emptying one jug in the process and leaving the other jug partially filled.

Let us return to part (a) and set up the problem algebraically as follows: Count $+1$ every time the 3 gallon jug is filled from the well and -1 every time it is emptied into the well. We thus obtain an integer x which represents that net number of fillings and emptyings of the 3 gallon jug from the well. (If the 3 gallon jug is filled more often than it is emptied, then x will be positive; if it is emptied more often than it is filled, then x will be negative.) In the same manner, we obtain a number y for the 5 gallon jug. Pouring from one jug to another, as in a type 3 move above, does not change x and y.

If we obtain 1 gallon by filling (or emptying) the 3 gallon jug x times and the 5 gallon jug y times, then we must have the equation

$$3x + 5y = 1,$$

where $3x + 5y$ is the net amount of water that has been removed from the well. (Note that, if one jug contains 1 gallon, then the other must be either empty or full. In the latter case, it may be emptied into the well, leaving only 1 gallon missing from the well.) This is a Diophantine equation, since x and y must be integers.

Because the coefficients of x and y are small, it is easy to find (by trial and error) solutions to this equation. One such solution is $x = 2$, $y = -1$. Another solution is $x = -3$, $y = 2$.

Consider the first solution, $x = 2$, $y = -1$. What does this mean?

According to our interpretation of x and y, it means we must fill the 3 gallon jug from the well twice and empty the 5 gallon jug into the well once. But how do we fill something twice without emptying it? And how do we empty something that has nothing in it to begin with? The answer is simple. We first fill the 3 gallon jug from the well and then empty it into the 5 gallon jug. (Since we are not emptying it into the well, this does not affect x.) We then can fill the 3 gallon jug a second time. We now want to empty the 5 gallon jug into the well. But it is not full; so we first fill it from the 3 gallon jug (leaving the desired 1 gallon in the 3 gallon jug), and then we empty the 5 gallon jug into the well. This is the same solution we found originally.

The other solution, $x = -3$, $y = 2$, can be interpreted similarly, where we start by filling the 5 gallon jug (see Figure 4.3).

In a similar manner, Problem 4.3(b) gives rise to the Diophantine equation $4x + 6y = 1$. However, we have seen that this equation does not have integer solutions (2 divides the left side of the equation but does not divide 1). So, as we suspected, Problem 4.3(b) has no solution.

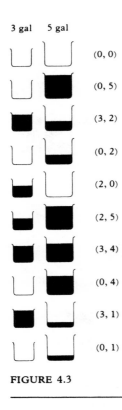

3 gal   5 gal

(0, 0)

(0, 5)

(3, 2)

(0, 2)

(2, 0)

(2, 5)

(3, 4)

(0, 4)

(3, 1)

(0, 1)

FIGURE 4.3

Now try Problem 4.4 (page 101) again, if you did not solve it the first time.

An algebraic attack may also be used for Problem 4.4. If x represents the number of dollars and y represents the number of cents on the check, then the check was in the amount of x dollars and y cents or, equivalently, $100x + y$ cents. By mistake, the teller gave out y dollars and x cents or $100y + x$ cents. He later gave 111 cents more, and then doubled the total disbursement to reach the correct figure. We thus obtain the following equation:

Solution of Problem 4.4

$$2(100y + x + 111) = 100x + y.$$

Collecting terms, this becomes

$$98x - 199y = 222.$$

This is a Diophantine equation, since x must represent a whole number of dollars and y must represent a whole number of cents. In fact, from the conditions of the problem, it is clear that both x and y must be greater than 0 or the interchange could not take place. Moreover, $y \leq 99$, since checks are not written with larger quantities of cents. By the same token, even a badly confused teller would have been forced to reexamine the situation if he found himself counting out a dollar's worth of cents or more; so x must be ≤ 99. Hence, we may assume that $0 \leq x \leq 99$ and $0 \leq y \leq 99$.

Since 98 and 199 (the coefficients of x and y in the preceding equation) are relatively large numbers, it is not so easy to solve this Diophantine equation by trial and error. We need more powerful mathematical techniques. (We know that solutions exist since $\gcd(98, 199) = 1$, and $1 \mid 222$.)

In general, there are several techniques for solving linear Diophantine equations. The most common makes use of the Euclidean algorithm. We do not present this method here. (You can find a discussion of the method in any book on elementary number theory—see, for example, [47], p. 7). Instead, we present a method that uses the notion of congruence (which will soon be discussed). This method often requires a little ingenuity but, if used thoughtfully, is generally a faster way of solving most recreational problems of this type.

DIVISION WITH REMAINDERS

Given two natural numbers n and d, we may divide n by d. If d does not divide n (evenly), then there will be a remainder r, which

will be less than d (otherwise we would continue dividing).

For example,

$$\begin{array}{r} 2 \\ 3{\overline{\smash{\big)}\,7}} \\ \underline{6} \\ 1 \end{array} \quad \text{or} \quad \begin{array}{r} 2\ r1 \\ 3{\overline{\smash{\big)}\,7}} \end{array} \quad \text{or} \quad 7 = 3 \cdot 2 + 1.$$

In this example, 1 is called the remainder and 2 is called the quotient (3 is the divisor and 7 is the dividend).

In general, we have the following result:

Theorem 4.7 (Division Algorithm) Given any two positive integers n and d, there exist integers q and r (respectively called the **quotient** and the **remainder**) such that

$$n = qd + r \quad \text{and} \quad 0 \le r < d.$$

Theorem 4.7 can be generalized for all integers n. For example, if $n = -5$ and $d = 3$, then

$$-5 = 3(-2) + 1.$$

Note that $r = 1$ is positive.

Similarly, if $n = -31$ and $d = 5$, then

$$-31 = 5(-7) + 4.$$

Again, $r = 4$ is positive.

PRACTICE PROBLEMS 4.H

1. [A] Find the quotient and remainder when 108 is divided by 3.
2. Find the quotient and remainder when 129 is divided by 7.
3. [A] Find q and r such that $87 = 7q + r$, where $0 \le r < 7$.
4. Find q and r such that $91 = 11q + r$, where $0 \le r < 11$.
5. [A] Find q and r such that $-93 = 5q + r$, where $0 \le r < 5$.
6. Find q and r such that $-41 = 8q + r$, where $0 \le r < 8$.

CONGRUENCE

Given a fixed divisor d, we can classify integers according to the remainder r obtained when the integer is divided by d. We can talk about a **remainder class modulo d**, consisting of all integers having the same remainder when divided by d. Since there are d possible remainders, $0, 1, 2, \ldots, d - 1$, there are d remainder classes. For example, if $d = 3$, then we get three remainder classes:

$$\bar{0} = \{\ldots, -9, -6, -3, 0, 3, 6, 9, \ldots\}$$
$$\bar{1} = \{\ldots, -8, -5, -2, 1, 4, 7, 10, \ldots\}$$
$$\bar{2} = \{\ldots, -7, -4, -1, 2, 5, 8, 11, \ldots\}$$

All numbers in remainder class $\bar{0}$ have remainder 0 when they are divided by 3. They are all divisible by 3 and can be expressed in the form $3k$.

For example, 231 is an element of $\bar{0}$, because

$$231 = 3 \cdot 77 + 0 = 3 \cdot 77.$$

Similarly, -9 is an element of $\bar{0}$:

$$-9 = 3(-3) + 0 = 3(-3).$$

All numbers in the remainder class $\bar{1}$ have remainder 1 when they are divided by 3. They are of the form $3k + 1$. For example, 241 and -23 are elements of $\bar{1}$, because

$$241 = 3 \cdot 80 + 1$$

and

$$-23 = 3(-8) + 1.$$

In a similar manner, the remainder class $\bar{2}$ contains all numbers that have remainder 2 when they are divided by 3. These numbers are of the form $3k + 2$.

Note that, once one element of a remainder class modulo 3 is known, other elements of the class may be obtained by adding 3's to or subtracting 3's from the known number.

---

1. $\boxed{A}$ Find the remainder classes when $d = 4$.
2. Find the remainder classes when $d = 5$.
3. $\boxed{A}$ How many remainder classes are there when $d = 12$? 35? n?

PRACTICE PROBLEMS 4.1

Definition 4.8 If a and b have the same remainder when divided by d, then we say that **a is congruent to b modulo d**, symbolically denoted by $a \equiv b \pmod{d}$.

In other words, $a \equiv b \pmod{d}$ if and only if a and b are in the same remainder class modulo d. For example,

$$241 \equiv -23 \equiv 1 \pmod{3},$$

as 241, -23, and 1 are all in the remainder class $\bar{1}$.

The concept of congruence is a powerful mathematical tool first formulated by Carl Friedrich Gauss in the 1790s.

From Definition 4.8, it is clear that

$$a \equiv b \pmod{d} \qquad \text{if and only if} \qquad b \equiv a \pmod{d}.$$

Also, for any number a,

$$a \equiv a \pmod{d}.$$

Finally,

$$\text{if } a \equiv b \pmod{d} \quad \text{and} \quad b \equiv c \pmod{d}, \quad \text{then } a \equiv c \pmod{d}.$$

Note also that

$$x \equiv r \pmod{d} \text{ if and only if } x = kd + r,$$

for some integer k.

In addition:

If $a \equiv b \pmod{d}$, then a and b belong to the same remainder class, say, $\bar{r}$. But then

$$a = qd + r \quad \text{and} \quad b = kd + r,$$

so

$$b - a = (kd + r) - (qd + r)$$

$$= (k - q)d;$$

thus

$$d \mid (b - a).$$

Conversely, one can show that if $d \mid (b - a)$ then $a \equiv b \pmod{d}$. In other words, $a \equiv b \pmod{d}$ if and only if $d \mid (b - a)$. This fact is often used as the definition of congruence.

$241 \equiv -23 \pmod{3}$
241 and -23 are both in remainder class $\bar{1}$

$$241 = 80 \cdot 3 + 1$$

and

$$-23 = (-8)3 + 1$$

$$241 - (-23) = (80 \cdot 3 + 1)$$
$$- [(-8)3 + 1]$$
$$= [80 - (-8)]3$$
$$= 88 \cdot 3$$

$$3 \mid [241 - (-23)],$$

that is, $\quad 3 \mid 264$.

In the other direction, since

$$29 - 17 = 12$$
$$3 \mid 12$$

therefore

$$29 \equiv 17 \pmod{3}$$

(note that both leave remainders of 2 modulo 3).

---

1. ⒜ Which of the following are true?
   (a) $7 \equiv 1 \pmod{3}$ (b) $7 \equiv -2 \pmod{3}$ (c) $7 \equiv 3 \pmod{3}$
   (d) $7 \equiv 7 \pmod{3}$ (e) $7 \equiv 147 \pmod{3}$ (f) $76 \equiv -152 \pmod{3}$.

2. Which of the following are true?
   (a) $8 \equiv 3 \pmod 5$ (b) $8 \equiv -3 \pmod 5$ (c) $8 \equiv -2 \pmod 5$
   (d) $8 \equiv 193 \pmod 5$ (e) $721 \equiv -484 \pmod 5$.

3. $\boxed{A}$ Find the smallest nonnegative integer to which 122 is congruent modulo 3.

4. Find the smallest nonnegative integer to which 122 is congruent modulo 7.

5. $\boxed{A}$ Find a number divisible by 5 to which 2 is congruent modulo 6.

6. Find a number divisible by 7 to which 2 is congruent modulo 9.

7. $\boxed{A}$ If $x \equiv 1 \pmod 3$, find an expression for x in the form $x = $ _____ .

8. If $x \equiv 3 \pmod 5$, find an expression for x in the form $x = $ _____ .

9. Prove that $a \equiv b \pmod d$ if and only if $d \mid (b - a)$.

The following theorem presents some important properties of congruence.

Theorem 4.8 If $a \equiv b \pmod d$ and $e \equiv f \pmod d$, then
 (i) $a + e \equiv b + f \pmod d$
 (ii) $a - e \equiv b - f \pmod d$
(iii) $ae \equiv bf \pmod d$.
We can add, subtract, and multiply congruences with the same modulus. In particular, since $c \equiv c \pmod d$, for any integer c, we obtain
(iv) $a \pm c \equiv b \pm c \pmod d$
 (v) $ac \equiv bc \pmod d$.

As a consequence of Theorem 4.8, *in any congruence modulo d, any number may be replaced by any other number in the same congruence class modulo d, and the congruence will be preserved.*
 For example,

if $207x \equiv 4 \pmod 5$, then $2x \equiv 4 \pmod 5$ since $207 \equiv 2 \pmod 5$.

That is, we can replace 207 by 2, since they are congruent modulo 5.
 As another example, suppose we wish to find the remainder when 3^{26} is divided by 7. This could be done by first evaluating 3^{26} and then dividing by 7. This would involve a lot of computation. A simpler approach is to express the problem in terms of congruence:

Find x such that $3^{26} \equiv x \pmod 7$.

Since $3 \cdot 3 = 9 \equiv 2 \pmod 7$,

$$x \equiv 3^{26} = (3 \cdot 3)^{13} \equiv 2^{13} \pmod 7.$$

Since $2 \cdot 2 \cdot 2 = 8 \equiv 1 \pmod 7$, we write 2^{13} as $(2 \cdot 2 \cdot 2)^4 \cdot 2$. Thus

$$x \equiv 2^{13} = (2 \cdot 2 \cdot 2)^4 \cdot 2 \equiv 1^4 \cdot 2 = 2 \pmod 7.$$

Therefore, when 3^{26} is divided by 7, the remainder is 2; that is, $3^{26} \equiv 2 \pmod 7$. We could also arrive at this result by observing that

$$3^3 = 27 \equiv -1 \pmod 7.$$

Then

$$x = 3^{26} = (3^3)^8 \cdot 3^2 \equiv (-1)^8 \cdot 9 = 9 \equiv 2 \pmod 7.$$

PRACTICE PROBLEMS 4.K

1. $\boxed{A}$ If $x \equiv 5 \pmod 8$ and $y \equiv 6 \pmod 8$, to what is $x + y$ congruent modulo 8? To what is xy congruent modulo 8?

2. If $x \equiv -2 \pmod 7$ and $y \equiv 3 \pmod 7$, to what is $x + y$ congruent modulo 7? To what is xy congruent modulo 7?

3. $\boxed{A}$ Simplify each of the following congruences by reducing the coefficients:
   (a) $214x \equiv -16 \pmod 5$ (b) $26x \equiv 413 \pmod 3$.

4. Simplify each of the following congruences by reducing the coefficients:
   (a) $39x \equiv 84 \pmod 7$ (b) $-13x \equiv 25 \pmod 3$.

5. $\boxed{A}$ Transform each of the following equations to congruences modulo 7:
   (a) $7x + 15y = 1$ (b) $-37x + 51y = 212$.

6. Transform each of the following equations to congruences modulo 13:
   (a) $42x + 27y = 182$ (b) $37x - 26y = 69$.

7. $\boxed{A}$ What is the remainder when 14^{30} is divided by 11?

8. What is the remainder when 5^{28} is divided by 3?

9. $\boxed{A}$ If $8^{79} \equiv x \pmod 5$ and if $0 \le x \le 4$, find x.

Another important theorem concerning congruences is the following:

Theorem 4.9 If $a \equiv b \pmod{d}$ and if $c \mid d$, then $a \equiv b \pmod{c}$. That is, if $d \mid (a - b)$ and $c \mid d$, then $c \mid (a - b)$.

For example, $22 \equiv 7 \pmod{15}$, so $22 \equiv 7 \pmod{5}$ since $5 \mid 15$, and $22 \equiv 7 \pmod{3}$ since $3 \mid 15$. Similarly, if $x \equiv 13 \pmod{15}$, then $x \equiv 13 \equiv 1 \pmod{3}$ and $x \equiv 13 \equiv 3 \pmod{5}$.

CASTING OUT NINES

The theory of congruences provides the explanation of a very interesting phenomenon.

Definition 4.9 The **digital root** of a natural number is the one-digit number obtained by adding all the digits of the original number to obtain a new number, then adding all the digits of the new number to obtain a third number, and so on until a one-digit result is obtained. (Observe that the sum of the digits of a number is less than the number, unless the number has only one digit, so that, by continuing the process above, a one-digit number must eventually be obtained.)

For example, the digital root of 2965184860 is 4:

$$2 + 9 + 6 + 5 + 1 + 8 + 4 + 8 + 6 + 0 = 49$$

$$4 + 9 = 13$$

$$1 + 3 = 4.$$

---

1. $\boxed{A}$ Find the digital roots of the following numbers:
   (a) 7349621 (b) 100000 (c) 4716318
2. Find the digital roots of the following numbers:
   (a) 8940361 (b) 180000 (c) 182736549

PRACTICE PROBLEMS 4.L

The interesting phenomenon mentioned above is that the remainder obtained when any natural number is divided by 9 is just the digital root of that number (unless the digital root is 9, in which case the remainder will be 0). In other words, any number is congruent modulo 9 to its digital root.

The justification of this phenomenon depends on Theorem 4.8 and on the fact that our number system uses the base ten (see Chapter 5). Specifically,

$$10 \equiv 1 \pmod{9} \qquad (10 = 1 \cdot 9 + 1),$$

hence, by Theorem 4.8,

$$10^2 \equiv 1^2 = 1 \text{ (mod 9)} \qquad (100 = 11 \cdot 9 + 1)$$

$$10^3 \equiv 1^3 = 1 \text{ (mod 9)} \qquad (1000 = 111 \cdot 9 + 1)$$

and so on,

$$10^n \equiv 1^n = 1 \text{ (mod 9)} \qquad (10^n = 11 \ldots 1 \cdot 9 + 1).$$

Any number may be expressed as a sum of multiples of powers of ten. For example:

The number written as ABCDEF in our number system may be expressed as	275314
$ABCDEF =$ $A \cdot 10^5 + B \cdot 10^4 + C \cdot 10^3 +$ $D \cdot 10^2 + E \cdot 10 + F \cdot 1$ $\Big\}(4.4)$	$= 200000 + 70000 + 5000$ $+ 300 + 10 + 4$ $= 2 \cdot 10^5 + 7 \cdot 10^4 + 5 \cdot 10^3$ $+ 3 \cdot 10^2 + 1 \cdot 10 + 4 \cdot 1$
(Recall, $10^0 = 1$.) By Theorem 4.8, this is congruent modulo 9 to	
$A \cdot 1 + B \cdot 1 + C \cdot 1$ $\qquad + D \cdot 1 + E \cdot 1 + F \cdot 1$ $= A + B + C + D + E + F.$	$\equiv 2 \cdot 1 + 7 \cdot 1 + 5 \cdot 1 + 3 \cdot 1$ $\qquad + 1 \cdot 1 + 4 \cdot 1 \text{ (mod 9)}$ $= 2 + 7 + 5 + 3 + 1 + 4$
If $A + B + C + D + E + F$ is not a one-digit number, we continue the process until a one-digit number is obtained.	$= 22$ $= 2 \cdot 10 + 2$ $\equiv 2 + 2 \text{ (mod 9)}$ $= 4$
	Thus
	$275314 \equiv 4 \text{ (mod 9)}.$

The general proof follows the same lines.

The fact that we are interested only in the result modulo 9 means that we can work step-by-step modulo 9 when we compute the digital root. That is, we may throw away multiples of 9. Hence the term "casting out nines." For example, in computing the digital root of 2965184860, we may throw away the 9, the 1 and the 8 (which add to 9), and the 4 and the 5, leaving $2 + 6 + 8 + 6 = 22; 2 + 2 = 4$.

The process of casting out nines provides a quick check (not foolproof) of addition and multiplication problems. If $d(m)$ and $d(n)$

denote the digital roots of m and n respectively, then $m \equiv d(m)$ (mod 9) and $n \equiv d(n)$ (mod 9), and so, by Theorem 4.8,

$$d(m + n) \equiv m + n \equiv d(m) + d(n) \quad \text{(mod 9)}$$

and

$$d(mn) \equiv mn \equiv d(m)d(n) \quad \text{(mod 9)}.$$

Thus, the digital root of a sum (product) of numbers should be the same as the digital root of the sum (product) of the digital roots of the numbers.

For example, is the following calculation correct?

$$269 \times 347 = 93543$$

The digital root of 269 is 8, that of 347 is 5, and that of 93543 is 6. Since $8 \times 5 = 40$, which has digital root 4 rather than 6, there is a mistake in the multiplication.

The process of casting out nines is also helpful in designing some mystifying number guessing tricks. (See Exercises 4.25 through 4.28 at the end of the chapter.)

---

1. $\boxed{A}$ Find the missing digit if 3_4103161 is divisible by 9.
2. Find the missing digit if 4731_29 leaves a remainder of 2 upon division by 9.
3. $\boxed{A}$ What is the remainder when 456823940 is divided by 9?
4. Use digital roots to find the mistake in the following:

$$
\begin{array}{r}
3641 \\
\times \quad 128 \\
\hline
29128 \\
7282 \\
3641 \\
\hline
468048
\end{array}
$$

5. $\boxed{S}$ Use equation 4.4 to find a rule that gives the remainder when ABCDEF is divided by 2; by 3; by 4; by 5; by 7; by 8; by 11.
6. What is the remainder when 24732151 is divided by 2? by 3? by 4? by 5? by 7? by 8? by 11?

PRACTICE
PROBLEMS
4.M

SOLVING LINEAR CONGRUENCES

Let us return to our goal of developing machinery for solving linear Diophantine equations.

The properties of congruence exhibited in Theorem 4.8 are almost identical to the properties of equality. However, there is one property of our number system that holds for equality but does not hold for congruence—namely, the cancellation law:
If $ax = ay$, and $a \neq 0$, then $x = y$.
An analog for congruence,

$$\text{if } ax \equiv ay \text{ (mod } d) \text{ and } a \not\equiv 0 \text{ (mod } d), \text{ then } x \equiv y \text{ (mod } d)$$

does not always hold. For example, putting $x = 3$, $y = 7$, $a = 6$, and $d = 8$, we see

$$6 \cdot 3 = 18 \equiv 42 = 6 \cdot 7 \text{ (mod 8).}$$

But $3 \not\equiv 7$ (mod 8), so we cannot cancel the 6.
There is a case, though, when the cancellation law for congruence multiplication does hold:

Theorem 4.10 If $ax \equiv ay$ (mod d) **and** if $\gcd(a, d) = 1$, then $x \equiv y$ (mod d).

Note the condition that $\gcd(a, d) = 1$; in this case we can cancel.
For example, if $3x \equiv 15$ (mod 34), then since $\gcd(3, 34) = 1$, $x \equiv 5$ (mod 34).
Actually, we can cancel in a more general setting, but we must change the modulus. That is, we have the following theorem:

Theorem 4.11 If $ax \equiv ay$ (mod d) and if $\gcd(a, d) = g$, then $x \equiv y$ (mod d/g).

(Note that Theorem 4.10 is a special case of Theorem 4.11, that is, $g = \gcd(a, d) = 1$.)
For example, if

$$6x \equiv 6y \text{ (mod 8),}$$

then

$$x \equiv y \text{ (mod 4)} \qquad (4 = 8/2 = 8/\gcd(6, 8)).$$

Thus,

$$6 \cdot 3 \equiv 6 \cdot 7 \text{ (mod 8) implies } 3 \equiv 7 \text{ (mod 4).}$$

Similarly, if

$$8x \equiv 12 \text{ (mod 20),}$$

then

$$2x \equiv 3 \text{ (mod 5)} \qquad (5 = 20/4 = 20/\gcd(8, 20)).$$

1. $\boxed{\text{A}}$ (a) If $7x \equiv 35 \pmod{81}$, solve for x.
   (b) If $7x \equiv 35 \pmod{84}$, solve for x.
2. Solve each of the following congruences for x:
   (a) $3x \equiv 27 \pmod{42}$.   (b) $-8x \equiv 16 \pmod{23}$.
   (c) $4x \equiv 24 \pmod{30}$.

PRACTICE PROBLEMS 4.N

Another important difference between equality and congruence is that we can always solve the linear equation $ax = b$, as long as $a \neq 0$ (just let $x = b/a$); but we cannot always solve linear congruences. For example, the congruence $2x \equiv 3 \pmod 4$ has no solutions. (When we deal with congruences, we are only interested in integers. For any integer x, $2x$ will be even and will leave a remainder of 0 or 2 when divided by 4; it will not leave a remainder of 3.)

In general, given the congruence $ax \equiv b \pmod d$, if $x = n$ is a solution, then any integer congruent to n modulo d is also a solution, since we have seen that any number in a congruence may be replaced by any other number in the same remainder class.

When we count the number of solutions of a congruence, we mean the number of distinct remainder classes in which solutions lie.

For example, $x = 3$ satisfies $2x \equiv 1 \pmod 5$; hence, $x = 8$ also satisfies it. $2(8) = 16 \equiv 1 \pmod 5$. Similarly, $x = \ldots, -7, -2, 3, 8, \ldots$ all satisfy this congruence. We write $x \equiv 3 \pmod 5$ is a solution.

One way of determining the solutions of a congruence modulo d is to substitute $0, 1, 2, \ldots, d - 1$ for x, testing when the congruence is satisfied. Thus, as we saw above, $2x \equiv 3 \pmod 4$ has no solutions since 0, 1, 2, and 3 do not satisfy this congruence.

Now consider $3x \equiv 1 \pmod 5$. Substituting

$$x = 0, \quad 3 \cdot 0 = 0 \not\equiv 1 \pmod 5$$

$$x = 1, \quad 3 \cdot 1 = 3 \not\equiv 1 \pmod 5$$

$$x = 2, \quad 3 \cdot 2 = 6 \equiv 1 \pmod 5$$

$$x = 3, \quad 3 \cdot 3 = 9 \not\equiv 1 \pmod 5$$

$$x = 4, \quad 3 \cdot 4 = 12 \not\equiv 1 \pmod 5,$$

we see that this congruence has a unique solution modulo 5: $x \equiv 2 \pmod 5$. Note that $x = 2, x = 7, x = 12, \ldots$ all satisfy the congruence.

As a further example, consider $2x \equiv 0 \pmod 4$. Substituting

$$x = 0, \quad 2 \cdot 0 = 0 \equiv 0 \pmod 4$$

$$x = 1, \quad 2 \cdot 1 = 2 \not\equiv 0 \pmod 4$$

$$x = 2, \quad 2 \cdot 2 = 4 \equiv 0 \ (\text{mod } 4)$$

$$x = 3, \quad 2 \cdot 3 = 6 \not\equiv 0 \ (\text{mod } 4),$$

we see that this congruence has two solutions modulo 4: $x \equiv 0 \ (\text{mod } 4)$, and $x \equiv 2 \ (\text{mod } 4)$.

The general situation is given in the following theorem:

Theorem 4.12 The congruence $ax \equiv b \ (\text{mod } d)$ is solvable if and only if $\gcd(a, d)$ divides b. If it is solvable, then there are $g = \gcd(a, d)$ solutions modulo d.

Note that, if $\gcd(a, d) = 1$, then, according to Theorem 4.12, $ax \equiv b \ (\text{mod } d)$ has a unique solution modulo d, since 1 certainly divides b.

PRACTICE PROBLEMS 4.0

1. $\boxed{A}$ How many solutions are there to each of the following congruences? That is, how many remainder classes give solutions?
   (a) $3x \equiv 6 \ (\text{mod } 8)$ (b) $3x \equiv 6 \ (\text{mod } 9)$
   (c) $3x \equiv 1 \ (\text{mod } 6)$ (d) $4x \equiv 26 \ (\text{mod } 98)$.

2. How many solutions are there to each of the following congruences? That is, how many remainder classes give solutions?
   (a) $4x \equiv 24 \ (\text{mod } 48)$ (b) $3x \equiv 1 \ (\text{mod } 17)$
   (c) $9x \equiv 3 \ (\text{mod } 6)$ (d) $6x \equiv 27 \ (\text{mod } 45)$.

3. $\boxed{A}$ Which of the following congruences can be solved? Solve, by trial, if possible:
   (a) $5x \equiv 7 \ (\text{mod } 12)$ (b) $3x \equiv 7 \ (\text{mod } 12)$
   (c) $12x \equiv 8 \ (\text{mod } 16)$ (d) $5x \equiv 8 \ (\text{mod } 3)$.

4. Which of the following congruences can be solved? Solve, by trial, if possible:
   (a) $5x \equiv 9 \ (\text{mod } 12)$ (b) $3x \equiv 9 \ (\text{mod } 12)$ (c) $3x \equiv 11 \ (\text{mod } 18)$.

The trial and error method of solving linear congruences works well if the modulus is small, but it is not practical for larger moduli.

Another method of solving congruences is as follows: We have $3x \equiv 1 \ (\text{mod } 5)$. We wish to replace 1 by a number that is congruent to it modulo 5 and that is divisible by 3. In this case, 6 works. (So does -9.) Thus $3x \equiv 1 \ (\text{mod } 5)$ has the same solutions as $3x \equiv 6 \ (\text{mod } 5)$. Since $\gcd(3, 5) = 1$, we can cancel 3 by Theorem 4.10. Thus $x \equiv 2 \ (\text{mod } 5)$. [If we had used -9 instead of 6, we would have obtained $x \equiv -3 \ (\text{mod } 5)$, which is the same solution since $-3 \equiv 2 \ (\text{mod } 5)$.]

In general, if we have $ax \equiv b \pmod{d}$ with $\gcd(a, d) = 1$ and if $a \nmid b$, then we wish to find a number that is divisible by a and is in the same remainder class as b. To do this we keep adding (or subtracting) d to b until we reach a number that is divisible by a.

$$7x \equiv 6 \pmod{37}$$

$$6 + 37 = 43 \; (7 \nmid 43)$$
$$43 + 37 = 80 \; (7 \nmid 80)$$
$$80 + 37 = 117 \; (7 \nmid 117)$$
$$117 + 37 = 154 \; (7 \mid 154)$$

so

$$7x \equiv 154 \pmod{37}$$

$$x \equiv 22 \pmod{37}$$

[Note $154 \equiv 6 \pmod{37}$.]

---

1. ⬛A⬛ Solve each of the following:
   (a) $5x \equiv 3 \pmod{12}$ (b) $5x \equiv 7 \pmod{12}$
   (c) $15x \equiv 39 \pmod{59}$.
2. Solve each of the following:
   (a) $7x \equiv 11 \pmod{16}$ (b) $4x \equiv -39 \pmod{11}$
   (c) $4x \equiv -38 \pmod{41}$.

PRACTICE PROBLEMS 4.P

SOLVING LINEAR DIOPHANTINE EQUATIONS

We are now ready to apply what we have learned about congruences to solve linear Diophantine equations of the type $ax + by = c$. As we saw above, there can be no solution unless $\gcd(a, b) \mid c$.

If $\gcd(a, b)$ does divide c, then we can divide through by $\gcd(a, b)$ to obtain

$$a^\star x + b^\star y = c^\star,$$

where

$$a^\star = a/\gcd(a, b),$$

$$b^\star = b/\gcd(a, b),$$

$$c^\star = c/\gcd(a, b).$$

This new equation has exactly the same solutions as the original equation. Furthermore, since we

$$66x + 48y = 30$$
$$\gcd(66, 48) = 6; \quad 6 \mid 30$$

$$11x + 8y = 5$$

have divided out the greatest common divisor of a and b, the resulting numbers, $a^\star$ and $b^\star$, can have no common divisor greater than 1. That is,

$$\gcd(a^\star, b^\star) = 1.$$

$$\gcd(11, 8) = 1.$$

Therefore, there is no loss in generality if we consider only Diophantine equations $ax + by = c$ for which $\gcd(a, b) = 1$.

We continue with the general method for solving $ax + by = c$ on the left side of the page, and with the example $11x + 8y = 5$ on the right.

Consider the new equation as a congruence modulo either a or b. (Usually, we choose the smaller of the two so that we deal with smaller numbers.) Working modulo b, we get

$$ax \equiv c \pmod{b}.$$

$$11x + 8y = 5$$
$$11x + 8y \equiv 5 \pmod{8}$$
Since $8 \equiv 0 \pmod{8}$,
$$11x + 0y \equiv 5 \pmod{8}$$

$$11x \equiv 5 \pmod{8}$$

Since $\gcd(a, b) = 1$, this congruence has a solution. If possible, we reduce a and c modulo b to obtain

$$a'x \equiv c' \pmod{b}.$$

$$11 \equiv 3 \pmod{8}, \text{ so}$$

$$3x \equiv 5 \pmod{8}$$

As indicated above, we can find the solution of this congruence by adding or subtracting multiples of b to c', until we obtain c'' which is divisible by a'. Then, the solution is

$$x \equiv d \pmod{b},$$

where $d = c''/a'$. Or, equivalently,

$$x = d + kb,$$

where $k = 0, \pm 1, \pm 2, \dots.$ For each value of k, we may substitute the corresponding

$$3x \equiv 5 - 8 \equiv -3 \pmod{8}$$
$$3x \equiv -3 \pmod{8}$$

$$x \equiv -1 \pmod{8}$$

$$(\text{where } -1 = -3/3)$$

$$x = -1 + 8k,$$

$$k = 0, \pm 1, \pm 2, \dots$$

value of x into $ax + by = c$ to find the corresponding value for y.

$$8y = 5 - 11x$$
$$= 5 - 11(-1 + 8k)$$
$$= 5 + 11 - 88k$$
$$= 16 - 88k$$
$$y = 2 - 11k$$

The solution is

$$x = -1 + 8k, \quad y = 2 - 11k,$$
$$k = 0, \pm 1, \pm 2, \dots.$$

Usually, the conditions of the problem limit the values that need to be considered for x, and hence limit the values for k that must be tried.

If the problem restricts x to be a number between 10 and 20, k must $= 2$, $x = 15$, $y = -20$.

---

1. $\boxed{A}$ Find all solutions for $17x - 19y = 1$, for which $50 \le x \le 100$.
2. Find all solutions for $17x - 19y = 3$.
3. $\boxed{A}$ Find all solutions for $37x - 53y = 11$.
4. Find all solutions for $44x + 23y = 13$.

PRACTICE PROBLEMS 4.Q

**We are now ready to consider Problem 4.4 (page 101).
If you have not already solved it, try it again now.**

Earlier we obtained the equation $98x - 199y = 222$, where x represents the number of dollars and y represents the number of cents on the check. Observe that $\gcd(98, 199) = 1$. Since $98 < 199$, we consider the equation as a congruence modulo 98:

$$98x - 199y \equiv 222 \pmod{98}.$$

Since $98 \equiv 0 \pmod{98}$, $199 \equiv 3 \pmod{98}$, and $222 \equiv 26 \pmod{98}$, this reduces (by Theorem 4.8) to

$$-3y \equiv 26 \pmod{98}.$$

Since 26 is not divisible by 3, we add 98, obtaining 124. [This doesn't change the congruence because $124 \equiv 26 \pmod{98}$.] As 124 is still not divisible by 3, we add 98 again, obtaining 222. Thus

$$-3y \equiv 222 \pmod{98}$$

Solution of Problem 4.4 (Continued)

or
$$-y \equiv 74 \ (\text{mod } 98)$$
or
$$y \equiv -74 \ (\text{mod } 98)$$
or
$$y = -74 + 98k.$$

Since we need $0 < y < 100$, we need only consider one value of k, $k = 1$. That is,

$$0 < -74 + 98k < 100$$

$$74 < 98k < 174$$

$$0 < 74/98 < k < 174/98 < 2$$

$$k = 1,$$

since k is an integer. Thus, $y = -74 + 98 \cdot 1 = 24$. Substituting this value of y into $98x - 199y = 222$, we obtain $98x = 222 + 199 \cdot 24 = 222 + 4776 = 4998$;

$$x = 51.$$

Thus, the amount of the check was \$51.24. (Of course, you should return to the original problem to check that this is the correct answer.)

There are times when a congruence is not solved so easily. For example, consider the Diophantine equation

$$285x + 391y = 72.$$

Transforming this to a congruence modulo 285,

$$285x + 391y \equiv 72 \ (\text{mod } 285).$$

Reducing modulo 285,

$$0x + 106y \equiv 72 \ (\text{mod } 285)$$

$$106y \equiv 72 \ (\text{mod } 285).$$

Since $\gcd(2, 285) = 1$, we can divide both sides of the congruence by 2:

$$53y \equiv 36 \ (\text{mod } 285). \tag{4.5}$$

If we now proceed as previously, we would start adding (or subtracting) 285 to 36 until we obtain a number that is divisible by 53:

$$36 + 285 = 321; \quad 53 \nmid 321$$
$$321 + 285 = 606; \quad 53 \nmid 606$$

etc.

Although we know that we must eventually reach a number divisible by 53 (this is guaranteed by the fact that gcd(53, 285) = 1), it might take a long time for this to happen. Furthermore, deciding whether or not a number is divisible by 53 is not easy. It therefore would be preferable for us to find an alternate approach.

Suppose we take the congruence 4.5 and write it as an equation:

$$53y = 36 + 285q, \tag{4.6}$$

for some integer q. This is again a Diophantine equation, and, as before, we can replace this by a congruence:

$$53y \equiv 36 + 285q \pmod{53} \tag{4.7}$$

$$0 \equiv 36 + 285q \pmod{53}$$

$$0 \equiv 36 + 20q \pmod{53}$$

$$-20q \equiv 36 \pmod{53}. \tag{4.8}$$

The important advantage of congruence 4.8 over 4.5 is that the modulus is smaller (53 as opposed to 285) and it should be easier to solve. Dividing by 4, (note gcd(4, 53) = 1),

$$-5q \equiv 9 \pmod{53} \tag{4.9}$$

we could now proceed by trying to add (or subtract) 53 to 9 until we obtain a number divisible by 5—since 5 is a smaller number, this would be a reasonable approach—or we could again replace congruence 4.9 by an equation

$$-5q = 9 + 53t \tag{4.10}$$

for some integer t. This equation can again be replaced by a congruence:

$$-5q \equiv 9 + 53t \pmod{5}$$

$$0 \equiv 4 + 3t \pmod{5}$$

$$-3t \equiv 4 \pmod{5}. \tag{4.11}$$

Observe that the modulus (5) of this congruence is smaller than previous moduli. In fact, 5 is small enough that congruence 4.11 can be easily solved by trial and error. The solution is

$$t \equiv 2 \pmod{5}.$$

We now wish to use this to solve congruence 4.5, so we can find y. Since knowing one solution of congruence 4.5 determines all solutions, we can take $t = 2$ rather than having to deal with $t \equiv 2 \pmod 5$. Substituting $t = 2$ into equation 4.10,

$$-5q = 9 + (53)(2) = 9 + 106 = 115$$

$$q = -23.$$

Substituting $q = -23$ into equation 4.6,

$$53y = 36 + 285(-23) = 36 - 6555 = -6519$$
$$y = -123.$$

Therefore the solution of congruence 4.5 is

$$y \equiv -123 \ (\text{mod } 285)$$

or, equivalently,

$$y = -123 + 285k.$$

Substituting into the Diophantine equation and solving for x, we obtain

$$285x + 391(-123 + 285k) = 72$$
$$285x + (285)(391k) = 72 + (391)(123)$$
$$285(x + 391k) = 72 + 48093 = 48165$$
$$(x + 391k) = 169$$
$$x = 169 - 391k.$$

Thus, the solution of the Diophantine equation is $x = 169 - 391k$, $y = -123 + 285k$. For each value of k, we get a pair of values for x and y.

In general, if we wish to solve

$$ax \equiv b \ (\text{mod } m),$$

with $\gcd(a, m) = 1$ and $0 < a < m$, we can replace the congruence by the Diophantine equation

$$ax + my = b,$$

which we in turn replace by the congruence

$$my \equiv b \ (\text{mod } a).$$

Since this congruence has a smaller modulus, it may be easier to solve by trial and error; if not, we can continue the procedure. As the modulus keeps getting smaller and smaller, the method must yield a solution in a finite number of steps. However, the trial method may be used at any point along the way to speed up the procedure.

We have presented two approaches to solving a congruence of the form $ax \equiv b \ (\text{mod } m)$. One is to keep adding (or subtracting) m to b until we obtain a number divisible by a, and the other is to replace the congruence by a Diophantine equation which is in turn replaced by a congruence, etc. Which of these two approaches is preferable? That will vary from problem to problem. If the coefficient a is reasonably small, we suggest trying the addition/subtraction method first; but if adding m to b one or two times and subtracting m from b one or two times doesn't result in a number divisible by a, then try the other approach. If the coefficient a is fairly large—say, greater than 20—then we suggest using the second approach immediately.

THE CHAPTER IN RETROSPECT

In this chapter you have met some more types of problems that have been of interest to recreational mathematicians for centuries. But this chapter has also given you a taste of some of the basic mathematical concepts in the field of number theory—a field that has been called the "Queen of Mathematics." We have only touched upon the subject; but many of the topics we have considered (divisibility, prime factorization, congruence) have analogs that are also important in other branches of mathematics.

This chapter is different from the first three in that it has formally presented definitions and theorems. Precise definitions are important in mathematics. We need them so that, when we refer to a particular term, we all know exactly what we mean and we all mean the same thing. For example, when we say that a number is prime, we should all understand that we are really saying that the number has exactly two divisors; we should also understand that we will be able to determine whether or not a number is prime if we can count its divisors.

Theorems are also important. They elaborate on the properties of and the relationships between the terms that are defined. They give us general information about all mathematical objects of some particular type, rather than just telling us about a particular example. For example, Theorem 4.12 gives us information about all congruences of the form $ax \equiv b \pmod{d}$. It tells us when such a congruence has a solution. The theorem is a general result. It tells us much more than that a particular congruence, such as $3x \equiv 1 \pmod{5}$, has a solution; it tells us that *all* congruences of the particular type which satisfy certain conditions have solutions.

Theorems must be proved before they are accepted as valid. Proofs are usually based on definitions, on the statements of other, previously proven theorems, and on primary assumptions called **axioms** which are accepted without proof.

This chapter has supplied very few actual proofs. Although most of the theorems that were presented are not difficult to prove, we have been more interested in discussing their applications rather than giving a rigorous presentation.

The theorems have been used to solve the sample problems. They also will be helpful for many of the exercises that follow. In general, problems of the type dealt with in this chapter can be stimulating; but they also can be frustrating. It is exciting to see how a little mathematical theory can be useful in solving such problems.

Exercises

Primes and Divisibility

4.1. A group of sorority sisters stopped for refreshments at the Soda Shoppe. Each ordered a Coke, and the total bill was $1.87, exclusive of tax. They were surprised to find out that a Coke was no longer $.15; prices are going up. How many were in the group, and what is the price of a Coke?

4.2. [A] A golden age group went on a bus tour. Each person paid his or her own exact fare with five coins. In totaling the fares, the bus driver obtained a sum of $21.83.

How many pennies did the bus driver receive?

4.3. [H] Using a desk calculator, a student was asked to obtain the complete factorization of 24,949,501. Dividing by successively increasing primes, he found the smallest prime divisor to be 499 with quotient 49,999. At this point, he quit.

Why didn't he continue his division? [52], Problem 16)

4.4. [A] When the accountants for Lose-a-digit Computer, Inc. had finished preparing their annual budget, they presented the final figures to the president, I. M. Smart.

"It looks like a good year," he exclaimed. "The amount of the budget just happens to

be the smallest number of cents (other than one cent) that is a perfect square, a perfect cube, and a perfect fifth power."

How much money has Lose-a-digit budgeted for the year?

4.5. [A] A census taker stopped at Hotel Sleep-Inn. The clerk at the desk was a mathematics student with a sense of humor. When asked how many guests were in the hotel, she replied, "The number is the smallest positive integer that has the following properties:

When divided by two, the result is a perfect square;
when divided by three, the result is a perfect cube."

The census taker, who had been calculating all day, slammed his book on the table. "I quit," he said.

Can you figure out how many guests were in the hotel?

4.6. [H] [A] If n is a positive integer, then $n!$ (read "n factorial") denotes the product of all positive integers that are less than or equal to n. For example,

$$5! = 5 \cdot 4 \cdot 3 \cdot 2 \cdot 1 = 120,$$

$$6! = 6 \cdot 5 \cdot 4 \cdot 3 \cdot 2 \cdot 1 = 720,$$

$$10! = 10 \cdot 9 \cdot 8 \cdot 7 \cdot 6 \cdot 5 \cdot 4 \cdot 3 \cdot 2 \cdot 1$$
$$= 3,628,800.$$

Note that 5! and 6! both end in one zero, and that 10! ends in two zeros. Without actually computing 50!, can you tell the number of zeros in which 50! ends? (Note: 50! is not equal to 5! · 10!)

★ **4.7.** [S] [A] The Wellington children wanted to buy an anniversary present for their parents. They each contributed a number of dollars equal to the number of boys in the family, and thereby collected a number of dollars that exceeded by 24 the number of years that their parents had been married.

The number of sons, the number of daughters, and the number of years that the parents had been married were three distinct prime numbers.

Find these numbers.

4.8. [H] A Christmas tree is decorated with 36 lights, which are numbered 1 through 36. Timers are set so that every 5 minutes a change occurs in the light pattern. The sequence of changes repeats every 3 hours.

Switches are set so that:

At the end of the first time interval, every light is turned on.

At the end of the second time interval, every second switch is reversed.

At the end of the third time interval, every third switch is reversed.

Etc.

(a) What is the state of the lights at the end of the first 3 hours (after the 36th time interval)?

(b) If the tree had n lights, what is the state after n time intervals?

★ **4.9.** [A] Three boys, Pedro, Quincy, and Ralph, and their sisters, Sandy, Trixie, and Vera (not necessarily in that order), had chickens for pets.

Last week was an unusual one. Each chicken laid as many eggs as its owner owned chickens. Quincy had three times as many chickens as his own sister, and had eight more chickens than Ralph's sister. Furthermore, by the end of the week, Quincy had collected 56 more eggs than Pedro; Ralph had collected 52 more than Sandy; and Pedro had collected as many eggs as Sandy and Trixie together.

How many chickens did each of the six people own? And, who was whose sister?

★ **4.10.** [S] [A] If $a^3 = b^2$ and $c^3 = d^2$, where a, b, c, and d are positive integers; and if $c - a = 25$, what are a, b, c, and d? ([60], Vol. 3, Problem 174)

4.11. [H] Find three positive integers whose sum is 25 and whose product is 540.

★ **4.12.** [H] [A] Once again, the census taker appears on the scene. He stops at a house, notes down the number on the door, and knocks. When a woman answers, he asks her age and notes the answer. Then he asks if anyone else lives at the house. She replies that three children live with her. Upon asking their ages, the census taker is given the reply that the sum of their ages equals the number on the door and that the product of their ages equals 36. The man does some quick computation and says that he needs another clue. He then asks if the youngest of the three is a twin. The woman replies that he is not; whereupon, the census taker is able to compute the ages of the three.

What are their ages?

4.13. [H] Find the largest number that leaves the same remainder when it is divided into 887, 959, 1007, and 1187.

Linear Diophantine Equations

4.14.   The annual dues for the Burbank Book Club are generally $23.00 per person. However, senior citizens pay only $17.00. If the total amount of dues collected this year was $1500.00, what is the smallest number of senior citizens that could belong to the club?

4.15. [A] Judy had absolutely no money for shopping, so she went to the bank to cash her tax refund check. Somehow, a $50 bill had become mixed in with the $20 bills in the teller's drawer, and the teller mistakenly gave this bill to Judy, thinking that it was a $20 bill.

Judy did not notice the mistake until after she had spent $2.40 in one store for some pairs of stockings and was about to purchase a dress in another store. The price of the dress (including tax) was exactly one-third of the money she then had with her. When she removed the "$20" bill from her wallet, she discovered that it was $50. She paid for the dress anyway, and observed that this left her with a number of cents equal to the amount of dollars of the original check and a number of dollars equal to the amount of cents of the check.

Being an honest person, Judy returned to the bank to correct the mistake.

How much was Judy's tax refund?

4.16.   A used car salesman purchased a number of very old and battered cars during the past month. He paid the same price, $89, for each. He has not yet sold all of the cars, but those that he has sold went for $225 apiece. He has already made a profit of $2,327 (total sale price minus outlay for all cars purchased).

If his car lot has room for no more than 50 cars at one time, how many cars remain?

4.17. [H] One hundred units of U.S. money were worth a total of $100. They included only half-dollars, $5 bills, and $10 bills. How many coins and bills of each denomination were there?

4.18. [A] At a meeting of the Coin Collectors of America, every member present had fifty U.S. coins (pennies, nickels, dimes, quarters, and half-dollars) in his pockets. Each person's coins added up to $1.00, and no two people brought exactly the same coins.

What is the largest number of people who could have attended the meeting?

4.19. [S] [A] Three freshmen stopped into their college book store for supplies. Garrett paid $6.95 for six notebooks, four pencils, and five pens. Beth bought two notebooks, five pencils, and three pens for a total of $2.85, commenting that the "ten cent pen" no longer existed.

Assuming that each notebook, each pencil, and each pen costs an exact number of cents, what did Wilma pay for one notebook, one pencil, and one pen?

4.20. [H] [A] The new saleswoman at Clocks Unlimited was unbelievable. During her first day on the job, she sold at least one of each of the three models that other salespeople

found most difficult to sell—the chartreuse kitchen clock that sells for $17, the $31 owl-shaped cuckoo clock that hoots instead of cuckooing, and the two-foot high grandfather clock that was on sale for $61. In all, the saleswoman collected $300 (excluding tax) from selling these models alone.

How many of each did she sell?

Congruence and Digital Roots

4.21. Locksher Mohles had finally decoded the message. It read, "The jewels are buried in St. Breka Cemetery under the 221st tombstone." The writing was followed by the picture in Figure 4.4.

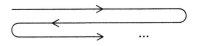

FIGURE 4.4

Mohles and his close friend, Dr. Swanto, reached the cemetery. There they found only 15 tombstones in the entire cemetery, and these were arranged in a straight line in front of the entrance.

"What do we do now, Mohles?" Swanto asked.

"Obviously, from the map, we must count the tombstones from 1 to 15 and then we must reverse direction, counting tombstone #14 as 16, #13 as 17 and so on, until we reach #1 which we count as 29. Then we reverse direction again, and so on," replied Mohles.

"But that will take a long time," snapped Swanto; "And it's spooky here."

"Don't worry. I know right now which tombstone to approach."

Do you know too? (No fair counting up to 221.)

4.22. What is the remainder when $3^{666,666}$ is divided by 7?

★ **4.23.** [H] Prove that, for any positive integer n, $3^{3n+1} + 2^{n+1}$ is always divisible by 5.

★ **4.24.** [H] [A] Magician: "Select any number from 1 to 12, but do not tell me what number you have selected. Now choose a second number from 1 to 12, but this time tell me the number."

Subject: "4."

Magician: "Fine. Now, starting with 4 and proceeding in a counterclockwise direction, I will tap the numbers on the face of my watch, one at a time. When I tap the 4, say to yourself the number you originally selected; with each successive tap I make, you add 1 to your number. Stop me when you reach 16. I will then be pointing at the number you selected."

As usual, the magician's prediction was realized.

Explain why this trick works. What would the magician have said if the second selected number had been a number other than 4?

4.25. [H] Pick a number. Multiply it by 3. If the result is odd, add 25; if it is even, add 34. The new result will be even; divide it by 2. Add 11. Now multiply by 6.

Add up the digits of your answer. If the number you obtain contains more than one digit, add the digits again. Continue doing so until you obtain a one-digit result.

Your answer will be 6.

How do we know?

4.26. [A] Giselda, the magician's wife, was learning the trade fast. She took two decks of cards from her pocket and placed them face down on the table. She then asked a volunteer from the audience to pick a number greater than 11 and smaller than 20. The volunteer did so. She then handed him one deck of cards and asked him to deal the cards face down on the table, placing each card on top of the previous one and counting one for each card, until he reached the number he selected. She then picked up the packet of cards that the volunteer had just dealt, and proceeded to arrange them like the numbers on a clock. That is, she placed the first card in the position of the number 1, the next card in the position of number 2, and so on, until 12 cards were so arranged. The remainder she returned to the deck.

The volunteer was then told to add the digits of the number he had selected and to note the card that was in the corresponding position. (For example, if the number he chose

had been 15, then he would note the card in position number 6.)

Giselda then handed the volunteer the second deck, asking him to announce his chosen card and to hold the deck so that the audience could see the face of the cards.

To everyone's amazement, all the cards the volunteer held were identical to the card he announced.

How was Giselda able to perform this trick without knowing in advance what number would be selected?

4.27. [H] The magician himself insisted on doing the next trick. He pointed to a young woman in the audience.

"Pick a number," he said. "Double it; add 7; multiply by 5; subtract the number you started with. Remove any nonzero digit from the answer; and tell me the remaining digits in any order."

"6 and 8," replied the woman.

"Then the digit you removed is a 3," announced the magician.

He was correct. How did he know?

4.28. [H] As another variation of the trick in Exercise 4.27, the magician handed a piece of paper and a pencil to another volunteer from the audience. He then gave the following instructions:

"Write down any number; scramble the digits to make another number; subtract the smaller of the two numbers from the larger; circle any nonzero digit in the result; add the remaining digits and tell me the sum."

"14," announced the volunteer.

"Then the number you circled is a 4."

Again, how did the magician know?

★ **4.29.** [H] [A] If the sum of the digits of 77^{77} is S, and the sum of the digits of S is T, find the sum of the digits of T.

4.30. [H] [A] Four boys were playing close to a bubble gum machine. Suddenly, Kevin knocked it over, the machine broke, and the bubble gum pieces rolled into a pile on the floor. The three older boys—Kevin, Esteban, and Tony—decided that they would split the gum four ways. They left Sean, the youngest, to watch over the pile while they went to get containers.

Kevin was the first to return. He counted the number of gum balls and found that, if the number were divided by four, there would be one left over. Feeling that he was entitled to the extra piece of gum (since, after all, it was he who knocked over the machine in the first place), Kevin took the extra piece plus one-fourth of the remaining gum and left.

Esteban, the oldest of the four, was the next to arrive. He counted the pieces of gum and again found there was one more than could be evenly divided among the boys. Sean was just about to tell him that Kevin had already taken his share, when Esteban said, "Since I'm the oldest, I get the extra piece. If you don't like it, I'll punch you in the nose." Sean decided to keep his mouth shut. So Esteban took the extra piece of gum and one-fourth of the rest and left.

When Tony arrived, the scene was essentially the same. He, too, found one piece of gum too many to be divided between the four boys and decided to keep the extra piece

for himself. He therefore took one gum ball and one-fourth of the remainder and left.

Sean then gathered up the remaining pieces of gum and went home.

What was the smallest number of gumballs that could have been in the machine? Who got the best of the deal? Who got the worst?

4.31. [S] [A] A farmer had eight baskets, some containing apples and the remainder containing plums. The baskets contained 67, 62, 35, 34, 30, 25, 19, and 17 fruits respectively. Each apple cost three times as much as each plum. The farmer's first customer purchased a number of baskets of apples and paid the farmer $11.88. The second customer purchased a number of baskets of plums, also paying $11.88.

This left the farmer with one basket of apples.

What was the remaining basket worth?

★ **4.32.** [H] According to the terms of the last will and testament of J. P. Moneybags, his entire fortune (which turned out to be worth $2,518,000) was to be divided equally (in exact multiples of $1000) among all of his direct descendants who were present at the conclusion of his funeral. The excess, if any, was to be donated to the Moneybags Home for Retired Mathematicians.

Mendel Giant, the director of the home, attended the funeral to determine how well the institution would make out.

Ten minutes before the funeral was scheduled to begin, only three heirs were present.

"I guess that means we'll get $1000," said Giant.

Just then, another heir entered.

"Good," exlaimed Giant. "That means $2000 for us."

One minute later, a fifth heir entered.

"$3000," Giant mumbled.

As the funeral was about to begin, three more heirs came in.

"Terrific," uttered Giant. "Now we'll get $6000."

During the funeral, three more heirs entered, one at a time. With each one, Giant mentioned a new figure. At the entrance of the 9th heir, Mendel said, "$7000"; the tenth heir, "$8000"; and the final heir, "$10,000."

How was Dr. Giant able to determine so quickly how much the home would receive? (That is, how could he tell the remainder when 2,518 was divided by each of the numbers above?)

★ **4.33.** [H] (a) [A] For each value of n, from 2 to 16, find the largest and smallest nine-digit numbers, containing each of the nine nonzero digits exactly once, that are exactly divisible by n.

(b) Do the same for ten-digit numbers containing each of the ten digits (including 0) exactly once. (Note that 0 may not be the first digit of a number.)

4.34. [H] Two digits of the nine-digit number

$$25\_37\_411$$

are missing. Find these digits if the number is divisible by 99.

4.35. [H] [A] In how many different ways can you fill in the blanks of

$$5\_3\_1\_672$$

with the digits 4, 9 and 8, in some order, so that the resulting number is divisible by 792?

★ **4.36.** The Gregorian calendar, which most nations use today, is based on a year containing 365 days. Every fourth year is a leap year (containing 366 days), except that years divisible by 100 are not leap years unless they are also divisible by 400, in which case they are leap years. Thus, for example, 1900 was not a leap year, but 2000 will be.

(a) [S] Show that every 400 years the calendar repeats itself (that is, the same dates fall on the same exact day of the week). Thus, for example, since January 1, 1976 was a Thursday, January 1, 2376 will also be a Thursday.

(b) Show that the calendar also repeats itself every 28 years that do not include the turn of a century.

(c) [S] [A] What day of the week was January 1, 1901?

(d) [H] [A] In what years of the 20th century does February have five Sundays?

(e) [H] [A] What day of the week, if any, can never be February 29?

(f) [H] [A] What day of the week, if any, can never be the first day of a new century?

(g) [H] [A] Which years of the 20th century contain 53 Sundays?

★ ★ (h) [A] How many times in the 20th century will the thirteenth of a month fall on a Friday?
(Note the 20th century runs from January 1, 1901 through December 31, 2000, inclusive.)
([60], Vol. 4, pp. 6, 34; [1], Vol. 40, p. 607. See also [47], p. 204)

5

More About Numbers:
Bases and
Cryptarithmetic

You have probably had an experience similar to the following: You are given a large pile of pennies to count. You begin counting. As you count each penny, it is taken from the original pile and placed in a "discard" pile. You reach, say, 1324 and you are interrupted. (A streaker enters the room, dances a jig, and exits down a fire escape.) You attempt to resume counting, but can no longer remember what number you were up to. Oh well, start counting all over again.

A little foresight could have saved you a lot of wasted effort. Suppose that, instead of placing all pennies that have been counted in the same discard pile, you placed them in piles of 10, combining ten of these piles into a pile of 100 when possible, and so on. When you were interrupted at 1324, you would have had before you an arrangement as in Figure 5.1.

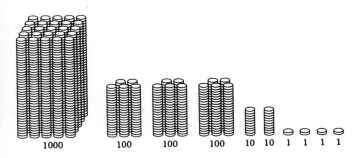

1000　　　100　　　100　　　100　　10　10　1　1　1　1

FIGURE 5.1

After the interruption, you would have had very little trouble resuming your count where you had left off.

This is the basic idea behind our **decimal**, or base ten, notation for numbers.

$$1324 = 1 \cdot 1000 + 3 \cdot 100 + 2 \cdot 10 + 4 \cdot 1.$$

In the decimal system, we express numbers in terms of powers of ten; hence the term "base ten." Historically, not all nations have used the base ten for their number systems. The ancient Babylonians used the base 60, and there is evidence that the bases 2, 3, 4, 5, 12, and 20 have also been used ([14], pp. 8–9).

In this chapter, we consider arithmetic in base ten and other bases. We also consider problems in which we are given only partial information about some arithmetical calculation. Some typical recreations of this type follow. Try them before reading further.

SAMPLE PROBLEMS

Problem 5.1

An assayer owns a balance scale that he uses to weigh ore samples. He always places the samples on the left pan of the balance and places weights on the right pan until a balance is achieved. If he wishes to be able to balance all possible samples weighing an integral number of grams between 1 and 127, what is the smallest number of weights he will need?

Problem 5.2

When the first Martian to visit Earth attended a high school algebra class, he watched the teacher show that the only solution of the equation $5x^2 - 50x + 125 = 0$ is $x = 5$.

"How strange," thought the Martian. "On Mars, $x = 5$ is a solution of this equation, but there is also another solution."

If Martians have more fingers than humans have, how many fingers do Martians have?* ([38], Chapter 7, Problem 9)

---

* Historically, at least part of the reason that we have adopted the base ten for our number system is that humans have ten fingers; the implication of Problem 5.2, then, is that the number of fingers that Martians have is the base of their number system.

Problem 5.3

A college student sent a postcard to her parents with the message

$$\begin{array}{r} \text{SEND} \\ + \text{MORE} \\ \hline \text{MONEY} \end{array}$$

If each letter represents a digit (0, 1, 2, 3, 4, 5, 6, 7, 8, or 9), with different letters representing different digits and the same letter representing the same digit each time it occurs, how much money is being requested? (SEND for the school tuition and MORE for other expenses.)

Problem 5.4

Reconstruct the division

$$\begin{array}{r} \text{C E C I} \\ \text{QUI)}\overline{\text{TROUVE}} \\ \underline{- - -} \\ \overline{- - -} \\ \underline{- - -} \\ - - - \\ \underline{- \, - \text{E}} \\ - - - \\ \underline{- - -} \\ 0 \end{array}$$

(*Qui trouve ceci?* is French for Who finds this?) (M. Pigeolet, [60], Vol. 2, p. 24)

POSITIONAL NOTATION

Our number system requires ten symbols—0, 1, 2, 3, 4, 5, 6, 7, 8, and 9—which we call digits. The position of each digit relative to the decimal point indicates the "place value" of the digit—that is, by what power of 10 it is multiplied. For example,

$$3 = 3 \cdot 1 = 3 \cdot 10^0$$

$$30 = 3 \cdot 10 = 3 \cdot 10^1$$

$$300 = 3 \cdot 100 = 3 \cdot 10^2$$

$$.3 = 3 \cdot \tfrac{1}{10} = 3 \cdot 10^{-1}$$

and so on. (Note that $10^0 = 1$. In fact, b^0 is defined to be 1 for any nonzero b.)

Thus the number 24 represents $2 \cdot 10 + 4 = 2 \cdot 10^1 + 4 \cdot 10^0$, while the number 204 represents $2 \cdot 100 + 0 \cdot 10 + 4 \cdot 1 = 2 \cdot 10^2 + 0 \cdot 10^1 + 4 \cdot 10^0$.

Note the importance of the digit 0 to help indicate position. This valuable digit was first introduced by the Hindus, prior to the year 800.

In general, it is possible to use any positive integer b greater than 1 as a base. We need b symbols, $0, 1, 2, \ldots, b-1$, and we still make use of positional notation.

Thus, in base two, we need two symbols, 0 and 1, and the number 1101 represents $1 \cdot 2^3 + 1 \cdot 2^2 + 0 \cdot 2^1 + 1 \cdot 2^0$. Similarly, in base three, we need three symbols, 0, 1, and 2, and the number 120 represents $1 \cdot 3^2 + 2 \cdot 3^1 + 0 \cdot 3^0$.

In the base twelve, we need twelve symbols, 0, 1, 2, 3, 4, 5, 6, 7, 8, 9, T, and E.

When more than one base may be involved, as in Problem 5.2, it is necessary to indicate the base with respect to which a number is being expressed. This is done by appending a subscript to the number.

For example, 110 may be ambiguous but the following are not:

$$(110)_{\text{two}} = 1 \cdot 2^2 + 1 \cdot 2^1 + 0 \cdot 2^0$$

$$(110)_{\text{three}} = 1 \cdot 3^2 + 1 \cdot 3^1 + 0 \cdot 3^0$$

$$(110)_{\text{ten}} = 1 \cdot 10^2 + 1 \cdot 10^1 + 0 \cdot 10^0.$$

If it is clear from a discussion what base is intended, the subscript may be omitted.

CHANGING BASES

Although every integer may be expressed with respect to any base, the representation of a number with respect to one base will usually differ from its representation with respect to another base. For example,

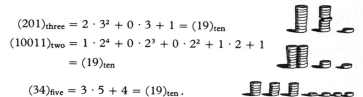

$$(201)_{\text{three}} = 2 \cdot 3^2 + 0 \cdot 3 + 1 = (19)_{\text{ten}}$$
$$(10011)_{\text{two}} = 1 \cdot 2^4 + 0 \cdot 2^3 + 0 \cdot 2^2 + 1 \cdot 2 + 1$$
$$= (19)_{\text{ten}}$$

and

$$(34)_{\text{five}} = 3 \cdot 5 + 4 = (19)_{\text{ten}}.$$

Thus, $(201)_{\text{three}}$, $(10011)_{\text{two}}$, $(34)_{\text{five}}$, and $(19)_{\text{ten}}$ are different representations of the same number—the number that we refer to as 19. In the first representation, the 19 objects have been grouped in terms of powers of three; in the second representation, powers of two have been used; and in the third, powers of five.

Since we are used to working in base ten, it is sometimes useful to be able to convert from the representation of a number in base b to its base ten representation, or vice versa.

To convert a number from base b to base ten is easy; it is simply a matter of expanding the number in terms of powers of b and then multiplying and adding. For example,

$$(1101101)_{two} = 1 \cdot 2^6 + 1 \cdot 2^5 + 0 \cdot 2^4 + 1 \cdot 2^3 + 1 \cdot 2^2 + 0 \cdot 2^1 + 1 \cdot 2^0$$

$$= 64 \quad + 32 \qquad\qquad + 8 \quad + 4 \qquad\qquad + 1$$

$$= (109)_{ten}.$$

---

1. $\boxed{A}$ Convert each of the following to base ten.
   (a) $(1101)_{two}$ (b) $(1001)_{seven}$ (c) $(3.5)_{six}$ (d) $(T81)_{eleven}$.
2. Convert each of the following to base ten.
   (a) $(10011101)_{two}$ (b) $(5403)_{six}$ (c) $(ET09)_{twelve}$ (d) $(31.2)_{seven}$.

PRACTICE PROBLEMS 5.A

To convert from base ten to base b is slightly more complicated because it requires division. We must express the given number as a sum of multiples of powers of b.

We illustrate by giving one general technique on the left and a corresponding example on the right:

If n is the given number that we wish to write in base b, we begin by finding the largest power of b that does not exceed n.	Suppose we want to write 185 $[= (185)_{ten}]$ in base 3.
	The powers of 3 are: $3^0 = 1$, $3^1 = 3$, $3^2 = 9$, $3^3 = 27$, $3^4 = 81$, $3^5 = 243$, ... The highest required power is $3^4 = 81$, since 243 is too big. That is,
Say this power is b^{a_1}.	$$3^4 = 81 \leq 185 < 243 = 3^5.$$
We next divide n by b^{a_1}, that is, we use the division algorithm (Theorem 4.7) to find q_1 and r_1 such that	$$\begin{array}{r} 2 \\ 81\overline{)185} \\ 162 \\ \hline 23 \end{array}$$
$$n = q_1 b^{a_1} + r_1,$$	so $q_1 = 2$, and $r_1 = 23$
with $0 \leq r_1 < b^{a_1}$. (Since $n < b^{a_1+1}$, it is easily seen that $q_1 \leq b - 1$.)	$$185 = 2 \cdot 81 + 23.$$

We now repeat the process using r_1 in place of n. That is, finding the highest power of b that does not exceed r_1 (say, b^{a_2}) we divide r_1 by this power, obtaining

$$r_1 = q_2 b^{a_2} + r_2$$

where $0 \le r_2 < b^{a_2}$.

Substituting in the expression for n, we get

$$n = q_1 b^{a_1} + q_2 b^{a_2} + r_2 .$$

(Note that $r_2 < b^{a_2} \le r_1 < b^{a_1}$ and, as above, $0 \le q_2 \le b - 1$.)

Continuing in this manner, we eventually have n expressed as a sum of multiples of powers of b:

$$n = q_1 b^{a_1} + q_2 b^{a_2} + \cdots + q_k b^{a_k}.$$

The process must terminate with a remainder of zero, since each remainder is smaller than the previous one.

Remembering to include the missing powers of b with coefficient zero, we can now express n in base b.

We continue, using 23:

Since $3^3 = 27$ is too big, we divide by $3^2 = 9$.

$$\begin{array}{r} 2 \\ 9\overline{)23} \\ 18 \\ \hline 5 \end{array}$$

$$23 = 2 \cdot 9 + 5$$

$$185 = 2 \cdot 81 + 2 \cdot 9 + 5.$$

Continuing, using 5, we get

$$5 = 1 \cdot 3 + 2$$

$$185 = 2 \cdot 81 + 2 \cdot 9 + 1 \cdot 3 + 2 \cdot 1$$

$$= 2 \cdot 3^4 + 2 \cdot 3^2 + 1 \cdot 3^1 + 2 \cdot 3^0.$$

Since the 3^3 term is missing, we add it in with coefficient 0:

$$185 = 2 \cdot 3^4 + 0 \cdot 3^3 + 2 \cdot 3^2 + 1 \cdot 3^1 + 2 \cdot 3^0$$

$$= (20212)_{\text{three}} .$$

PRACTICE PROBLEMS 5.B

1. $\boxed{A}$ Convert $(2087)_{\text{ten}}$ into each of the following bases:
   (a) 2 (b) 3 (c) 6 (d) 7 (e) 12.
2. Convert $(17854)_{\text{ten}}$ into each of the following bases:
   (a) 2 (b) 3 (c) 5 (d) 12 (e) 16.
3. $\boxed{A}$ (a) Express two in base two.
   (b) Express three in base three.
   (c) If b is a positive integer greater than 1, express b in base b.

Aside from the historical significance, other bases are used in various ways today. For example, many retailers count inventory in terms of dozen (12) and gross (12^2); they are essentially using base twelve. Thus, for example, 4 gross 3 dozen and 11 items could be expressed as $(43E)_{twelve}$. For another example, we may consider time as being measured in base sixty (a remnant from the Babylonian system). Thus,

$$3 \text{ hours } 12 \text{ minutes and } 23 \text{ seconds}$$

is really

$$3 \cdot 60^2 + 12 \cdot 60 + 23 \text{ seconds.}$$

Probably the most important system today other than the decimal system is the **binary system**—base two. Since an electric switch may be placed in one of two states [off (0) or on (1)], it is possible to represent a number in the binary system by turning a sequence of switches on or off. Thus, operations with numbers may be carried out electronically. This is the way in which some computers do calculations. The binary system also has many applications to mathematical puzzles. Problem 5.1 is one example.

**If you have not already solved it, try Problem 5.1 (page 146)
now before reading on.**

Problem 5.1 is attributed to Bachet de Méziriac (1581–1638). The problem essentially says: What is the smallest number of weights that can serve to counterbalance each integral number of grams from 1 to some fixed number n?

The answer lies in the binary system.

Recall that every positive integer n may be expressed in the binary system by first expressing n as a sum of multiples of powers of 2. Since the base is 2, each coefficient is either 0 or 1. That is,

$$n = a_0 \cdot 1 + a_1 \cdot 2 + a_2 \cdot 2^2 + a_3 \cdot 2^3 + \cdots + a_k \cdot 2^k,$$

where each a_i is either 0 or 1.

Suppose we had weights in the denominations of 1 gram, 2 grams, 2^2 grams, and so on. Then, to balance an object weighing n grams, we need only select those weights corresponding to powers of 2 that have coefficient 1 in the expansion of n.

For example, the binary expansion of 109 is

$$109 = 1 \cdot 2^6 + 1 \cdot 2^5 + 0 \cdot 2^4 + 1 \cdot 2^3 + 1 \cdot 2^2 + 0 \cdot 2^1 + 1 \cdot 2^0.$$

Thus, an object weighing 109 grams can be balanced by placing 64, 32, 8, 4, and 1 gram weights in the other pan.

Solution of Problem 5.1

The fact that the coefficients in the binary expansion of a number are at most 1 means that we need only one 2^i gram weight for each i.

In the version of the problem at hand, the assayer wanted to be able to balance all weights up to 127 grams. He therefore may select 1, 2, 4, 8, 16, 32, and 64 gram weights—seven weights in all. (In general, to be able to balance all weights up to n, we may choose weights of 1, 2, 4, ..., 2^k, where $2^k \leq n < 2^{k+1}$.)

The discussion above shows that the weights 1, 2, 4, ... will suffice, but the question remains as to whether it might be possible to use still fewer weights.

We demonstrate that this is not possible by showing that, if a set of k weights suffices to balance every integral weight up to n, then n must be less than 2^k. In other words, if $2^k \leq n < 2^{k+1}$, then we need at least $k + 1$ weights. (And since 1, 2, ..., 2^k will do, the minimum number of weights needed is exactly $k + 1$.)

Suppose now that we have k weights which can balance every integral weight (in grams) up to n.

Since we can balance 1 gram, one of the weights must weigh 1 gram. Call this weight W_1.

Since we can balance 2 grams, we must have a second weight, W_2, which does not exceed 2 grams, that is, $W_2 = 1$ or 2. (If every weight other than W_1 weighed more than 2 grams, we could not balance an object weighing exactly 2 grams.)

W_1 and W_2 together weigh at most $1 + 2 = 3$ grams. Hence, to be able to balance an object weighing 4 grams, we need a third weight, W_3, weighing no more than 4 grams.

Since W_1, W_2, and W_3 together weigh at most $1 + 2 + 4 = 7$ grams, we need a fourth weight, W_4, not exceeding $2^3 = 8$ grams, in order to be able to balance an object weighing 8 grams.

And so on. For each r, we must have another weight, W_r, that does not exceed 2^{r-1} grams.

Thus, the sum of all the weights W_1, W_2, ..., W_k cannot exceed $1 + 2 + 2^2 + \cdots + 2^{k-1} = 2^k - 1$ grams.*

Thus, n cannot exceed $2^k - 1$; that is, n is less than 2^k.

The argument above gives the outline of a proof, but the lack of precision in the "and so on" step leaves something to be desired. The proof can, with very little change, be formalized through the use of mathematical induction. (A discussion of mathematical induction appears in Appendix B.)

---

* If $1 + 2 + \cdots + 2^{k-1} = x$, then $2x = 2 + 4 + \cdots + 2^k$, so $x = 2x - x = (2 + 4 + \cdots + 2^k) - (1 + 2 + \cdots + 2^{k-1}) = 2^k - 1$ (all other terms cancel). Or see Appendix B for a proof by mathematical induction.

Before leaving this version of Bachet's weight problem, one more comment is in order. Suppose we wish to be able to balance all weights up to 100 grams. From our discussion above, we know that at least seven weights will be needed; and, in fact, the seven weights 1, 2, 4, 8, 16, 32, and 64 will do. However, since we wish to go up only to 100, there are other sets of seven weights that will also work, for example, 1, 2, 4, 8, 16, 32, and 63; but no set of fewer than seven weights will do. The first set will actually enable us to go up to 127, while the second set will enable us to go up only to 126, but both sets will handle weights up to 100.

Another version of the weight problem is also of interest—weights may be placed on both pans of the balance. We leave this for an exercise at the end of the chapter (see Exercise 5.25.)

ADDITION AND MULTIPLICATION IN OTHER BASES

Just as we can add and multiply numbers in base ten, we can do so in any base. Consider the following addition in base ten:

$$28 \\ + 57 \\ + 39$$

To obtain the sum, we say 8 plus 7 is 15, plus 9 is 24; so we write 4 under the 8–7–9 column and carry 2. Then we say $2 + 2$ is 4, plus 5 is 9, plus 3 is 12. We write 12 and thus obtain 124 as our answer.

The justification for this process is as follows:

$$28 + 57 + 39 = (2 \cdot 10 + 8) + (5 \cdot 10 + 7) + (3 \cdot 10 + 9)$$

$$= 2 \cdot 10 + 5 \cdot 10 + 3 \cdot 10 + 8 + 7 + 9$$

$$= (2 + 5 + 3) \cdot 10 + (8 + 7 + 9)$$

$$= (2 + 5 + 3) \cdot 10 + 24$$

$$= (2 + 5 + 3) \cdot 10 + 2 \cdot 10 + 4$$

$$= (2 + 5 + 3 + 2) \cdot 10 + 4 \cdot 1 \quad \text{(notice 2 being carried)}$$

$$= 12 \cdot 10 + 4 \cdot 1$$

$$= (10 + 2) \cdot 10 + 4 \cdot 1$$

$$= 1 \cdot 10^2 + 2 \cdot 10 + 4 \cdot 1$$

$$= (124)_{\text{ten}}.$$

Base ten arithmetic should be second nature to us, because the addition and multiplication tables were pounded into our heads in the early grades at school. Base b arithmetic would be just as easy if we knew the base b tables. These tables can easily be formed by using base ten equivalents. For example, in base five, we first write the base five equivalents of the base ten numbers from 0 to 16★ :

base ten	0	1	2	3	4	5	6	7	8	9	10	11	12	13	14	15	16
base five	0	1	2	3	4	10	11	12	13	14	20	21	22	23	24	30	31

This table can be used to add or multiply any two base five numbers from 0 to 4, by doing the addition or multiplication in base ten and then converting back to base five. For example,

$$3_{five} + 3_{five} = 3_{ten} + 3_{ten} = 6_{ten} = (11)_{five}$$

or

$$(3 \cdot 4)_{five} = 3_{ten} \cdot 4_{ten} = (12)_{ten} = (22)_{five} .$$

We can now construct the addition and multiplication tables for base five (although it is no longer really necessary to do so). See Figure 5.2.

+	0	1	2	3	4
0	0	1	2	3	4
1	1	2	3	4	10
2	2	3	4	10	11
3	3	4	10	11	12
4	4	10	11	12	13

·	0	1	2	3	4
0	0	0	0	0	0
1	0	1	2	3	4
2	0	2	4	11	13
3	0	3	11	14	22
4	0	4	13	22	31

FIGURE 5.2

To add and multiply larger numbers in base five, we make use of positional notation in following the same procedure as in base ten, carrying and borrowing where necessary.

For example, in base five,

$$\begin{array}{r} 34 \\ + 23 \\ \hline 112 \end{array}$$

since

$$(4 + 3)_{five} = (12)_{five} \qquad \text{(write 2, carry 1)}$$

$$(1 + 3 + 2)_{five} = (11)_{five} \qquad \text{(write 1, carry 1);}$$

★ We stop at 16 because, in base ten, $16 = 4 \cdot 4$. In general, we must go up to $(b - 1) \cdot (b - 1)$.

and

$$\begin{array}{r} 34 \\ \times\ 3 \\ \hline 212 \end{array}$$

since

$$(3 \cdot 4)_{\text{five}} = (22)_{\text{five}} \qquad \text{(write 2, carry 2)}$$

$$(3 \cdot 3)_{\text{five}} + 2_{\text{five}} = (21)_{\text{five}}.$$

The justification of this procedure is similar to that for base ten. For example, in base five,

$$\begin{aligned} 34 + 23 &= (3 \cdot 5 + 4) + (2 \cdot 5 + 3) \\ &= (3 + 2) \cdot 5 + (4 + 3) \\ &= 5 \cdot 5 + (12)_{\text{five}} \\ &= 1 \cdot 5^2 + 1 \cdot 5 + 2 \\ &= (112)_{\text{five}}. \end{aligned}$$

PRACTICE PROBLEMS 5.C

1. $\boxed{\text{A}}$ Construct the addition and multiplication tables for the base eight.

2. Construct the addition and multiplication tables for each of the following bases: (a) two (b) three (c) seven.

3. $\boxed{\text{A}}$ Add: (a) $(2012)_{\text{three}} + (2122)_{\text{three}}$. (b) $(222)_{\text{three}} + (120)_{\text{three}}$.

4. Add: (a) $(5621)_{\text{seven}} + (3106)_{\text{seven}}$. (b) $(4613)_{\text{seven}} + (2536)_{\text{seven}}$.

5. $\boxed{\text{A}}$ Subtract: (a) $(101101101)_{\text{two}} - (1111111)_{\text{two}}$. (b) $(1111100)_{\text{two}} - (1011001)_{\text{two}}$.

6. Subtract: (a) $(2713)_{\text{eight}} - (1746)_{\text{eight}}$. (b) $(4531)_{\text{eight}} - (2617)_{\text{eight}}$.

7. $\boxed{\text{A}}$ Multiply: (a) $(2012)_{\text{three}} \cdot (201)_{\text{three}}$. (b) $(201)_{\text{three}} \cdot (201)_{\text{three}}$.

8. Multiply: (a) $(5621)_{\text{seven}} \cdot (3106)_{\text{seven}}$. (b) $(2463)_{\text{seven}} \cdot (253)_{\text{seven}}$.

9. Use binary system addition to show that $1 + 2 + 2^2 + \cdots + 2^k = 2^{k+1} - 1$.

We are now ready to solve Problem 5.2 (page 146). If you haven't already solved it, try again now.

Solution of Problem 5.2

Consider the coefficients of the equation as being in some base, **b**. The equation is

$$(5)_\mathbf{b}\, x^2 - (50)_\mathbf{b}\, x + (125)_\mathbf{b} = 0.$$

Converting to base ten, this becomes

$$5x^2 - (5\mathbf{b} + 0)x + (\mathbf{b}^2 + 2\mathbf{b} + 5) = 0.$$

We are told that $x = 5$ is a solution. Thus,

$$5 \cdot 5^2 - (5\mathbf{b} + 0) \cdot 5 + (\mathbf{b}^2 + 2\mathbf{b} + 5) = 0.$$

$$125 - 25\mathbf{b} + \mathbf{b}^2 + 2\mathbf{b} + 5 = 0.$$

$$\mathbf{b}^2 - 23\mathbf{b} + 130 = 0.$$

$$(\mathbf{b} - 13)(\mathbf{b} - 10) = 0.$$

$$\mathbf{b} = 13 \quad \text{or} \quad \mathbf{b} = 10.$$

Therefore, since Martians have more fingers than Earthlings have, they must have 13 fingers. (We should check that in base thirteen, the equation $5x^2 - 50x + 125 = 0$ has two solutions:

$$(5)_{\text{thirteen}}\, x^2 - (50)_{\text{thirteen}}\, x + (125)_{\text{thirteen}} = 0.$$

Converting to base ten, we get

$$5x^2 - 65x + 200 = 0$$

$$x^2 - 13x + 40 = 0$$

$$(x - 5)(x - 8) = 0$$

$x = 5$, $x = 8$ in base ten, which also gives $x = 5$, $x = 8$ in base thirteen.)

CRYPTARITHMETIC

The remaining problems discussed in this chapter fall into the general area of recreational mathematics known as cryptarithmetic. This type of problem was popularized during the 1930s in *Sphinx*, a Belgian journal of recreational mathematics. Each problem is an arithmetic calculation in which digits are either coded or partially missing. Our task is to reconstruct the original calculation.

In many problems of this type, positional considerations are often relevant. For example, if

$$\begin{array}{r} AB \\ + CD \\ \hline EFG \end{array}$$

then E must be 1. To see why, remember that the first written

digit of a number is never zero, so E is at least 1. Since neither AB
nor CD can be larger than 99, their sum is no larger than
$99 + 99 = 198$. This shows that E is no larger than 1, so E = 1.

Similarly, if

$$\begin{array}{r} ABCD \\ -\ \ EFG \\ \hline HI \end{array}$$

then A must be 1, B must be 0, and E must be 9 (otherwise the
difference between ABCD and EFG would be at least 100 and we would
have a three digit difference rather than just HI.)

**We are now ready to attack Problem 5.3 (page 147). If you
have not already solved it, try it again.**

Cryptarithmetic problems in which letters form meaningful words are
also called Alphametics, a term coined by J. A. Hunter [31]. Problem
5.3, one of the most famous cryptarithmetic problems, is such a
problem. All the information of the problem is given in code. (Note
the similarity with cryptograms, except that here each distinct letter
stands for a specific digit rather than for another letter.) The tools
needed to solve the problem are just the basic properties of base ten
arithmetic and a little ingenuity.

To describe our reasoning, we label the columns starting from the
right. That is, the D–E–Y column is the first column, and so on.
Observe that the M in MONEY must result from a "carryover," since
there are no other numbers in the fifth column. Hence, as we observed
at the beginning of this section, M must be 1 (SEND $\leq$ 9999 and
MORE $\leq$ 9999 so MONEY $\leq$ 19998.) Substituting 1 for M every-
where M occurs, we obtain

Solution of Problem 5.3

$$\begin{array}{r} SEND \\ 1ORE \\ \hline 1ONEY \end{array}.$$

Since END $\leq$ 999 and ORE $\leq$ 999, at most 1 can be carried over from
the addition in the third column.

Consider the fourth column. There are two possible cases:

Case 1	*Case 2*
If nothing was carried from the third column, then $S + 1 = 10 + O$. Since $S \leq 9$, this gives $S = 9$ and $O = 0$.	If 1 was carried from the third column, then $S + 2 = 10 + O$, or, equivalently, $S = 8 + O$. Thus, $S = 8$ and $O = 0$, or $S = 9$ and $O = 1$. But $M = 1$, so O cannot be 1. Hence $S = 8$, $O = 0$.

So $O = 0$ in either case, and $S = 9$ if there is no carryover from the third column, and $S = 8$ if there is a carryover.

Substituting,

$$\begin{array}{r} \text{SEND} \\ \underline{1\,\text{ORE}} \\ \text{10NEY} \end{array}$$

Look now at the third column. Since at most 1 can be carried from the second column ($99 + 99 \leq 198$),

$$E + 0\ (+1) = N\ (+\ 10).$$

The $(+1)$ indicates the possibility that 1 was carried from the second column, and the $(+10)$ indicates the possibility that the sum of the third column might be 10 or more. But the only way that $E + 0\ (+1)$ could be $N + 10$ is if $E = 9$ and $N = 0$, which cannot be ($O = 0$). Hence, there is nothing carried from the third column. Thus, $S = 9$ and $E\ (+1) = N$.

Since $E \neq N$, we must have a carryover from the second column. Thus,

$$E + 1 = N$$

and

$$N + R\ (+1) = 10 + E.$$

Substituting $E + 1$ for N in this last equation,

$$E + 1 + R\ (+1) = 10 + E$$

$$R\ (+1) = 9.$$

Since $S = 9$, we must have $R = 8$ and there must be a carryover from the first column. Thus

$$\begin{array}{r} \text{9 END} \\ \underline{1\ 0\ 8\ \text{E}} \\ \text{10NEY.} \end{array}$$

with

$$E + 1 = N$$

and

$$D + E = 10 + Y.$$

Since D is at most 7 ($R = 8$, $S = 9$) and Y is at least 2 ($O = 0$, $M = 1$), E must be at least 5, since $D + E = 10 + Y$.

Also, since $E + 1 = N \leq 7$, then $E \leq 6$.

If $E = 6$, then $N = 7$, and so D can be at most 5; but then $D + E \leq 11$, which cannot be.

Hence, E = 5, N = 6, D = 7, and Y = 2. That is,

$$
\begin{array}{r}
9567 \\
+\ 1085 \\
\hline
10652.
\end{array}
$$

The student is asking for \$10,652!

---

1. $\boxed{A}$ Consider the following cryptarithmetic problem (distinct letters represent distinct values):

$$
\begin{array}{r}
CF \\
AB)\overline{CDEF} \\
\underline{AB} \\
BGF \\
\underline{BGF}
\end{array}
$$

(a) What are the possible values for C?
(b) Which is larger, B or F? (Note F · AB = BGF.)
(c) What are the possible values for B and F? (Note F · AB ends in F.)
(d) For each possible set of values for B and F, what are the possible values for A? (Note F · AB = BGF.)
(e) Which of the above cases actually lead to a solution? What is the solution?

2. Consider the following cryptarithmetic problem (distinct letters represent distinct values):

$$
\begin{array}{r}
ABC \\
\times\ \ BC \\
\hline
EDC \\
FEB \\
\hline
CDAC
\end{array}
$$

(a) What are the possible values for C? (Note, C · C ends in C.)
(b) For each possible value for C, what are the possibilities for B? (B · C ends in B.)
(c) Eliminating those values of C for which no values of B exist, what is the unique remaining possibility for C?
(d) What can F be equal to? (Note addition in the fourth column.)
(e) For which of the remaining possible values for B is it possible to determine a value for A? (ABC · B = FEB.)
(f) Complete the solution.

It is now time to try Problem 5.4 (page 147) again, if you have not already solved it.

Solution of Problem 5.4

Problem 5.4 is a partial ghost—we are told which positions are occupied by numbers, but we are given only partial information about what each letter stands for. (As before, different letters represent different numbers, and each time a letter appears it represents the same number.)

To facilitate referring to specific entries, we label the rows of the division as follows:

$$\begin{array}{r}
\text{quotient} \\
\overline{\text{divisor})\text{dividend}} \\
\text{row 1} \\
\hline
\text{row 2} \\
\hline
\text{row 3} \\
\hline
\text{row 4} \\
\hline
\text{row 5} \\
\hline
\text{row 6} \\
\hline
\text{row 7}
\end{array}$$

We begin the solution by observing that the last entries in rows 2, 4, and 6 respectively must be U, V, and E, since these are brought down from the dividend.

$$\begin{array}{r}
\text{C E C I} \\
\overline{\text{QUI})\text{TROUVE}} \\
\underline{\text{-- -- --}} \\
\text{-- --U} \\
\underline{\text{-- -- --}} \\
\text{-- --V} \\
\underline{\text{-- --E}} \\
\text{-- --E} \\
\underline{\text{-- -- --}} \\
\text{0}
\end{array}$$

We next observe that the final entry of row 7 must be E, since there is no remainder. Since the final entry in row 5 is also E, we conclude that $C \cdot I$ ends in E and I^2 also ends in E. Making a chart of the possible values of I and recalling that C, E, and I are distinct, we are left with only three possible cases:

I	E	C
0	impossible	
1	impossible	
2	4	7
3	9	impossible
4	6	9
5	impossible	
6	impossible	
7	9	impossible
8	4	3
9	1	impossible

The second of the possible cases may be eliminated because, if $C = 9$, then row 1 = $C \cdot QUI \geq 900$, which forces T to be 9; but T and C cannot both be 9.

In the remaining two cases, since $C \cdot QUI$ = row 1 and $I \cdot QUI$ = row 7 are three-digit numbers, Q must be 1 and we have the possibilities for U shown in Figure 5.3.

I	E	C	Q	U
2	4	7	1	0
2	4	7	1	3
8	4	3	1	0

($U \neq 1, 2,$ or 4, and $152 \cdot 7$ is four digits.)

($U \neq 1,$ and $8 \cdot 128$ is four digits.)

FIGURE 5.3

Since there is no remainder, row 6 must equal row 7 = $I \cdot QUI$. But row 6 is obtained by subtracting row 5 from row 4; hence, the second entry in row 6 is V (+ 10) − E (here the (+ 10) indicates the possibility of borrowing). This enables us to compute V. We do so in each case (Figure 5.4):

I	E	C	Q	U	row 6	V	
2	4	7	1	0	204	4	impossible V = E = 4
2	4	7	1	3	264	0	
8	4	3	1	0	864	0	impossible V = U = O

FIGURE 5.4

In the first case, $V = E = 4$; and, in the third case, $V = U = 0$. Hence, these cases may be eliminated.

We are left with $Q = 1$, $U = 3$, $I = 2$, $C = 7$, $E = 4$, and $V = 0$.

It is now a simple matter to complete the problem. We first find rows 1, 3, 5, and 7 (and hence 6) by multiplying C · QUI, E · QUI, C · QUI, and I · QUI respectively. This gives

$$
\begin{array}{r}
7472 \\
132)\overline{\text{T R O 3 0 4}} \\
924 \\
\hline
\_\_3 \\
528 \\
\hline
\_\_0 \\
924 \\
\hline
264 \\
264 \\
\hline
\end{array}
$$

Then, working from the bottom up, we can fill in row 4, then row 2, and finally the dividend. We obtain

$$
\begin{array}{r}
7472 \\
132)\overline{986304} \\
924 \\
\hline
623 \\
528 \\
\hline
950 \\
924 \\
\hline
264 \\
264 \\
\hline
\end{array}
$$

Thus, $T = 9$, $R = 8$, $O = 6$ and, as we already found, $Q = 1$, $U = 3$, $I = 2$, $C = 7$, $E = 4$, $V = 0$.

Problems 5.3 and 5.4 involve base ten arithmetic, as do most of the cryptarithmetic exercises at the end of this chapter. It is also possible, however, to consider cryptarithmetic problems in other bases—see, for example, Exercises 5.57–5.64.

THE CHAPTER IN RETROSPECT

This chapter first considered some aspects of base ten (decimal) arithmetic, then generalized to arithmetic in an arbitrary base b. (Mathematicians love to generalize.) We found that no matter what base we work in, the operations are essentially the same.

The problems in the text have stressed some recreational aspects of numerical computation. The exercises at the end of this chapter continue along the same lines. Some of the games we will encounter in Chapters 7 and 8 will also make use of the concepts we have considered here.

Exercises

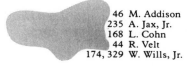

Some of the concepts discussed in Chapter 4 can be helpful in the solutions of some of the problems in this chapter. These exercises are denoted by #.

Decimal Notation

In Exercises 5.1 through 5.14, all numbers are assumed to be expressed in the base ten.

5.1. Ⓐ In his autobiography, Count Burr Turr tells about an occasion during his childhood when he decided to write all the counting numbers from one to one million. It was a noble undertaking, but his arm gave out after he had written only 31,676 digits.

Assuming that the Count was using decimal notation, what was the last digit he wrote before his arm grew numb?

 # 5.2. (a) Ⓗ Show that the last digit of the square of any integer must be 0, 1, 4, 5, 6, or 9.

(b) Ⓐ Which digits could possibly be the last digit of the cube of a positive integer?

(c) Which digits could possibly be the last digit of the fourth power of an integer?

 # 5.3. Ⓢ Show that the next to last digit of a perfect square (a number that is the square of a positive integer) is odd if and only if the last digit is 6.

 # 5.4. Ⓗ Ⓐ For the first time in the history of Weapons, Armaments, and Rockets, Inc., there was a real battle over the presidency of the company. Each of the five candidates had the backing of more than 100,000 shares of stock. The final tabulation was recorded on a piece of paper; but, before the results could be read at the board meeting, someone spilled some coffee. The resulting situation is shown in Figure 5.5.

46	M. Addison
235	A. Jax, Jr.
168	L. Cohn
44	R. Velt
174, 329	W. Wills, Jr.

FIGURE 5.5

Fortunately, someone had noted earlier that the number of votes received by the winner was a perfect square.

Who won the election?

5.5. An automobile riding along Route 101 passed a sign showing the distance to Los Angeles. This distance, in miles, was a three-digit number, the middle digit of which was zero. Exactly one hour later, the car passed another sign showing the distance to Los Angeles. This time, the distance was a two-digit number, containing the same two nonzero

digits as the original sign, but in the reverse order. Exactly one hour after that a sign indicated the distance to Los Angeles as a two-digit number containing the original two nonzero digits in their original order.

Assuming that the car maintained a uniform speed, how long after passing the third sign did the car arrive in Los Angeles? ([60], Vol. 5, p. 139)

5.6. [S] [A] Find all two-digit numbers that are divisible by the product of their digits.

5.7. Find all two-digit numbers that are divisible by the sum of their digits.

5.8. [H] Find all two-digit numbers that have two distinct nonzero digits, such that the difference between the number and its reversal (the same number with the order of the digits reversed) is divisible by the sum of the digits of the number.

5.9. (a) Find all two-digit numbers that, when added to their reversals, give perfect squares. ([45], Problem 125e)

(b) [S] [A] Find all two-digit numbers that, when subtracted from their reversals, give perfect squares.

(c) [A] Find all two-digit numbers that, when subtracted from their reversals, give perfect cubes.

5.10. Find all two-digit numbers with nonzero digits, that, when added to their reversals, give a result divisible by 66.

5.11. [H] [A] Find two-digit numbers with distinct digits such that the difference between the square of the number and the square of the reversal of the number is itself a perfect square. ([60], Vol. 1, p. 85)

★ **5.12.** Show that if any three-digit number is subtracted from its reversal, then the result is divisible by 99.

5.13. [S] [A] The magician removed a sealed envelope from his pocket and handed it to a member of the audience. He then asked another member of the audience to select a three-digit number with distinct digits, to reverse the order of the digits, and to subtract the smaller three-digit number from the larger. The magician then asked his subject whether the resulting number had three digits. When he was told that it did, he asked the subject to reverse the order of the digits in the result and to add the number thus obtained to the previous result. The subject was then told to announce the sum to the audience. After he did so, the sealed envelope was opened; inside it was an index card on which the subject's result was correctly predicted.

How did the magician do the trick?

5.14. [A] The magician asked the subject to do the following:

1. Represent the month of your birth as a number. For example, January = 1, February = 2, and so on.
2. Multiply that number by 5.
3. Add 17.
4. Double the result.
5. Subtract 13.
6. Multiply the result by 5.
7. Subtract 8.

8. Double the result.

9. Add 9.

10. Add the day of the month on which you were born. For example, if your birthday is August 23, then, in step 1, start with 8 and, in step 10, add 23.

11. Tell me the result.

The subject announced: 1332.

"Oh, I see your birthday was this past Thursday," said the magician.

If today is Tuesday, what is today's date?

Binary System

5.15. ⑤ Ⓐ Tweedledee and Tweedledum are playing a game. Tweedledee thinks of a number between 1 and 1000, and Tweedledum tries to guess the number. Tweedledum is allowed to ask a limited number of questions of the form "Is the number greater than (less than) so and so?"

How many questions must he be allowed to ask in order for him to be sure of getting the right number on his first guess if he proceeds properly? How should he proceed?

5.16. Ⓗ Ⓐ Now it is Tweedledum's turn to think of a number; but the rules have changed slightly. Tweedledee is allowed eight guesses. With each incorrect guess, Tweedledum says either "higher" or "lower," whichever is correct.

What is the largest number that Tweedle-

dum should be allowed to pick if Tweedledee is to be assured of success provided that he proceeds properly? How should Tweedledee proceed?

5.17. Once again the rules have changed (see Exercises 5.15 and 5.16). This time, Tweedledee may choose any positive integer whatsoever, and Tweedledum is allowed as many "yes or no" questions as he likes; but, if he receives two "no" answers, then Tweedledum loses.

How should Tweedledum proceed to be sure of success?

5.18. Ⓗ Ⓐ The magician selected three volunteers from the audience. He instructed each to think of a number between 1 and 31. He then displayed five cards containing numbers as indicated in Figure 5.6. He asked each volunteer to determine which cards contain the number he or she selected.

Card A	Card B	Card C
1 9 17 25	2 10 18 26	4 12 20 28
3 11 19 27	3 11 19 27	5 13 21 29
5 13 21 29	6 14 22 30	6 14 22 30
7 15 23 31	7 15 23 31	7 15 23 31

Card D	Card E
8 12 24 28	16 20 24 28
9 13 25 29	17 21 25 29
10 14 26 30	18 22 26 30
11 15 27 31	19 23 27 31

FIGURE 5.6

"A, C, and E," announced the first volunteer.

"Then your number is 21," replied the magician quickly.

"My number is only on card D," said the second volunteer.

"Then your number is 8."

"My number is on cards B, C, D, and E," proclaimed the third volunteer.

"Then your number is 30," the magician announced triumphantly.

How was the magician able to determine the selected number so quickly? And why does the trick work?

5.19. The magician displayed a packet of 16 cards, each bearing a number from 0 to 15. The pack was arranged, from top to bottom, in numerical order. He then dealt two piles face up on the table as follows: First he dealt the 0 face up and then 1 face up in another pile. Then he dealt 2 face up on top of 0, 3 face up on top of 1, 4 face up on top of 2, and so on. When he finished dealing the 16 cards, he selected the pile into which the last card was dealt, placed it face down in his left hand, and then placed the remaining pile face down on top of the one in his hand. He thus again held the 16 cards face down in his hand, although they were no longer in their original order. He then repeated the entire process, alternately dealing to two piles of face up cards and then picking up the two piles in the prescribed manner. After he had completed this process a total of four times, he displayed the cards and they were back in their original order.

(a) $\boxed{\text{S}}$ Why does this work?

(b) $\boxed{\text{A}}$ What would happen if he went through the above process four times, dealing the cards face down rather than face up? Explain why.

(c) $\boxed{\text{A}}$ What would happen if he dealt the cards face down, but placed the pile into which the last card is dealt on top of the other pile rather than below it?

(d) For each of the three methods of dealing described above, how many times would the process have to be repeated for the cards to return to their original order, if the packet contains 32 cards rather than 16?

Other Bases

★ **5.20.** $\boxed{\text{H}}$ The magician placed an apple, a banana, a peach, and 25 matchsticks on the table. He then selected three volunteers from the audience. Their names were Alvin, Julia, and Melonie. The magician then handed one of the matchsticks to Alvin and two matchsticks to Julia, but he did not give any to Melonie. He was then blindfolded, and each of the volunteers was asked to remove one of the fruits from the table.

"Whichever of you removed the apple should now remove a number of matchsticks equal to the number I handed you a moment ago. Whoever removed the banana should remove three times as many matches as you were originally given. Finally, the remaining person should remove nine times as many sticks as I gave to him or her."

"Now, tell me how many sticks remain on the table."

"Seven," was the reply.

"Then Alvin took the peach, Julia took the banana, and Melonie took the apple," declared the magician.

Naturally, he was correct.

Explain how this trick works.

5.21. The magician held up a packet of 27 cards, marked, in numerical order from top to bottom, with the numerals from 0 to 26.

He then dealt them face up into three piles as follows: First he put 0, 1, and 2 in three separate piles face up on the table. Then he put 3 on 0, 4 on 1, 5 on 2, 6 on 3, 7 on 4, and so on.

When he finished dealing, he picked up the

pile into which the last card was dealt and placed it face down in his left hand. Then he placed the next pile face down on top of the pile in his hand; and, finally, he placed the third pile face down on top of the other two.

He then repeated the entire process two more times. When he was done, the cards were back in their original order.

Explain why this happens. (See also Exercise 5.19.)

5.22. The magician removed a packet of 27 cards from an ordinary deck of playing cards. He handed the packet to a volunteer from the audience and asked her to select a card, look at it, and then mix it up in the packet. The magician then asked the audience for a number between 1 and 27.

"Eight," someone called out.

Next the magician dealt the 27 card packet face up into three piles of nine cards each and asked the volunteer to indicate which pile contained the selected card. The subject pointed, and the magician picked up the cards, placing them face down in his hand with the indicated pile between the other two. He then again dealt three face up piles, alternating, one card at a time, between the piles (as in Exercise 5.21). The subject again indicated which pile contained the selected card, and the magician picked up the cards, this time placing the selected pile face down below the other two piles.

Once more the magician dealt three piles as before, and this time, when he picked the cards up, he placed the indicated pile face down on top of the other two.

"Now," he said, "the audience has selected the number eight."

He counted seven cards off the top of the packet, and turned over the eighth card, saying, "This is the card you selected." As usual, he was correct.

(a) What would the magician have done differently if the audience had selected the number twenty-three?

(b) [S] Explain why the trick works.

5.23. At a party celebrating the end of the math course, a student who had decided to major in magic asked one of her friends to think of a number between 1 and 26.

She then placed the following cards on the table and asked the friend which cards contained the selected number.

Card A	Card B	Card C
1 10 19	2 11 20	3 12 21
4 13 22	5 14 23	4 13 22
7 16 25	8 17 26	5 14 23

Card D	Card E	Card F
6 15 24	9 12 15	18 21 24
7 16 25	10 13 16	19 22 25
8 17 26	11 14 17	20 23 26

"B, C, and E," was the reply.

"Then your number is 14."

(a) [A] What would the student have said if her friend had announced "A, C, and F"?

(b) Explain why the trick works. ([60], Vol. 7, p. 96) (See also Exercise 5.18.)

★ **5.24.** [S] Another student at the same celebration produced three cards from his pocket.

Card A					Card B				
1	4	7	10	13	2	3	4	11	12
16	19	22	25	2*	13	20	21	22	5*
5*	8*	11*	14*		6*	7*	14*	15*	
17*	20*	23*	26*		16*	23*	24*	25*	

Card C
5   6   7   8   9
10   11   12   13   14*
15*   16*   17*   18*
19*   20*   21*   22*

He asked his subject to select a number from 1 to 26 and to indicate those cards on which the selected number appeared. In addition, for each appearance of the selected number, the subject was to say whether or not the number was followed by an asterisk.

"Card A with an asterisk and Card C without one."

"Your number is 8."

Explain how the trick works. ([60], Vol. 7, p. 97)

★ **5.25.** [H] [A] In Problem 5.1 at the beginning of the chapter, the assayer always placed his ore sample on the left pan and weights on the right pan. Normally, however, we may place weights on both pans of the balance. Thus, for example, with weights of 1 gram and 3 grams, we may weigh an ore sample weighing 2 grams—simply place the sample and the 1 gram weight in one pan and the 3 gram weight in the other pan.

Assuming that the assayer uses both pans in this manner, what is the smallest number of weights he can possess and still be able to weigh all possible ore samples weighing an integral number of grams between 1 and 13? Between 1 and 40? Between 1 and 121? Generalize.

5.26. According to the will of G. Barefoot Tiptoe, the value of his entire estate was to be divided among his heirs (in a given order of priority) according to the following rules:

    1. Each heir is to receive $1, $5, or a number of dollars equal to some other power of 5.

    2. No more than four people are to receive the same amount of money.

When Mr. Tiptoe died, his estate was valued at $1,013,652.

How many of Mr. Tiptoe's heirs actually received money, and how much did each receive?

5.27. [A] If 1/6 of 30 is 4, what is 1/4 of 10? ([41], Vol. 31, p. 178)

★ **5.28.** [H] In what base is 11111 a perfect square? ([46], Vol. 15, Problem 149)

★ **5.29.** [H] [A] In what base is 297 a factor of 792? ([41], Vol. 38, Problem 124)

5.30. [A] An integer is said to be even if it is divisible by two, and is said to be odd otherwise.

    (a) If ABCDE is the base b representation of a number n, under what circumstances is n odd if

      (i) $b =$ ten   (ii) $b =$ two   (iii) $b =$ twelve
      (iv) $b =$ three (v) $b =$ five?

    (b) Generalize. That is, given the base b representation of an integer n, how can you quickly tell whether or not n is odd? ([46], Vol. 12, Problem 197)

5.31.   (a) [S] Prove that $(121)_b$ is a perfect square for every base $b > 2$.

    (b) Prove that $(1331)_b$ is a perfect cube for every base $b > 3$. ([35], p. 52)

★★ **5.32.**   (a) [S] Prove that for any base $b > 2$, the base b representation of the numbers $2(b - 1)$ and $(b - 1)^2$ are the reversals of each other.

    (b) For any base $b > 3$, if the digits of the representation of $3(b - 1)$ are reversed and the digit $(b - 1)$ is added at the right, prove that the resulting three-digit number is the representation of $(b - 1)^3$. ([60], Vol. 2, Problem 79)

Cryptarithmetic

5.33. ☐A☐ If $I^2 = ME$, but $I^3 = YOU$, what is the reversal of the difference between ME and I?

5.34. ☐A☐ Find MAID if $O^3 = DAD$ and $(IM)^2 = MOM$.

5.35. ☐H☐ ☐A☐ Albert M. Adic was brilliant in all matters concerning numbers, but he was sometimes careless about his spelling. Once he wrote "to" instead of "two" in answer to an age problem. His sister, who was always teasing her brother about his spelling and his love for problem solving, delighted at the opportunity to correct him.

"Can't you spell? It's age 'two' not 'to.' All you ever do is math problems. You're a real square."

"Oh yeah," replied Albert after a moment's thought. "If AGE, TWO, NOT, and TO are perfect squares, then do you know what you are? You are a TWO + TO + TOO."

What was Albert calling his sister?

5.36. An archeologist discovered the following addition and multiplication problems engraved on the wall of a Roman ruin.

$$\begin{array}{r} LIX \\ + LVI \\ \hline CXV \end{array} \qquad X^2 = C$$

Being a fan of cryptarithmetic problems, the archeologist decided to consider the engraving in that light. She found there is a unique solution. What is it? ([38], Chapter 7, Problem 29)

5.37. ☐A☐ Solve:

$$\begin{array}{r} AHBCF \\ + EBFAF \\ \hline AGGHFE \end{array}$$

5.38. Shortly after Scott Troupe became the new scoutleader in Camptown, his wife gave birth to a baby boy. The scouts threw a party in the baby's honor, with music provided by the scout brass band. The local newspaper's coverage of the gala proceedings was headlined as follows:

$$\begin{array}{c} TROOP \\ TOOTS \\ NEWSON \end{array}$$

When Scott noticed the article, he realized that it makes an interesting cryptarithmetic problem: "TROOP + TOOTS = NEWSON."

Can you solve it?

5.39. ☐S☐ ☐A☐ Solve:

$$\begin{array}{r} LET \\ \times\ NO \\ \hline SOT \\ NOT \\ \hline FRET \end{array}$$

([35], Problem 5, p. 82)

5.40. ☐H☐ ☐A☐ Solve:

$$\begin{array}{r} ABC \\ \times\ DEC \\ \hline FBGC \\ FHCE \\ GDE \\ \hline HBHIC \end{array}$$

5.41. Solve:

$$\begin{array}{r} ABC \\ \times\ DEC \\ \hline FGH \\ IAC \\ ACAE \\ \hline AFAGCH \end{array}$$

5.42. Solve:

```
        A B C
  ×     A B C
      ─────────
      D E F C
      D A G H
    D H A C
    ─────────
    D D I H F C
```

5.43. [H] Fill in the missing digits:

```
        _ _ 3 _
    ×   _ _ _
    ─────────
    _ _ _ 0 _
    _ _ _ 7 _
    _ 2 3 _
    ─────────
    _ 4 _ _ _ 5 _
```

5.44. [H] If all the X's represent the same digit, find X and fill in the other digits of the following multiplication problem:

```
        _ X _
    ×   _ X
    ─────────
    _ X X _
    _ _ _
    ─────────
    _ _ _ X _
```

★ **5.45.** [H] [A] If all the X's represent the same digit, find X and fill in the other digits of the following multiplication problem:

```
        _ _
    ×   _ _
    ─────────
    X X X
    _ _
    ─────────
    _ 0 _ _
```

★ **5.46.** [H] [A] Solve:

```
        A B
    ×   C D
    ─────────
      E E E
      F A G
    ─────────
    B F E E
```

★ **5.47.** [H] [A] In 1978, the authors composed the following problem:

```
        C H E I N
        1 9 7 8
      ─────────────
      _ _ _ _ _
      _ _ _ 4 4 _
      _ _ R _ _ _
    ───────────────
    A V E _ B A C H
```

Solve it.

5.48. [H] Fill in the missing digits:

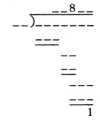

([35], p. 81)

5.49. Fill in the missing digits:

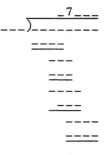

([60], Vol. 1, p. 52)

5.50. A Solve:

```
              GEABF
      ABC)ACDEEDFB
         ACFD
           CED
           ABC
           HHF
          I HD
           GGB
           GGB
```

5.51. Solve:

```
              FDAG
      ABC)DECFGA I
         DF J C
           BDG
           FB I
          KAGA
          KKE C
           KCE I
           KCE I
```

★ **5.52.** H A Find the missing digits (all the 0's and 1's are shown):

```
             _____1
      __1)101010101
           _1_
           1__1
           1___
            __0
            ___
            1__1
            1___
             __0
             _1_
              __1
              __1
               0
```

([60], Vol. 1, p. 108)

★ ★ **5.53.** H A Fill in the missing digits (all the 4's are shown):

```
            _____
      ___)__4___4
          ___
          ____
          __4_
           _4__
           _4_4
            __4
            __4
```

★ ★ **5.54.** Don Dakota loved cryptic division problems, so he asked a friend to take any long division problem and to replace each digit by a dash. The result was

```
            _____
      ___)_____
          ___
          ___
          ___
          ___
          ___
          ___
           ___
           ___
            ___
            ___
```

Don's efforts to recapture the original problem are doomed to failure because there is not enough information to determine a unique problem from which the ghost could have come. Let us aid Don a little:

```
              D O OM E D
      DON)D I V I S I ON
          _ _ _
           _ _ _
          _ _ _
           _ _ _
          _ _ _
            _ _ _
           _ _ 4
          _ _ _
           _ _ _
```

Now the problem has a unique solution. Find it.

★ 5.55. [A] Fill in all the digits, given that the quotient in the first division is the dividend in the second.

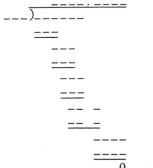

([11], Problem 150)

 # ★★ 5.56. [H] [A] Fill in the missing digits:

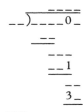

([11], Problem 145)

Cryptarithmetic in Other Bases

★ 5.57. [H] Given that, in base six,

$$GALON = (GOO)^2$$

and

$$ALONG = (OOG)^2$$

find the numerical value of each letter. (A. Colago, [60], Vol. 4, p. 136)

 # ★ 5.58. [S] [A] In base seven, find all squares of the form ABCABC. ([60], Vol. 4, p. 152)

 # ★ 5.59. [H] Show that there are no squares of the form ABCABC in base five, if A, B, and C represent distinct digits. ([60], Vol. 4, p. 152)

★ 5.60. [A] In what base(s) does the following addition have a unique solution?

$$\begin{array}{r} TEN \\ +NOT \\ \hline NINE \end{array}$$

★★ 5.61. [H] [A] In what base(s) does the following addition have a unique solution?

$$\begin{array}{r} ONE \\ +TWO \\ \hline NEXT \end{array}$$

★ 5.62. If, in base nine,

$$\begin{array}{r} TO \\ -BE \\ \hline OR \end{array} \quad \text{and} \quad \begin{array}{r} NOT \\ -TO \\ \hline BE \end{array}$$

find all solutions.

★★ 5.63. [A] In what base does the following division have a solution? Find the solution.

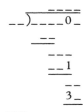

([63], Problem 173)

★ 5.64. Prove that for all bases $b \geq 3$, $(ONE)_{b+1} - (ONE)_b = (ON)_{2b+1}$, no matter what digits O, N, and E represent, $O \neq 0$.

6

Solve It With Networks: An Introduction to Graph Theory

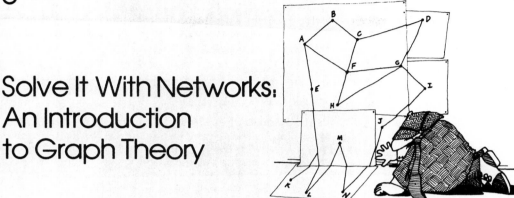

There are areas of mathematics whose development was motivated by the desire to solve problems originally posed as recreations. As the theory in these areas was further developed, important applications to other fields were found.

One such area of mathematics is graph theory, which today has wide applications in the design and programming of computers, and in life sciences, electrical networks, and business, as well as other fields.

Although graph theory did not come into its own until the middle 1800s, the groundwork of this theory was laid by Leonhard Euler (pronounced "oiler") in 1735. He had visited the town of Königsberg, Prussia, where he encountered the following problem that has become known as the Königsberg Bridge Problem:

In Königsberg, there are two islands surrounded by the two branches of the Pregel River. There are seven bridges crossing the river in various locations in the city (see Figure 6.1). Can a person plan a walk in the city so that he or she will cross each bridge exactly once?

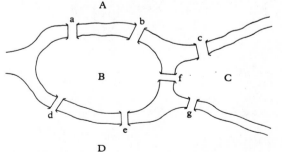

FIGURE 6.1

We use this problem as well as the remaining sample problems below to introduce some of the basics of graph theory. As usual, we suggest that you try them before reading on.

SAMPLE PROBLEMS

Problem 6.1

(a) Solve the Königsberg Bridge Problem stated above.

(b) Suppose that, owing to severe rusting, bridge b collapsed, leaving the situation pictured in Figure 6.2.

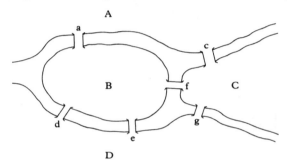

FIGURE 6.2

Can a person plan a walk so that he or she will start at D and cross each of the bridges once but not more than once? What if the individual starts at B? Can a walk be planned that crosses each bridge exactly once and terminates at the same place at which it starts?

(c) Suppose that the town planners want to build a new bridge to replace the one that collapsed. However, instead of building it in the same location as before, they wish the bridge to be situated in such a way that, after it is completed, it will be possible for a person, starting anywhere in town, to plan a walk that will cross each bridge exactly once and return to the starting point.

Where should the new bridge be built?

Problem 6.2

Figure 6.3 shows the floor plan of a house having two exterior doors.

Is it possible to enter the house through door *a*, travel through the house, passing through each doorway exactly once, and then exit through door *i*?

Problem 6.3

Starting at point A in Figure 6.4, travel along the lines of the figure so that each labeled point is passed through exactly once and you end your trip back at A.

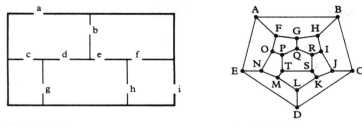

FIGURE 6.3   FIGURE 6.4

Problem 6.4

Three missionaries were conducting three cannibals to the mission school. On the way, they came to a piranha-filled river, which had to be crossed. There was a canoe that could hold only two people, but only one of the missionaries and one of the cannibals could be trusted to paddle without tipping it. In addition, as the cannibals had not yet been shown the light, the missionaries were afraid to create a situation in which cannibals would outnumber missionaries on either shore.

How should they cross the river?

GRAPHS

Euler's approach to the Königsberg Bridge Problem was to realize that the sizes of the land masses and bridges are not significant; and so there is no loss of generality if we shrink the land masses to points and represent the bridges as lines or curves connecting these points. Doing this with Figure 6.1, we get Figure 6.5a. Similarly, Figure 6.2 becomes Figure 6.5b.

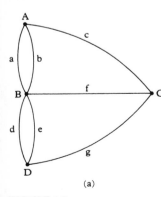
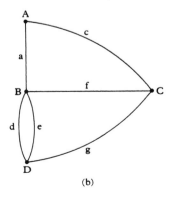

(a)   (b)

FIGURE 6.5

The new diagrams are called **graphs** or **networks**. The word graph is being used here in a sense that may be unfamiliar to you. As used in this chapter, a **graph** $\mathscr{G}$ consists of a finite set $\mathscr{V}$ of points, called **vertices** (**vertex** is the singular) or **nodes**, and a finite set $\mathscr{E}$ of line segments or curves, called **edges, branches,** or **roads**. Each edge either connects two distinct vertices (⌢) or else it connects a vertex to itself (◊). In the latter case, the edge is called a **loop**.★

Figure 6.6 gives some examples of graphs and networks—highway maps, diagrams representing chemical compounds, and electrical circuits. The tree diagrams discussed in Chapter 1 are further examples.

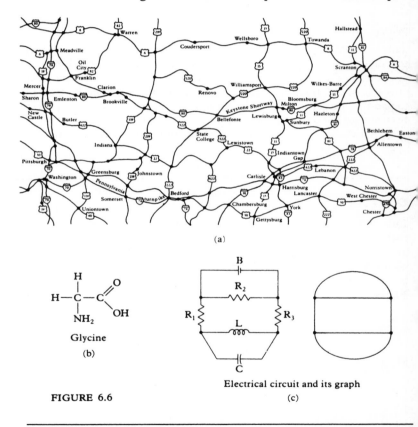

(a)

Glycine

(b)

Electrical circuit and its graph

(c)

FIGURE 6.6

★ Present practice favors using the term graph for structures with at most one edge between any two vertices and not containing any loops. The term multigraph is used for structures that contain multiple edges (⟬⟭) or loops. The term network is used for graphs whose edges are labeled with numbers representing physical quantities such as distances or resistances.

Nevertheless, the use of these terms is not uniform, and in the interests of simplicity and clarity, we use the terms graph and network synonymously, as defined above.

Many graphs, such as those in Figures 6.5a and 6.5b, can be drawn on a piece of paper—that is, in the plane—so that edges intersect only at vertices. A graph that can be drawn in this manner is called **planar**.

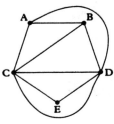

There are also **nonplanar** graphs. For example, consider a graph having five vertices, each pair of which is connected by exactly one edge. (This graph is called the **complete graph on five vertices**.) This graph is nonplanar. If we try to draw this graph on paper—in the plane—we may start as in Figure 6.7.

FIGURE 6.7

Now, there is no way to connect B to E without intersecting one of the edges already drawn. We might try to start over again but, no matter how we begin, we will not succeed in drawing the entire graph in the plane without introducing points of intersection other than the vertices. On the other hand, if we consider the graph in three dimensions (3-space), we can connect B to E by an edge that passes over edge CD and does not intersect it.

Similarly, a graph consisting of two sets of three vertices each, with edges connecting each vertex in the first set with each vertex in the second set, cannot be drawn in the plane without introducing new points of intersection. (For example, in Figure 6.8, C cannot be connected to X in the plane, but may be connected to X in 3-space.)

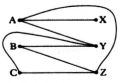

FIGURE 6.8

This graph is called the **complete bipartite graph on two sets of three vertices**. In general, a **bipartite** graph is one in which the vertices can be divided into two sets so that each edge of the graph connects a vertex in one set to a vertex in the other.

Although not all graphs are planar, it can be shown that all graphs can be constructed in 3-space so that edges intersect only at vertices. On the other hand, because all pictures have to be drawn on a two-dimensional page, we must somehow represent nonplanar graphs in the plane. In Figure 6.9 we illustrate two ways in which this can be done

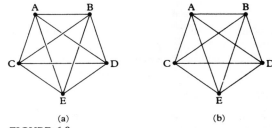

(a)   (b)

FIGURE 6.9

for the complete graph on five vertices. This book uses the approach of Figure 6.9b. Note that, in this figure, the only vertices are A, B, C, D, and E. The other points at which lines intersect are not considered to be vertices or intersection points; they are only "apparent points of intersection." With this in mind, the complete bipartite graph on two sets of three vertices may be drawn as in Figure 6.10.

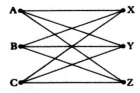

FIGURE 6.10

EULERIAN PATHS AND CIRCUITS

If we start at a vertex in a graph, follow an edge from that vertex to a second vertex, then follow an edge from that vertex, and so on; and if no edge is followed more than once in this process, then the resulting sequence of vertices and edges is called a **path**.

We may denote a path by listing the vertices and edges in the order in which they are encountered. For example, the path in Figure 6.11, starting at A and ending at C, can be denoted by

FIGURE 6.11

$$AaBeDcC.$$

If there is only one edge connecting two vertices, then omitting specific mention of the edge should create no confusion. That is, the path above could also be denoted by

$$ABeDC.$$

A path that terminates at the same vertex at which it starts is called a **circuit**. A path that traverses every edge of a graph exactly once is called an **Eulerian** (pronounced "oil air′ ee en") **path**; an **Eulerian circuit** is defined analogously; it is a circuit in which every edge of the graph is traversed exactly once. Thus, for example, DEABCFEBFDC is an Eulerian path for Figure 6.12a, and ABCDA is an Eulerian circuit for Figure 6.12b.

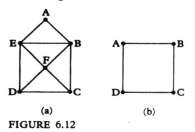

FIGURE 6.12

A graph possessing an Eulerian path is said to be **traceable**: By following the path, we can trace the graph without repeating any edge and without lifting our pencil from the paper.

Problem 6.1 (page 174) may now be rephrased as follows:

(a) Is the graph in Figure 6.5a traceable?

(b) Is there an Eulerian path starting with D for the graph in Figure 6.5b? What about starting at B? Does the graph in question have an Eulerian circuit?

(c) Can one edge be added to the graph in Figure 6.5b so that the resulting graph will have an Eulerian circuit?

We can answer these questions if we can answer the following, more general one: Under what circumstances does a graph contain an Eulerian path (circuit) starting at a particular vertex?

Before we attempt to answer that question, let us consider some further examples.

The graph in Figure 6.13a is clearly not traceable, because no path can contain both edge e and edge f. The problem with this graph is that it is not connected—it consists of two disjoint components. (Mathematically speaking, a graph is said to be **connected** if, for every possible pair of vertices, U and W, there is a path from U to W.) To avoid problems of this type, we restrict our attention to connected graphs.

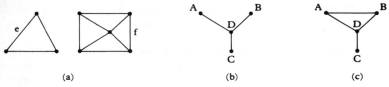

| (a) | (b) | (c) |

FIGURE 6.13

The graph in Figure 6.13b is connected, yet it is still not traceable. There are three roads leading to D but because A, B, and C are all dead ends, at most two of the roads to D can be traveled—one to reach D and one to leave.

The addition of an edge from A to B in Figure 6.13c allows us to find Eulerian paths: CDABD and CDBAD. That is, reaching D from C, we can travel to A or B and still be able to traverse the third road to D. As the reversal of an Eulerian path is also an Eulerian path, we also have DBADC and DABDC.

All four of these paths begin at C or D and end at C or D. Are there also Eulerian paths starting at A or B? The answer is no. There are still three roads to D; and, if we start at A or B, we can reach D by one road, leave by a second, but can never get to travel the third road. Thus, there are Eulerian paths starting and ending at C and D, but none that starts or ends at A or B.

What is different about A and B that prevents Eulerian paths from starting or ending there? Maybe we should look at it from the other point of view: What is different about C and D that forces Eulerian paths to begin and end there? Clearly, the fact that C is a dead end is important. Once we reach C, we cannot leave; so, unless we start at C, we must end there. But, what about D? The fact that there are three roads to D seems to have played a role in our argument. If we do not start at D, then, possibly, we can reach D by one road, leave by a second and, if we are able to travel the third road to D, this third road leaves us stranded at D. Thus, if we do not start at D, we must end there.

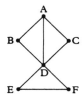

Can we generalize? We first consider one more example:

Playing around with Figure 6.14, we find several Eulerian paths (for example, ABDCADEFD and ADEFDBACD), all of which seem to begin and end at A and D.

What is special about A and D?

To help answer this question, we make some definitions.

FIGURE 6.14

ODD AND EVEN VERTICES

If the edge e connects the vertices U and W, then e is said to be **incident** to U and to W.

The **degree** of a vertex U is the number of edges that are incident to U. Each loop through U is counted twice since it has two ends at U. If the degree of U is odd, then U is said to be an **odd vertex**; otherwise, U is an **even vertex**.

For example, in Figure 6.14, A and D are odd vertices (of degrees three and five respectively) and B, C, E, and F are even (each of degree two). In Figure 6.13c, C and D are odd (of degrees one and three respectively), and A and B are even. In Figure 6.13b, A, B, C, and D are all odd—A, B, and C are of degree one, D is of degree three.

PRACTICE PROBLEMS 6.A

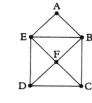

1. [A] In each of the graphs in Figure 6.15, find the degree of each vertex.

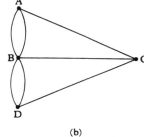

FIGURE 6.15   (a)   (b)

2. In each of the graphs in Figure 6.16, find the degree of each vertex.

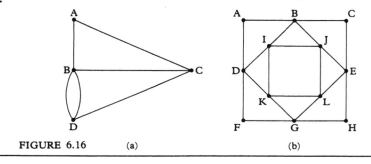

FIGURE 6.16   (a)   (b)

We can now answer the question: What is special about the vertices A and D in Figure 6.14, and the vertices C and D in Figure 6.13c? The answer: They are odd, and all other vertices in these figures are even. The following theorem applies in general.

Theorem 6.1   If a graph, $\mathscr{G}$, containing an odd vertex, U, has an Eulerian path, then each Eulerian path for $\mathscr{G}$ must either begin or end at U.

To prove this, we use an argument very similar to the one used above to show that every Eulerian path in the graph of Figure 6.13c must either begin or end at D. Namely, if we do not begin at U, then each time we visit and leave U we use up two edges, one to reach U and the other to leave U. Because each edge must be used exactly once, and because there are an odd number of edges incident to U, the edges incident to U must be used two at a time until, eventually, we have left U and only one unused edge incident to U remains. When we use this final edge incident to U, we will be stranded at U, and so our path must end there.

Since a path can start at only one vertex and end at only one vertex, Theorem 6.1 implies the following corollaries:

Corollary 6.1   If a graph, $\mathscr{G}$, containing two odd vertices U and V, has an Eulerian path, then each such path must start at U and end at V or vice versa.

Corollary 6.2   A graph that has more than two odd vertices cannot have an Eulerian path.

Since the graph in Figure 6.13b has four odd vertices, it can have no Eulerian path.

We are now able to answer Problems 6.1(a) and 6.1(b) (page 174). If you have not yet solved them, try them again.

Solution of Problems 6.1(a) and 6.1(b)

The graph of the problem, Figure 6.5a, has four odd vertices (B is of degree five, and A, C, and D are of degree three); hence it can have no Eulerian path. That is, a person cannot plan a walk as required.

On the other hand, Figure 6.5b—the graph of Problem 6.1(b)—has only two odd vertices, C and D. Thus, it is still conceivable that an Eulerian path for this graph exists. However, because C and D are of odd degree, such a path—if one exists—must begin and end at C and D. Thus, a walk starting at B cannot be planned, and neither can a walk that terminates at its starting point.

To plan a walk that starts at D is not very difficult; it is merely a matter of trial and error. For example, one such path in Figure 6.5b is

$$DdBCABeDC.$$

This completes Problem 6.1(b).

Corollary 6.2 tells us about graphs with more than two odd vertices. But what about graphs having two or fewer? Is it always possible to find an Eulerian path for a graph having no more than two odd vertices? Because graphs containing only even vertices are included among those having no more than two odd vertices, let us consider what happens when an Eulerian path begins at an even vertex. By an argument very similar to that used to prove Theorem 6.1, we can prove the following theorem.

Theorem 6.2 If a graph, $\mathscr{G}$, has an Eulerian path that begins at an even vertex, U, then the path must also end at U; and, hence, the path is an Eulerian circuit.

Proof Suppose the path begins at U and continues to W via an edge e. Then the remainder of the path gives an Eulerian path, starting at W, for the graph $\mathscr{G}'$ obtained by omitting the edge e from the graph $\mathscr{G}$. (For example, see Figure 6.17.) But, in $\mathscr{G}'$, U is of odd degree (one edge incident to U in $\mathscr{G}$ has been omitted); hence, every Eulerian path for $\mathscr{G}'$ that starts at W (does not start at U) must end at U. Therefore, the Eulerian path for $\mathscr{G}'$, and hence for $\mathscr{G}$, ends at U.

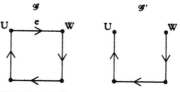

FIGURE 6.17

We are now ready to prove that graphs having no more than two odd vertices always do have Eulerian paths. We begin with the following theorem:

Theorem 6.3 If a connected graph, $\mathscr{G}$, has no odd vertices, then $\mathscr{G}$ has an Eulerian circuit.

The proof of this theorem actually gives a method, or algorithm, for finding an Eulerian circuit. We outline the proof on the left below, and on the right we apply the method to find an Eulerian circuit for the graph in Figure 6.18a.

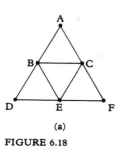

(a)

FIGURE 6.18

Begin at any vertex you like, say, U. Start tracing a path and, without retracing any edges, continue until you get stuck. Where will you be when this happens? Each time you enter and leave a vertex, you use up two edges at that vertex, so that, since all vertices are even, each time you enter a vertex other than U, there must remain at least one edge by which to leave that vertex. Thus, you cannot get stuck at a vertex other than U, and so you must be stuck at U. That is, you have completed a circuit Γ, beginning and ending at U.

If Γ is an Eulerian circuit, you are done. If not, then—since 𝒢 is connected—there must be at least one vertex, W on Γ, for which not all edges incident to W have been used. (If e is an unused edge incident to a vertex R not on Γ, then—since 𝒢 is connected—there must be a path from U to R. The first place in which this path uses an edge not on Γ yields the desired vertex W.) Starting with W, take a side excursion, not using any edges used in Γ, until you get stuck. Since Γ uses an even number of edges at each vertex, the number of unused edges at each vertex must still be even; and so, by the argument above, the only place you can get stuck is at W. But then you can create a larger circuit Γ', starting at U, as follows: Follow Γ until you reach W; then take the side excursion, returning to W; then continue on Γ back to U. If Γ'

Start at A:

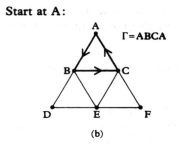

(b)

We are now stuck at A.

Choose a vertex on the circuit for which unused edges remain.

For example, choose **B and take a side excursion (Figure 6.18c).**

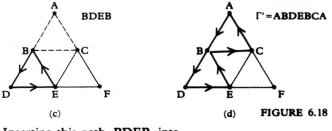

(c)   (d)   **FIGURE 6.18**

Inserting this path, BDEB, into the original circuit at B, we get a new circuit Γ' = **ABDEBCA** (Figure 6.18d).

is not the desired Eulerian circuit, you can continue as above until the desired Eulerian circuit is found. (Since the number of edges is finite, the process must terminate.)

This is still not the desired Eulerian circuit, so we take another side excursion, say at E (see Figure 6.18e).

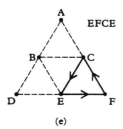

(e)

Now, inserting this path into Γ' at E, we have the Eulerian circuit shown in Figure 6.18f on the left:

ABDEFCEBCA

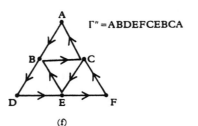

FIGURE 6.18   (f)

Γ" = ABDEFCEBCA

PRACTICE PROBLEMS 6.B

1. [A] Find an Eulerian circuit (if one exists) for each of the graphs in Figure 6.19.

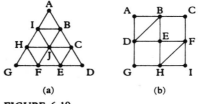

(a)   (b)   (c)

FIGURE 6.19

2. Find an Eulerian circuit (if one exists) for each of the graphs in Figure 6.20.

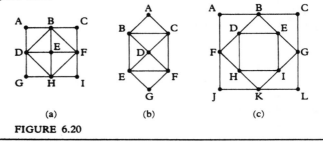

(a)   (b)   (c)

FIGURE 6.20

As a simple consequence of Theorem 6.3, we have the following theorem:

Theorem 6.4 If a connected graph $\mathscr{G}$ has exactly two odd vertices, U and W, then there is an Eulerian path for $\mathscr{G}$, starting at U and ending at W.

Proof Add a new edge e connecting U to W. (We may have to leave the plane to do this, but it always can be done.) This results in a new graph $\mathscr{G}'$ containing only even vertices. By the proof of Theorem 6.3, $\mathscr{G}'$ has an Eulerian circuit Γ, starting by traveling via e from W to U and then eventually returning to W. Omitting e from Γ gives the desired Eulerian path for $\mathscr{G}$.

---

1. Find an Eulerian path (if one exists) for each of the graphs in Figure 6.21.

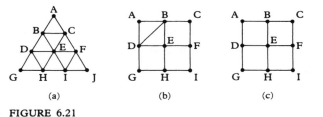

(a)   (b)   (c)

FIGURE 6.21

2. Find an Eulerian path (if one exists) for each of the graphs in Figure 6.22.

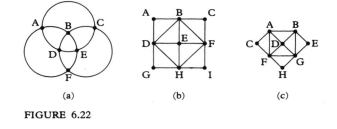

(a)   (b)   (c)

FIGURE 6.22

Theorems 6.3 and 6.4 take care of graphs with either zero or two odd vertices. All that remains to consider, then, are graphs having only one odd vertex. Actually, however, there are no such graphs. This follows from the following theorem.

Theorem 6.5 Every graph has an even number of vertices of odd degree. (And hence there cannot be only one odd vertex.)

Proof Observe that if S is the sum of the degrees of all the vertices, and if E is the number of edges, then $S = 2E$, since each edge contributes exactly two in the count of S—one at each vertex to which the edge is incident. Thus, S is always even. For example, the graph in Figure 6.12a (page 178) has ten edges and six vertices—A, B, C, D, E, and F—of degrees 2, 4, 3, 3, 4, and 4 respectively. Note that

$$S = 2 + 4 + 3 + 3 + 4 + 4 = 20 = 2 \cdot 10 = 2E.$$

Similarly, if S_e is the sum of the degrees of all the even vertices, then S_e is a sum of even numbers and hence is even.

Finally, if S_o is the sum of the degrees of all odd vertices, then $S = S_e + S_o$. Thus $S_o = S - S_e$, which is even. But S_o is a sum of odd numbers. The only way a sum of odd numbers could be even is if there are an even number of them. Hence, there must be an even number of odd vertices.

Corollary 6.2 and Theorems 6.3, 6.4, and 6.5 tell us the whole story about Eulerian paths. Summing up, we have the following:

1. A connected graph has an Eulerian circuit if and only if all vertices are even.

2. A connected graph has an Eulerian path that is not a circuit if and only if the graph contains exactly two odd vertices.

We are now able to complete Problem 6.1(c) (page 174). If you have not already solved it, try again now.

Solution of Problem 6.1(c)

Because part (c) of the problem requires us to find an Eulerian circuit, we must construct the new bridge so that all vertices are even. As C and D are the only odd vertices in Figure 6.5b, adding a new edge that joins them will make all vertices even. Thus, the new bridge must be constructed from C to D (see Figure 6.23).

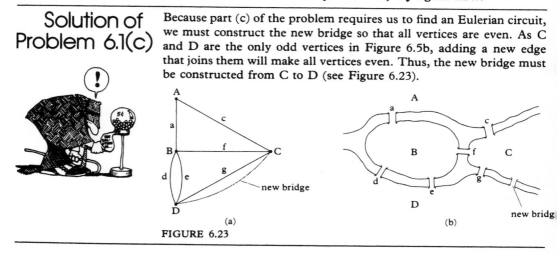

FIGURE 6.23

1. Plan a walk for a person starting at A in Figure 6.23, so that he or she can cross each bridge exactly once and end up again at A.

✩ MORE THAN TWO ODD VERTICES

The problem of finding an Eulerian path for a graph is sometimes referred to as the "highway inspector's problem," because we try to inspect every road once without unnecessarily retraveling any road.

Corollary 6.2 tells us that one inspector cannot accomplish this task if there are more than two odd vertices; but several inspectors might be able to succeed without duplicating any of the work. In fact, having a different inspector for each stretch of road would certainly succeed; but then more people would be hired than are really needed.

We now turn our attention to the question: How many inspectors really are needed? That is, given a connected graph $\mathscr{G}$ with more than two odd vertices, what is the smallest number of edge-disjoint paths by which the graph can be covered? **Edge-disjoint paths** are paths that have no edges in common with each other. For example, the graph in Figure 6.24a has four odd vertices (B, C, D, and G) so it has no Eulerian path. It can be covered, though, by two edge-disjoint paths, DABEDGECB and CFIHFEHG (see Figure 6.24b).

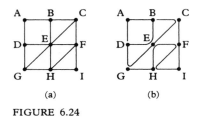

(a)   (b)

FIGURE 6.24

By Theorem 6.5, we may assume that the graph has $2k$ odd vertices. Then the answer is given in the following theorem.

Theorem 6.6 A connected graph, $\mathscr{G}$, having $2k$ odd vertices may be covered by k edge-disjoint paths.

Proof From our previous discussions, it is clear that fewer than k paths will not work, because each path can remove the oddness of at most two vertices. On the other hand, k paths always suffice.

On the left below, we show that k paths are always enough, while we simultaneously consider an example on the right.

We begin by pairing off the odd vertices. Say they are $U_1, \dot{W}_1, U_2, W_2, \ldots, U_k, W_k$.	The graph in Figure 6.25a has four odd vertices. We label these $U_1, W_1, U_2,$ and W_2 (and label the other vertices A, B, C, D, and E), to get Figure 6.25b.

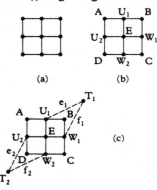

(a)   (b)

We now form a new graph $\mathscr{G}'$ by first adding k new vertices, $T_1, T_2, \ldots, T_k$ and then adding $2k$ new edges $e_1, e_2, \ldots, e_k,$ $f_1, f_2, \ldots, f_k$, where, for each i, e_i connects U_i and T_i, and f_i connects W_i and T_i. (Again, we may have to leave the plane to do this.)

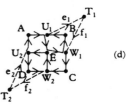

(c)

Since one edge has been added to each of the odd vertices of $\mathscr{G}$, and all new vertices are of degree two, $\mathscr{G}'$ has only even vertices.

Hence, $\mathscr{G}'$ has an Eulerian circuit. Since T_i can be reached only from U_i or W_i, each of the segments $e_i T_i f_i$ (or $f_i T_i e_i$) must appear in the circuit.

We obtain an Eulerian circuit for Figure 6.25c, starting with $U_1 e_1 T_1 f_1 W_1$. For example, $U_1 e_1 T_1 f_1 W_1 C W_2 E U_2 D W_2 f_2 T_2$-$e_2 U_2 A U_1 E W_1 B U_1$ (Figure 6.25d).

(d)

Removing these cuts the circuit into k pieces—the k paths required to solve the original problem.	Omitting the segments $e_1 T_1 f_1$ and $f_2 T_2 e_2$, we obtain the two paths $W_1 C W_2 E U_2 D W_2$ and $U_2 A U_1 E W_1 B U_1$, which give Figure 6.25e, an edge-disjoint covering of the graph in Figure 6.25a.

FIGURE 6.25   (e)

Instead of hiring several highway inspectors to cover a given network, we may allow a highway inspector to retrace some roads, if necessary. We can now ask the following question: How can one highway inspector cover the entire network while keeping the number of re-traced edges to a minimum?

Retracing an edge is essentially the same as adding a new edge that is incident to the same two vertices as the edge that is to be retraced. Our aim is to add these new edges so that all but two of the vertices become even and therefore an Eulerian path exists.

For example, to cover the graph in Figure 6.26 with a minimal amount of retracing, we must eliminate two odd vertices.

We do this by adding an edge (or retracing an edge) between two of the odd vertices, say, U_1 and W_1. Actually, we must retrace two edges, for example, U_1BW_1 or U_1EW_1. We then get an Eulerian path, such as

$$U_2AU_1BW_1EU_1BW_1CW_2DU_2EW_2$$

from U_2 to W_2.

FIGURE 6.26

If we want to start at a particular point and end at that same point, then we need an Eulerian circuit and hence must eliminate all odd vertices. In the above example, this can be done by retracing two more edges (U_2DW_2, for example).

---

1. [A] Completely cover each of the graphs in Figure 6.27 by the fewest possible edge-disjoint paths.

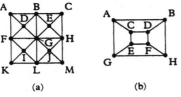

(a)   (b)

FIGURE 6.27

2. Completely cover each of the graphs in Figure 6.28 by the fewest possible edge-disjoint paths.

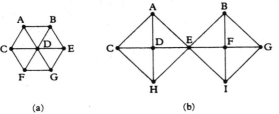

(a)   (b)

FIGURE 6.28

3. [A] Trace each of the figures in Problem 1 above without lifting your pencil from the page, in such a way that you retrace as few edges as possible.

4. Trace each of the figures in Problem 2 above without lifting your pencil from the page, in such a way that you retrace as few edges as possible.

5. [A] Starting at A and ending at A, trace each of the figures in Problem 1 so that you retrace as few paths as possible.

6. Starting at A and ending at A, trace each of the figures in Problem 2 so that you retrace as few of the paths as possible.

☆ DIRECTED GRAPHS

Another interesting variation of the Eulerian path problem occurs if we assume that each edge can be traveled in only one direction, like a one-way street. The resulting geometric structure is called a **directed graph** or **digraph**.

As before, we want to know when we can traverse every edge of the graph exactly once; but this time we must obey the "one-way" signs. Although this new problem seems more difficult, its solution is actually similar to that of the original problem.

We define the **in-degree** of a vertex as the number of edges incident to that vertex which are directed toward the vertex. We define the **out-degree** of a vertex analogously—the number of incident edges which are directed away from the vertex. For example, in Figure 6.29, the in-degrees and out-degrees are:

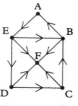

FIGURE 6.29

	A	B	C	D	E	F
in-degree	1	2	1	1	1	4
out-degree	1	2	2	2	3	0

We now have the following theorem.

Theorem 6.7   A connected digraph $\mathscr{G}$ has an **Eulerian dicircuit** (that is, an Eulerian circuit which obeys directional signs) if and only if the in-degree of each vertex is equal to the out-degree. (Note that this condition implies that every vertex is of even degree.)

We leave the details for you to supply; see how similar they are to the discussion above concerning graphs in general.

1. By analogy, define what is meant by an **Eulerian dipath**.

2. $\boxed{A}$ If possible, find an Eulerian dicircuit or Eulerian dipath for each of the graphs in Figure 6.30. If it is not possible, explain why not.

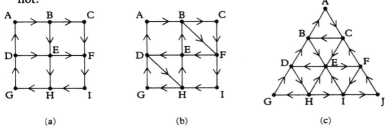

(a)   (b)   (c)

FIGURE 6.30

3. If possible, find an Eulerian dicircuit or Eulerian dipath for each of the graphs in Figure 6.31. If it is not possible, explain why not.

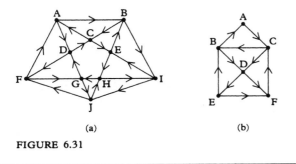

(a)   (b)

FIGURE 6.31

Before ending this discussion, we should point out that problems involving Eulerian paths are occasionally disguised in various ways.

Problem 6.2 (page 174) is such a problem. Try it again now if you have not already solved it.

Solution of Problem 6.2

If we represent the rooms and exterior by letters, then we see that the problem really requires us to find an Eulerian path on the graph in Figure 6.32b.

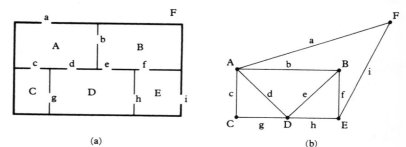

(a)   (b)

FIGURE 6.32

Because this graph has two odd vertices, B and E, there exist Eulerian paths beginning at B and ending at E or vice versa. However, it is not possible to find such a path that begins and ends at F, the exterior of the house.

Thus, the answer to Problem 6.2 is "No. It is not possible."

HAMILTONIAN CIRCUITS

Problem 6.3 (page 174) is a two-dimensional representation of a game called "Around the World," which was invented by Sir William Rowan Hamilton in 1857. The original game consisted of a solid regular dodecahedron (one of the five regular solids) with each of the 20 corners labeled with the name of a city—see Figure 6.33b. The object of the game was to travel around the world, along the edges of the dodecahedron, visiting each of the 20 cities exactly once.

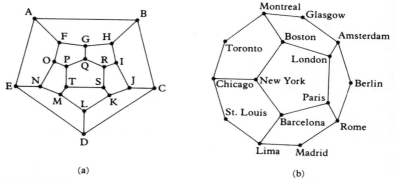

(a)

(b)

FIGURE 6.33

Although Problem 6.3 seems to be closely related to our discussion of Eulerian circuits, it is actually quite different. With Eulerian circuits, each edge has to be traversed exactly once; here, some edges need not be traversed at all. With Eulerian circuits, we could visit a vertex as often as necessary (as long as we use different roads each time to reach and leave the vertex); here, each vertex must be visited exactly once.

Given a graph, a path that visits each vertex exactly once is called a **Hamiltonian path**; and, if it returns to its starting point, it is called a **Hamiltonian circuit**.

As with Eulerian paths, a graph must be connected in order for a Hamiltonian path to exist; but the similarity ends there. Although there are some theorems that give conditions under which Hamiltonian paths will exist, there are no known criteria that completely characterize those graphs which have Hamiltonian paths; nor is there any simple algorithm that enables us to find a Hamiltonian path if one exists. Each graph must be examined individually, and finding a Hamiltonian path—if one exists—is a matter of luck, insight, and reasoned trial and error.

If you could not solve Problem 6.3 (page 174) before,
try it again now.

With Problem 6.3, finding a Hamiltonian path is not too difficult. With a little experimentation, we find, in fact, a Hamiltonian circuit:

Solution of Problem 6.3

ABCDENMLKJIHGQRSTPOFA

There are many other circuits that also work.

Although there is no simple algorithm that tells us how to find a Hamiltonian path if one exists, two observations are worth noting:

1. If there is a vertex of degree one, then any Hamiltonian path must start or end there.

2. If there is a vertex of degree two, then either the path must start or end there, or else both edges incident to the vertex must be traveled.

These observations will be helpful in attacking some of the exercises at the end of the chapter. An example of how they can be applied is given in the discussion that follows.

THE KNIGHT'S TOUR

As with Eulerian path problems, Hamiltonian circuit problems often appear in disguise. One well-known example is the **knight's tour** of the checkerboard.

The problem may be stated as follows:

> Place a knight somewhere on the checkerboard; and then move it so that it visits each square of the board exactly once.

If the square from which the knight started can be reached in one knight's move from the final square, then the tour is said to be **reentrant**.

In order to discuss the knight's tour problem, we must first be familiar with the knight's move in chess. We may describe this move as follows: A knight moves horizontally or vertically two spaces and then turns left or right for one space (an "L" in essence, as Figure 6.34a shows). Thus the knight—indicated by N in Figure 6.34b—could, on its next move, move to any of the squares indicated by an X.

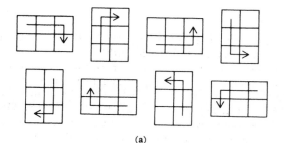

(a)

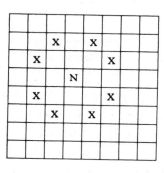

(b)

FIGURE 6.34

An example of a reentrant knight's tour on a 5 × 6 checkerboard is illustrated in Figure 6.35. The knight starts in the upper lefthand corner, then moves to the square labeled 2, then 3, and so on.

But what does this have to do with graphs? We can think of the squares of the checkerboard as the vertices of a graph, with edges connecting two vertices if and only if a knight could move directly from one to the other. For example, for the 3 × 4 case, we get the graph shown in Figure 6.36. Note that only the 12 labeled vertices are considered as points of intersection of the edges.

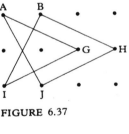

1	22	27	12	7	16
28	11	30	15	26	13
21	2	23	8	17	6
10	29	4	19	14	25
3	20	9	24	5	18

FIGURE 6.35

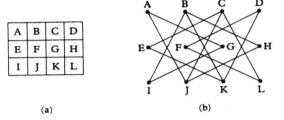

(a)   (b)

FIGURE 6.36

The knight's tour requires that he visit each vertex exactly once—that is, we must find a Hamiltonian path.

For the graph in Figure 6.36, such a path can be found fairly easily as a result of our observation pertaining to vertices of degree two. If a vertex is of degree two, then both edges incident to that vertex must be traveled in any Hamiltonian circuit or in any Hamiltonian path which does not start or end at that vertex. Applying this to Figure 6.36, we see that vertices A, G, H, and I are of degree two (as are D, E, F, and L). Hence, a Hamiltonian path must contain all of the edges shown in Figure 6.37, with the possible exception of one incident to the starting or ending point of the path.

Similarly, the other six vertices give rise to the edges pictured in Figure 6.38. The only connections between these two circuits are the edges B–K and C–J. Thus, to find a Hamiltonian path, we must traverse the first six vertices (A, B, G, H, I, and J) ending on B or J, and then we must continue with a traversal of the other six vertices, beginning at K or C. One path that works is:

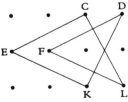

FIGURE 6.37

AGIBHJCEKDFL

Others may be found similarly.

There is no Hamiltonian circuit for the Figure 6.36 graph—that is, a reentrant knight's tour on a 3 × 4 checkerboard is not possible.

With larger boards, graph theory can still be used to represent the

FIGURE 6.38

knight's tour problem. But the lack of a simple algorithm for finding Hamiltonian paths often causes the graph-theoretical representation to be of only limited value.

OTHER APPLICATIONS

So far, we have considered problems in which the object was to travel along all edges or through all vertices of a graph. Frequently, however, we are satisfied with much less. Suppose, for instance, that there is a system which may be in any of several "states" (conditions, situations), and the object is to get from one particular state to another one. Then the problem may be represented graphically as follows: We consider the possible states of the system as the vertices of a graph, and connect two states by an edge if it is possible to go directly from one state to the other. The problem then becomes one of finding any path between the vertices corresponding to the particular states in question.

"Crossing problems" are a class for which this approach is sometimes helpful. Problem 6.4 is a classical example.

If you have not already solved Problem 6.4 (page 175), try it again now.

Solution of Problem 6.4

In Problem 6.4, let M represent the missionary who can paddle, C the cannibal who can paddle, m_1 and m_2 the other missionaries, and c_1 and c_2 the other cannibals. Then, according to the conditions of the problem, the only sets of people who can be on the original side of the river together with the canoe are:

$$Mm_1m_2\, Cc_1c_2, \ Mm_1m_2\, Cc_i, \ Mm_1m_2\, c_1c_2, \ Mm_1m_2\, C, \ Mm_1m_2\, c_i,$$

$$Mm_i\, Cc_j, \ Mm_i\, c_1c_2, \ m_1m_2\, Cc_i, \ Mm_1m_2,$$

$$Cc_1c_2, \ Cc_i, \ MC, \ Mc_i, \ m_i\, C, \ or \ C,$$

(where $i = 1$ or 2, and $j = 1$ or 2). All other states are prohibited, either because cannibals would outnumber missionaries on one of the sides of the river, or because there is no way the boat could have returned (no one on the side with the boat can paddle).

We represent the possible states as the vertices of a graph, and connect two vertices by an edge if it is possible to go directly from one of the states or situations to the other. Note that we consider only states with the canoe at the original bank, so a "move" consists of someone paddling over and someone paddling back. For example, we can go from the state $Mm_1m_2\, C$ to the state $Mm_1\, Cc_2$ by having M row m_2 over and bring c_2 back. The total graph is indicated in Figure 6.39. Note that, in getting from one state to the next, we must consider

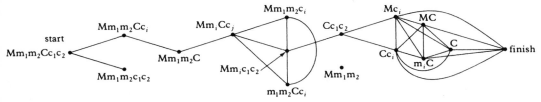

FIGURE 6.39

what the situation would be when the canoe is brought to the second side of the river. Thus, for example, from $Mm_1m_2c_1c_2$, we cannot have Mm_i paddle over, for that would leave two cannibals and one missionary on the near shore; nor can we have Mc_i paddle over, because, when they arrive, the cannibals on the second shore will outnumber the missionary.

From the graph, it is now easy to find the solution to the problem. All that is needed is a path from start to finish. It is not only easy to find such a path but, in fact, we can find one that minimizes the distance to be paddled. We exhibit the solution in a table (Figure 6.40).

	in the boat	on the near shore	on the far shore
start		$Mm_1m_2Cc_1c_2$	
trip 1.	out Mc_1	$m_1m_2Cc_2$	
	return M	$m_1m_2Cc_2$	c_1
trip 2.	out Cc_2	Mm_1m_2	c_1
	return C	Mm_1m_2	c_1c_2
trip 3.	out Mm_1	m_2C	c_1c_2
	return Mc_2	m_2C	m_1c_1
trip 4.	out MC	m_2c_2	m_1c_1
	return Mc_1	m_2c_2	m_1C
trip 5.	out Mm_2	c_1c_2	m_1C
	return C	c_1c_2	Mm_1m_2
trip 6.	out Cc_1	c_2	Mm_1m_2
	return C	c_2	$Mm_1m_2c_1$
trip 7.	Cc_2		$Mm_1m_2c_1$
finish			$Mm_1m_2Cc_1c_2$

FIGURE 6.40

The same general approach used in Problem 6.4 may also be used in many other types of problems. The main difficulty is that, if the number of possible states is large, the graph will have many vertices and will become time consuming to construct and difficult to read.

✩ COLORING GRAPHS AND MAPS

No chapter on modern graph theory would be complete without mention of the recent proof of the Four Color Conjecture, which deals with the coloring of maps in the plane. According to the evidence available, in 1852 Francis Guthrie first raised the following question*: Given any map, can it be colored in at most four colors so that countries that share a common border are colored in different colors? Guthrie conjectured that the answer was yes. What makes this conjecture so noteworthy is that, although many of the best mathematical minds of the nineteenth and twentieth century worked very hard to try to prove (or disprove) the conjecture, the problem remained unsolved until 1976, when Kenneth Appel and Wolfgang Haken completed a proof. We can now refer to the result as the Four Color Theorem.

The theorem may be stated very simply as follows:

Theorem 6.8 (The Four Color Theorem) Every planar map may be colored in at most four distinct colors, so that no two regions which share a common border have the same color.

The statement of the theorem requires some explanation. Any planar graph divides the plane into a number of regions. For example, the graph in Figure 6.41 divides the plane into three regions. (Note that what might be called the exterior of the graph is counted as a region— region 3.)

The graph together with these regions is called a **planar map**, provided that each edge of the graph is on the boundary of two distinct regions.

Two regions of the map are said to share a common border if some edge of the graph is part of the boundary of each region. Two regions of a map that meet only in one point or a finite number of points are not considered to share a common border. For example, in Figure 6.42, region 3 shares a common border with 2, 4, and 5, but does not share a common border with 1. Similarly, 2 shares a common border with 1 and 3 but does not share one with 4 or 5.

---

* See [2] and [7] pp. 90–93 for fuller discussions of the history of the conjecture.

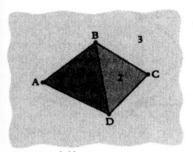

FIGURE 6.41

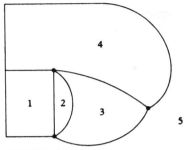

FIGURE 6.42

It should be pointed out that a region or country of the map must be connected. That is, we would have to consider the United States and Alaska as different regions because they are separated by Canada. These two regions need not receive the same color.

The proof of the Four Color Theorem is lengthy and beyond the scope of this book. It depends on an involved case analysis that is carried out with the help of computers.

On the other hand, it is interesting to note that a weaker Five Color Theorem was proved in 1870, over a century ago. This theorem is as follows:

Theorem 6.9 (The Five Color Theorem) Any planar map may be colored in at most five distinct colors so that no two countries that share a common border have the same color.

Of course, now that the Four Color Theorem has been proved, the Five Color Theorem is no longer of great interest. However, it does suggest an interesting question: Might it not be possible to improve the Four Color Theorem to a Three Color Theorem? In fact, it is not possible to do so, as the map in Figure 6.43 shows.

In this map the four numbered regions all touch each other, so at least four colors are required to color this map.

On the other hand, there are some maps that may be colored in fewer than four—or even three—colors. For example, we have the following theorem:

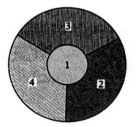

FIGURE 6.43

Theorem 6.10 A planar map may be colored in two colors if and only if each vertex is of even degree.

Given any planar map, if we shrink each region to a point or vertex and consider each edge between two regions as a bridge or edge connecting the vertices (as we did in the Königsberg Bridge Problem), then we get a new graph called the dual of the original map. For example, the graph in Figure 6.44b is the dual of the map in 6.44a. The vertices of this graph are the regions of the map.

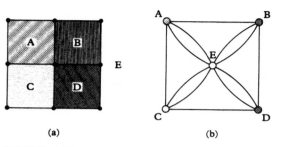

(a)　　　　　　　　　　(b)

FIGURE 6.44

If the original map is colored, then the vertices of the graph we obtain will also be colored. In fact, no two different vertices that are connected by an edge will have the same color.

This notion may be applied to define a coloring for any graph—color the vertices so that no two distinct vertices that are connected by an edge receive the same color. It may be shown that theorems about map coloring can be translated into theorems about coloring graphs. In particular, any planar graph may be colored (in the above sense) in at most four colors.

Instead of coloring the vertices of a graph, we may color the edges so that no two edges incident to the same vertex receive the same color. With this notion of coloring, there can be no analog of the Four Color Theorem, because a vertex of degree n will necessitate at least n colors (see Figure 6.45).

FIGURE 6.45

PRACTICE PROBLEMS 6.G

1. **A** Excluding the exterior regions, what is the fewest number of colors needed for each of the maps in Figure 6.46 so that no two regions which share a common border receive the same color?

(a)

(b)

FIGURE 6.46

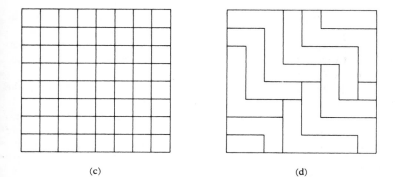

(c)   (d)

FIGURE 6.46

2. Including the exterior regions, what is the fewest possible colors needed for each of the maps in Figure 6.46 so that no two regions which share a common border receive the same color?

3. [A] If you were to color the edges of each of the graphs in Figure 6.47 so that no two edges which are incident to the same vertex receive the same color, what is the minimum number of colors you need?

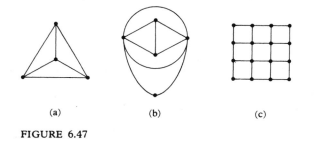

(a)   (b)   (c)

FIGURE 6.47

THE CHAPTER IN RETROSPECT

This chapter has presented a modern topic in mathematics which is different in flavor from the basically algebraic and arithmetic topics of Chapters 3, 4, and 5.

The chapter has also pointed out how a mathematical theory can be developed from an apparently simple question: Is it possible to take a certain walk in Königsberg without retracing any part of the trip? By abstracting the basic elements of the problem and defining certain terms—graph, vertex, edge, path, and so on—we can restate the question simply as, "Does the graph representing the problem have an Eulerian path or circuit?"

Our next step was to try a few problems, and we discovered that the parity of the vertices was somehow involved. These discoveries led to the development of some conjectures, which became theorems when we finally proved them. The proofs were based on definitions, previously proved theorems, and logical reasoning.

The theorems we finally proved enabled us to answer not only the original question, but also related ones.

Much mathematical theory develops along the above lines, although many years may pass between conjecture and proof. For example, as we saw above, it took over a hundred years for the Four Color Theorem finally to be proved. And, as we saw in Chapter 4, Fermat's famous conjecture is still not proved today.

Once a mathematical theory has been developed in response to a question, other applications frequently appear. For example, the Königsberg problem is of interest in terms of highway inspection and its analogs—police patroling streets, trash collection, and others. The question is relevant and practical, even though it initially arose in a recreational setting.

The exercises at the end of this chapter can all be stated or solved graph-theoretically. However, in some cases, the graph-theoretic approach is not practical without the aid of a computer—there are too many possibilities to be handled efficiently. Nevertheless, we include these problems so that you will be aware of their graph-theoretical connection.

Exercises

Eulerian Paths and Circuits

6.1. E. Sterner of Philadelphia, Pennsylvania, is planning her first trip through the part of the United States pictured in Figure 6.48. She intends to fly to Omaha, Nebraska, beginning her tour there. The remainder of her trip is to be made by car, along such a route that she crosses the border between each pair of neighboring states exactly once. (That is, she crosses the Nebraska–Wyoming border, the Nebraska–South Dakota border, the Wyoming–South Dakota border, etc.)

Can you help Ms. Sterner plan her trip?

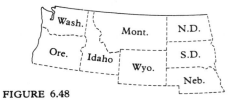

FIGURE 6.48

6.2. [A] Ms. Sterner (see Exercise 6.1) decided that she has enough time to add four more states to her itinerary (see Figure 6.49). She still wants to cross the border between each pair of neighboring states exactly once.

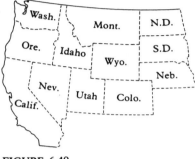

FIGURE 6.49

If she flies to Omaha, Nebraska, in what state must she catch a return flight in order for her to be able to complete such a trip?

6.3. [H] Gia Graphy has a map showing a cluster of states in the United States. She has labeled each state with a number equal to the number of states bordering it on the map.

Prove that the number of states which have been labeled with odd numbers is even.

6.4. [H] Six famous numismatists, Messrs. Dollar, Pound, Frank, Mark, Lira, and Yen, met to trade coins with each other. Each trade that took place was between only two people. After the meeting, each of the six was asked how many people he had traded with. The answers were 5, 4, 2, 1, 3, and 2 respectively.

Prove that at least one person is mistaken.

6.5. [H] [A] Two cells of a chessboard are said to be separated by a knight's move if it is possible for a knight to move from one of them to the other in one move.

(a) Is it possible for a knight to travel over a 4 × 4 chessboard in such a way that it moves exactly once between each pair of cells which are separated by a knight's move? (If it moves from cell X to cell Y, then it cannot also move from cell Y to cell X.)

(b) For what values of n is it possible for a knight to make such a trip on an $n \times n$ board?

(c) Considering a move and its reversal as two different moves, plan a trip for a knight starting on cell A of the 3 × 4 board pictured in Figure 6.50 so that every possible knight's move is made exactly once.

A	B	C	D
E	F	G	H
I	J	K	L

FIGURE 6.50

6.6. [H] (a) If possible, plan a trip for a bishop on the 7 × 7 chessboard in Figure 6.51 so that

it moves exactly once between each diagonally adjacent pair of black cells. If such a trip is not possible, explain why not.

(b) If possible, plan such a trip on the white cells. If no such trip is possible, explain why not. (Use cordinate notation to represent the cells. For example, G–1 is the lower right hand corner, A–7 is the upper left hand corner, D–4 is the center square.)

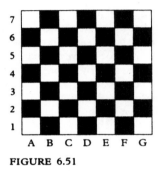

FIGURE 6.51

6.7. **S** **A** Figure 6.52 shows the layout of Woodlawn Gardens. Each line represents a path along which flowers are growing, and each path is exactly 100 meters long. The entrance to the gardens is at the point labeled A, and the exit is at point B.

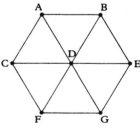

FIGURE 6.52

Mr. E. Ficiency was hired to conduct a tour of the entire gardens in such a way that the necessary amount of walking would be held to a minimum.

What route should he follow? What is the total distance walked?

6.8. **H** Maggie Zehan is working her way through college by selling subscriptions to periodicals. Today, she is planning to try her luck at a housing development, a map of which is shown in Figure 6.53. Each street has houses on only one side. Maggie wants to cover every street, but she also wants to keep her walking to a minimum.

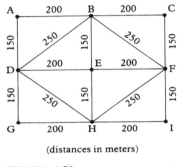

(distances in meters)

FIGURE 6.53

(a) **A** What is the minimum distance she will have to walk if she can start and end at any points she chooses? What should her route be?

(b) If the only parking lot in the development is at point A (that is, she must start and end at A), what is the minimum distance she will have to walk? Describe her route.

6.9. The Ramada County Department of Highways has just resurfaced the county roads, and now the yellow stripe down the middle of each road must be repainted. The truck used for this purpose is very inefficient as far as gas consumption is concerned, and so the Department would like to have the truck travel the shortest distance possible. A road map of the county is shown in Figure 6.54.

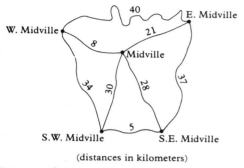

(distances in kilometers)

FIGURE 6.54

The county truck is garaged in Midville, and it must return there when the job is done.

What route should it follow?

6.10. [H] The magician placed a standard 28-piece set of dominoes on the table, removed the doubles (double zero through double six) and turned his back. He then asked the volunteer to remove any one of the dominoes and to arrange all the remaining dominoes in a chain, in accordance with the rules of the game of dominoes (touching parts of two different dominoes must both contain the same number; no side branches are allowed).

When the chain was completed, the magician turned around, took a very quick glance at the arrangement on the table, and said, "The domino you removed is the two–five."

How did he know? Explain why the trick works.

★ **6.11.** [S] [A] There are 27 different three-digit numbers that can be made from the digits 1, 2, and 3—111, 121, 123, 312, etc.

Use graph theory to determine how to place nine 1's, nine 2's, and nine 3's in a circular arrangement so that each of the 27 numbers appears exactly once when all sets of three consecutive digits around the circle are read in a clockwise direction. That is, the sequence shown in Figure 6.55 would give 121, 213, 132, etc. ([64], Problem 23c, p. 110)

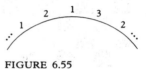

FIGURE 6.55

6.12. [H] Place four 0's and four 1's in a circular arrangement, so that each of the triples (0, 0, 0), (0, 0, 1), (0, 1, 0), (1, 0, 0), (0, 1, 1), (1, 1, 0), (1, 0, 1), and (1, 1, 1) appears exactly once when all sets of three consecutive digits are read around the circle in a clockwise direction.

Hamiltonian Circuits

6.13. [H] [A] A saleswoman, starting at town A on the map in Figure 6.56, wishes to visit every town on the map exactly once, returning to A at the end of her trip.

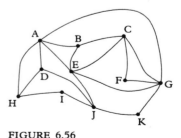

FIGURE 6.56

How many different possible routes could she plan that would satisfy these conditions?

6.14. ☐H☐ A salesman wishes to visit all the cities on the map in Figure 6.57 and to return to his starting point.

What route should he follow if he does not wish to visit any city more than once?

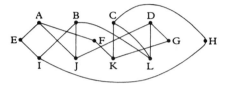

FIGURE 6.57

6.15. ☐H☐ Albatross Airlines serves the nine cities on the map in Figure 6.58, with the flights indicated (two points are connected by an edge if and only if there is a direct Albatross flight between the corresponding cities).

Show that a vacationer wanting to visit each of these cities exactly once would not be able to end her trip at her starting point if she uses only Albatross Airlines for transportation.

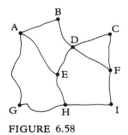

FIGURE 6.58

6.16. ☐A☐ On the graph in Figure 6.59, start at the vertex labeled start, and travel along the graph so that you visit every vertex exactly once, ending at a vertex labeled T. So far, most people who have seen this puzzle have said, " This is too difficult."

What do you say?

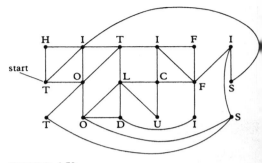

FIGURE 6.59

★ 6.17. ☐S☐ ☐A☐ Each of the 22 vertices on the graph in Figure 6.60 represents a shop that serves as a front for a bookie. The edges joining these vertices represent the only streets connecting these shops. The police have placed stakeouts outside each shop with orders to arrest anyone passing the same shop twice.

Can a numbers runner make the rounds of all the shops without being arrested? ([38], Chapter 1, Problem 30)

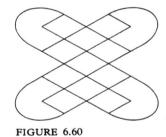

FIGURE 6.60

6.18. ☐S☐ The lighting technician at the Footlight Theater has three colored filters (red, green, and blue) that can be placed over the main spotlight. By varying which filters are on and off he can obtain eight different lighting effects. The filters may be changed one at a time either by adding a filter not being used or by removing one that was just used.

Assuming that he starts and ends with no filters, how can he test all eight lighting effects, without repeating any effect except the final one?

(c) Show that a bishop's tour is not possible on the cells of either color of an $n \times n$ board, where n is any even integer greater than two.

(d) Show that a reentrant bishop's tour (one that ends where it starts) is not possible on the cells of either color of an $n \times n$ board, where n is any integer greater than five.

6.21. [H] Describe a knight's tour on the 3×7 chessboard pictured in Figure 6.62.

A	B	C	D	E	F	G
H	I	J	K	L	M	N
O	P	Q	R	S	T	U

FIGURE 6.62

★ **6.22.** [H] Describe a reentrant knight's tour on the 3×10 chessboard pictured in Figure 6.63.

1	2	3	4	5	6	7	8	9	10
11	12	13	14	15	16	17	18	19	20
21	22	23	24	25	26	27	28	29	30

FIGURE 6.63

6.23. [H] King Arthur has hired a consultant, Sir Cumference, to plan a seating arrangement for Arthur and nine of his knights to sit around the round table. This would not be a difficult task, were it not for the jealousies and petty rivalries that exist among the knights.

6.19. [A] (a) Is it possible for a rook to start on a corner cell of a 6×6 chessboard and to visit every cell of the board exactly once in such a way that it ends on its starting cell? (Note that a rook may move horizontally or vertically but not diagonally. A cell is considered visited if the rook passes over that cell or lands there.)

(b) [H] What happens on a 5×5 chessboard?

(c) Generalize.

6.20. [H] Consider the 7×7 chessboard in Figure 6.61. Recall that a bishop may move diagonally but not horizontally or vertically.

(a) Describe a bishop's tour of the black cells of the chessboard (each black cell is visited once).

(b) Show that a bishop's tour of the white cells of the board is not possible.

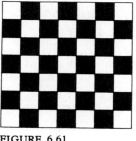

FIGURE 6.61

Arthur insists that Lancelot should sit on his right and that Kay should occupy the seat on Arthur's left; Bedivere refuses to sit next to anyone but Lionel and Tristan; Gawain won't sit next to Tristan, Lancelot, or Lionel; Gareth won't sit next to Galahad, Lancelot, or Kay; Perceval objects to sitting next to Galahad, Lancelot, or Lionel; Tristan refuses to sit next to Lancelot, Perceval or Kay; Galahad will sit next to anyone except Gawain and Kay; and Lionel will not sit next to Galahad. The other two knights are not particular about whom they sit next to.

Help Sir Cumference find a suitable seating arrangement.

Crossing Problems

6.24. A farmer owns a wolf, a goat, and a cabbage, and he wishes to transport them to the other side of a river. Unfortunately, the only available boat is large enough to hold only the farmer and one of his possessions at one time. (It is a giant cabbage!) The farmer cannot afford to leave the wolf and the goat together unchaperoned, for the former would eat the latter. Similarly, the goat and cabbage may also not be left together unattended.

How can the farmer safely transport all three of his possessions across the river?

6.25. [H] Three married couples want to cross a river. The only boat available is capable of holding only two people at a time. This would present no difficulty were it not for the fact that the women are all very jealous, so that each woman refuses to allow her husband to be in the presence of another woman unless she herself is also present.

How should they cross the river with the least amount of rowing?

6.26. [H] Five couples want to cross a river. There is a canoe that holds three people. No man will allow his wife to be in the presence of another man unless he himself is present.

How can the crossing be made with the least amount of rowing?

6.27. How can n couples cross a river in a boat that holds four people, so that no man is in the presence of a woman unless both his wife and her husband are present?

The following problems are similar to the previous ones and may be represented graph-theoretically, but the graph-theoretic approach is not always a reasonable method of attacking them.

★ **6.28.** Four couples want to cross a river. There is a kayak that will hold only two people. No man will allow his wife to be in the presence of another man unless he himself is also present.

(a) [S] Show that the crossing is not possible under the stated conditions.

(b) [A] Show how the crossing can be made if there is an island in the middle of the river.

6.29. [A] Five men come to a river and wish to cross. The only boat they can see is a very small one which is being used by two boys. In fact, the boat is so small that, although the boys can both fit in it, it can hold only one adult and cannot even accommodate one man together with one child.

Enlisting the aid of the boys, what is the

smallest number of trips required for the men to cross, with the boys ending up in the boat?

6.30. [H] [A] Jon, Val, and John plan to steal a treasure from a castle tower, with the aid of John's trained dog, Hugo. The men will have to escape through the tower window, by means of a tackle consisting of a long rope with a basket at each end (outside the window) and a pulley. However, they must be careful. If the difference in weight between the contents of the two baskets is more than 20 lb, then the heavier basket will descend too rapidly for the safety of any of the robbers or of Hugo; however, the treasure chest would not be harmed by the shock of the impact.

John weighs 210 lb, Jon weighs 100 lb, Val weighs 120 lb, and Hugo weighs 20 lb. The treasure is in a chest weighing 60 lb, to which the men will steal the key (which weighs 1 lb).

How can the men escape from the tower so that no one is hurt and no one is double crossed? (The treasure chest is worthless without the key.)

6.31. [H] When Jon, Val, and John (see Exercise 6.30) actually carried out the theft, everything went smoothly during the first part of the operation. After they were all safely down from the tower, they opened the chest and each took a share of the treasure inside.

Each person placed his share in a bag he had brought for the occasion. Since Jon masterminded the entire operation, he received the lion's share of the loot—$30,000 worth. John received $20,000 worth, since his dog had been vital to the success of the plan. Val received the remaining treasure, which was worth $10,000. On the way home from the castle, the men came to a river. The only boat available could hold only two men at a time; or it could hold one man and the dog; or else, one man and one bag of loot.

How can they cross the river so that no man is ever alone (ignore Hugo) with more than his share and no two men are ever alone with more than their combined shares? (In case you were wondering, Hugo, the trained dog, does not know how to row.)

Miscellaneous Graph-Related Problems

6.32. [S] Show that if there are more books in a library than there are pages in any one book, then at least two of the books must contain the same number of pages.

6.33. Show that in a room full of people (more than one person), there are at least two people who have the same number of friends in the room. (Assume that if B is A's friend, then A is also B's friend.)

6.34. [H] Place a marker on a dot in the diagram in Figure 6.64, then slide it along

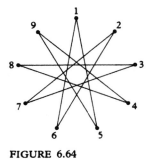

FIGURE 6.64

one of the two lines from that dot to an adjacent dot. (For example, place it on 7 and slide it to either 2 or 3.) Leave the marker on this new dot. Place a second marker on an empty dot and slide it to a vacant adjacent dot. Continue in this manner until you are no longer able to do so. Your goal is to place eight markers.

How should you do this?

6.35. [H] [A] Move the knights in the diagram in Figure 6.65 so that the white knights change places with the black knights.

How can this be accomplished in the fewest moves?

FIGURE 6.65

6.36. [H] [A] Place a white marker on each of the numbers 1, 2, and 3 in the Figure 6.66 diagram, and place a blue marker on 10, 11, and 12. The object of this puzzle is to move the markers until the white ones occupy the spaces originally occupied by the blue ones, and vice versa. A move consists of moving one marker from one vertex of the graph to an adjacent vertex (along an edge of the graph). However, at no time may a white marker and a

blue marker occupy graphically adjacent edges. That is, if a white marker is on 4, then no blue marker may be on 3, 9, or 11.

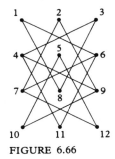

FIGURE 6.66

What is the fewest moves in which the transfer may be made?

6.37. [H] [A] On the graph in Figure 6.67, find five disjoint paths (no two paths have any edge or vertex in common) so that the first connects the two points labeled A to each other, the second connects the points labeled B, and so on.

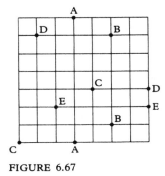

FIGURE 6.67

6.38. [S] [A] A builder is constructing three houses on adjacent lots. He wishes to draw a plan showing how to connect each house to the gas company, the electric company, and the water authority. (See Figure 6.68.)

Can he do this without any of the nine lines crossing each other or passing through any house or utility?

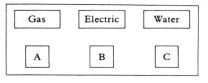

FIGURE 6.68

★ **6.39.** [S] If there are six people at a party, prove that at least three are mutual acquaintances (they all know each other) or else three are mutual non-acquaintances (none of the three knows either of the other two). ([63], Problem 151)

6.40. [H] [A] Amos, Burl, Crawford, Dirk, and Everett were seated, in that order, around a circular table, playing poker. They took a break to watch the fights on TV. When they resumed playing, they changed their seats, so that no one was now sitting next to a person next to whom he had sat before.

What was the new seating arrangement?

6.41. [H] Seven women, Flora, Gail, Hala, Iris, Jaclyn, Kitty, and Leona, were seated, in that order, around a circular table, playing poker. They took a break to watch the fights on TV. Then they resumed playing. At midnight, they interrupted their game again, for a snack. When they resumed playing again, it was observed that each woman had new neighbors. In fact, each pair of women had been neighbors during one of the three segments of the card game.

Find seating arrangements that would satisfy the stated conditions.

6.42. Monique, Nelson, Olivia, Patrick, Queenie, Rosa, Sharim, Theo, and Una plan to set out in teams of three to canvass a neighborhood for a polling organization. They intend to repeat this procedure for a total of four consecutive days (counting the first). They wish to arrange the teams so that no two

people will be in the same threesome on more than one day.

Can you help them?

★ **6.43.** [H] [A] The eight queens on the chessboard in Figure 6.69 dominate the board in the sense that every cell not occupied by a queen is under attack by at least one of the queens. On the other hand, no two of the queens attack each other. The problem is to move exactly three of the queens so that, in the final position, the queens will again dominate the board and no two queens will attack each other. (A queen may move any distance horizontally, vertically, or diagonally.)

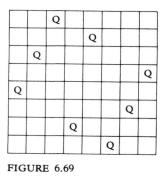

FIGURE 6.69

6.44. [H] It is possible to place five queens on an 8 × 8 chessboard so that they dominate the board (see Exercise 6.43). In the diagram in Figure 6.70, four of these queens have already been placed. Place the fifth one, and show that the answer is unique. ([48], pp. 206–207)

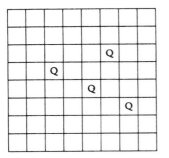

FIGURE 6.70

6.45. [H] (a) Prove that on an $n \times n$ chessboard it is not possible to place more than n queens so that they are independent (that is, so that no two queens attack each other).

(b) On the board in Figure 6.71, four queens have already been placed. Place four more so that the resulting eight queens are independent. ([48], p. 211)

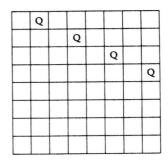

FIGURE 6.71

6.46. [H] [A] What is the largest number of knights that can be placed on a 6×6 chessboard so that they are independent? (See Exercise 6.45.)

7

Games of Strategy
for Two Players

The word "game" is derived from the Old English word, *gamen*, which means sport or amusement. In this sense the recreational problems we have considered so far are games—games with one player, you, the solver.

In this chapter we consider games involving two players. Some such games were played prior to 2000 BCE. Many old games have weathered time or, in some cases, the rules have been altered. For example, the game of chess, in one form, was known to the Hindus and Persians in the sixth century. However, it did not come into its present form until the sixteenth century.

On the other hand, new games are constantly being developed. Probably the best sources of information in regard to recent games, as well as old ones, are Martin Gardner's column, "Mathematical Games," which appears in *Scientific American*, and his books ([18] to [25]). In addition, several important new books on the subject have recently appeared (for example, see [6] and [9]).

The determining factor in solving the problems we have encountered in earlier chapters has been our ability to reason. In a two-person game, another factor seems to be introduced—the reasoning ability of an opponent. In some games, such as bridge, this is of great importance. There is a class of games, however, for which the opponent's ability may (at least theoretically) be neglected completely. The games that we will investigate fall into this category. Some of these games will be easy to analyze (that is, determine whether either player can force a win), while others are mind boggling.

Many games of this type can now be played against a computer.

Your analyses of these games will tell you whether or not you can beat the computer, and, if so, what your strategy should be. Also, the logic of the analyses can be quite useful in programming the computer to play.

The following problems present a few games of the type we consider in this chapter. See if you can analyze them on your own; then read on for a discussion of some of the principles involved in conducting an analysis. We suggest that if you find the analysis difficult at first, you should play the game several times and, toward the end of each round, start asking questions such as "What would have happened if I had made a different move?"

SAMPLE PROBLEMS

Problem 7.1

Two players, Thinker and Patzer, are about to play the following game. From a given pile of matchsticks, each player, in turn, selects any number of sticks up to some fixed maximum, thus gradually depleting the pile. The player who selects the last stick is the winner.

If Thinker has the choice of whether or not to go first, tell which he should choose and what strategy he should follow if

(a) there are 28 matchsticks in the pile and each player may select 1 or 2 sticks at each turn;

(b) there are 28 matchsticks in the pile and each player may select 1, 2, or 3 sticks at each turn;

(c) there are n matchsticks in the pile and each player may select $1, 2, \ldots, m - 1$, or m sticks at each turn.

Problem 7.2

Consider the following game played on a 3×3 checkerboard. Two players alternate turns placing checkers on the board. Each player, in turn, may place checkers on as many of the vacant squares as he or she wants to, provided that all checkers placed during a given turn lie in the same horizontal row or the same vertical column. The winner is the player who places the final (ninth) checker.

Is it better to go first or second? ([38], Chapter 1, Problem 39)

Problem 7.3

The game of Nim is played as follows: It begins with three (or more) piles containing arbitrarily specified (usually unequal) numbers of matchsticks. The players alternate turns. Each player, in turn, may select as many sticks (up to the limit of the pile) as desired from any one pile in which there remain any sticks. He or she may not select from more than one pile on any one turn. The player who removes the final stick from the final pile is the winner.

(a) One commercial version of the game begins with piles of 3, 5, and 7 sticks respectively. In this version of the game, is it best to go first or second, and what strategy should you follow?

(b) Generalize to the game beginning with three piles containing p, q, and r sticks respectively.

Problem 7.4

Given 15 dots on a line,

. . . .

two players take turns placing an X through one of the dots. The first player to mark off a dot so that three consecutive dots are marked is the winner.

Which player should win—the one who goes first or the one who goes second?

For each game that we consider in this chapter, there will be two players, Alfie and Bette, whom we will generally refer to as A and B. They play or "move" alternately, A moving first.

In most games, at his or her turn, a player may have to choose one of several moves. A **strategy** is a rule or decisionmaking formula that tells the player which choice to make at each turn. If this strategy enables the player to win no matter what moves his or her opponent makes, we will call it a **winning strategy** for that player. If the game can end in a draw, we can also speak of a **drawing strategy**—a strategy which, although it does not guarantee a win for the particular player, guarantees that he or she does not lose.

For each game we ask the following question: "Does A have a winning strategy?" If not, we ask whether B has one. If the answer to either question is yes, we try to find this winning strategy. It is not possible for both players to have winning strategies. And, of course, it might be that neither has.

CHANCE-FREE DECISIONMAKING

In this chapter we ignore games in which "Lady Luck" plays a role (poker, backgammon—games in general in which a player's fortune depends on the roll of dice or the deal of cards). The analyses of games such as these are far too complicated, involving lengthy probabilistic

arguments, and, even if successfully completed, yield only strategies by which one can optimize the chances of winning but cannot guarantee a win. We will investigate only games in which each player at each turn is free to choose any legal move—games in which decisions are not made by chance.

GAMES OF PERFECT INFORMATION

As a second restriction, we are interested here only in games of **perfect information**—games in which both players are aware at all times of all aspects of the structure of the game. Each knows, at any point in the game, what moves have been made prior to that point as well as what moves the opponent will be able to make in response to any possible move. Consider Scrabble for example. Aside from being a game of chance (chance determines what letters you pick), it is also a game of imperfect information, for you do not know what letters your opponent holds. As a result, you might be reluctant to place certain letters on the board for fear that doing so would open excellent scoring opportunities for your opponent, opportunities of which he or she might in fact not be able to take advantage. In general, not having full information about a game introduces an element of chance. As a result, it may not be possible for either player to develop a strategy that guarantees a win—although it may be possible to find a strategy that optimizes one's chances.

Consider, for example, the following game. Player A secretly selects three numbers from 1 to 9. B then announces two numbers from 1 to 9. If the sum of at least one of B's numbers with at least one of A's numbers is 10, then B wins. Otherwise, A wins. (For instance, if A picks 3, 5, and 7 and B picks 1 and 4, then B wins because $1 + 4 + 5 = 10$; similarly, B wins if she chooses 1 and 2 because $2 + 3 + 5 = 10$.) Although chance is not involved in the decision-making process of this game (each player is free to choose whatever number he or she likes), the lack of information about the opponent's move makes luck a factor in the game. Neither player can adopt a strategy that will guarantee a win.

FINITENESS

There is one other requirement we place on the games we consider here—they must be finite. A **finite game** is one that must necessarily terminate in a finite number of moves. Many board games, such as backgammon and Monopoly, are not finite; they could conceivably continue forever. As another example of an infinite game, consider the following: A and B alternate turns at rolling a die. With each roll, the player who rolled adds the resulting number to his or her previous total. The game ends when one player's total exceeds the other's total by 20 points or more. This game could conceivably end after A's fourth turn; or it could continue indefinitely. Frequently, we can make a game finite by the imposition of a rule that limits play. For example, in the game above, we could add a rule which says that if no player has won by the end of 1000 rolls, then B is the winner.

A more interesting example of a rule that forces a game to be finite occurs in chess. The rule says that if 50 consecutive moves are made without a piece being captured or a pawn being moved, then the game is a draw. Since each side can have at most a finite number $(8 \cdot 6 = 48)$ of pawn moves and can make at most a finite number (15) of captures, the game must terminate in a finite number of moves—at most, $2(48 + 15) \cdot 50 = 6300$, since the game ends immediately if only the two kings remain.

If there is a number, n, such that the game cannot last for more than n moves, then we say that the game is **bounded**. If n is the smallest such number, then we say that the game is of **length** n. Although there are finite games that are not bounded (for example, player A chooses a number and then A and B alternate turns rolling a die until the total exceeds the chosen number), we do not consider such games here. All bounded games have some fixed (maximal) length n, although it may be difficult to say what that length is.

1. [A] Consider each of the following games: Is it chance-free; is it a game of perfect information; is it finite?
   (a) There are 6 matchsticks on the table. Each player at a turn takes 1 or 2 matchsticks. The one who picks the last stick wins. To see who goes first, the players flip a coin.
   (b) The first player, A, writes any positive integer on a piece of paper, then B writes one. If the sum of the two numbers is odd, A wins. Otherwise, B wins.
   (c) Backgammon.
   (d) Gin rummy.
   (e) Chinese checkers.
   (f) Tic Tac Toe.
   (g) Parcheesi.

PRACTICE PROBLEMS 7.A

THE EXISTENCE OF WINNING STRATEGIES

If we now impose the above restrictions on a game, we have the following result:

Theorem 7.1 In any finite two-person game of perfect information in which the players move alternately and in which chance does not affect the decisionmaking process, either

(a) one of the two players must have a winning strategy

or

(b) the game is a theoretical draw (both players must have drawing strategies).

To see this, observe that, since the game is finite, it must end in a win for one player or in a draw. If A does not have a winning strategy, then, whatever play A makes, B must have a counterplay to prevent A from winning (that is, if there is a play that A can make which cannot be countered by B, then that play would be part of a winning strategy for A); thus, B has at least a drawing strategy. It may be that B actually has a winning strategy. If not, then it must be possible for A to play in such a way that, no matter what B does, B does not win; in other words, A has a drawing strategy.

Thus, one of the players has a winning strategy or both players have drawing strategies.

Note that, if the game cannot end in a draw, then one of the two players must have a winning strategy—preventing your opponent from winning is equivalent to winning yourself.

A formal proof of Theorem 7.1 for games of finite length can be given, using induction on the length of the game. We do not present the complete details of the proof, but we illustrate the inductive step by the following example. (For a discussion of mathematical induction, see Appendix B.)

Consider a game in which A selects a number, either 3 or 4; B adds either 3 or 4 to A's number; A then adds 3 or 4 to the total; etc. If either player brings the total to 13, he or she wins; if the player brings the total to 14, the game is a draw; if the player brings the total over 14, that player loses.

The tree diagram of the game is shown in Figure 7.1. Since the game ends in at most five moves, the game is of length 5.

Consider the nodes from which all moves end the game immediately (in Figure 7.1, these nodes are circled). If such a node is reached in the course of play, it should be clear who will win. Specifically, if it is A's turn at such a node and if one of A's possible choices results in a win for A, then obviously A will make that choice and will win. If none of A's choices result in a win for A but one of them results in a draw, then A will make that choice and the game will be drawn. If A cannot make a choice that results in a win or a draw, then B will win if that node is reached. The same principles apply if B is to make the choice. Since the outcome of the game is strategically determined when a node of the type in question is reached, we can

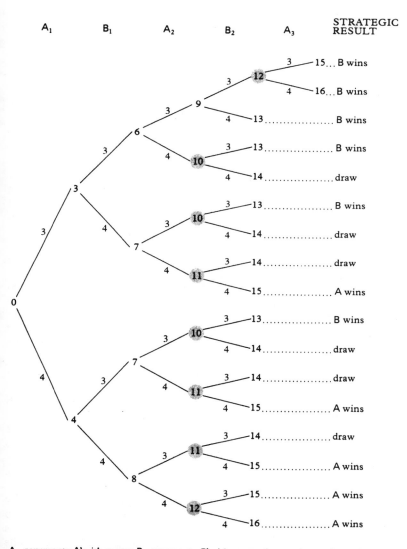

A$_i$ represents A's ith move; B$_i$ represents B's ith move; the number at the node indicates the total at that point; the number above each line indicates the number selected at that turn.

FIGURE 7.1

replace such nodes by the associated strategic results. Doing this for the tree in Figure 7.1, we obtain Figure 7.2.

This figure represents a game of length 4. In general, the length will

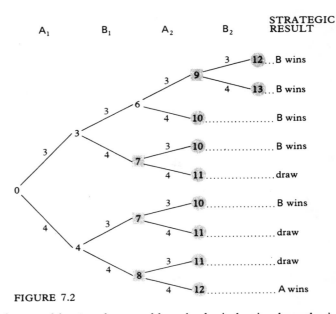

FIGURE 7.2

be decreased by 1 and we could apply the induction hypothesis; that is, if we assume that a suitable game of length n satisfies the conclusion of Theorem 7.1, we would know that each such game of length $n + 1$ is reducible to a game of length n, and therefore also satisfies the conclusion of the theorem. The circles in Figure 7.2 indicate where nodes were eliminated from the previous diagram.

Although it would not be necessary to do so in the inductive proof, let us continue the process for our example: We consider next the nodes indicated by boxes in Figure 7.2. Replacing them by their strategic equivalents, we obtain Figure 7.3a.

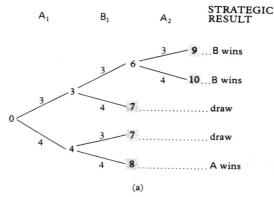

(a)

FIGURE 7.3

Continuing this process further, we obtain **Figure 7.3b,**

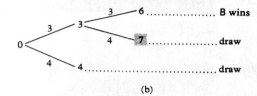

(b)

then Figure 7.3c,

(c)

and finally Figure 7.3d.

0 . . . . draw

(d)

Thus, the game is a theoretical draw.

In the analysis of the game, we have associated a value with each node. Returning to the original tree diagram, we fill in these values, obtaining Figure 7.4 (next page).

From this diagram, each player's strategy should be clear. A must always play to a node whose strategic value is "A wins" or "draw," and B must always play to a node whose value is "B wins" or "draw." If a player deviates from this strategy, he or she may lose.

A word of caution: In general, drawing a tree diagram is not always a reasonable approach to finding strategies—the diagrams are usually too immense. We will return later to the question of how to go about trying to find a winning strategy.

Theorem 7.1 says that in every game of the type under consideration, either one player has a winning strategy or both players have drawing strategies. But determining which of these cases applies may not be easy. For some games, such as checkers and chess, the determination has never been made. In these games, Theorem 7.1 implies that either the first player has a theoretical (forced) win, or the second player has a theoretical win, or the game is a theoretical draw; but the vast number of possible sequences of moves in these games has enabled them to withstand complete analysis.

A₁   B₁   A₂   B₂   A₃

3 — 15 — B wins
3
12 — B wins
4 — 16 — B wins
3
9 — B wins
4 — 13 — B wins
3
6 — B wins
3 — 13 — B wins
4
10 — B wins
4 — 14 — draw
3
3 — B wins
3 — 13 — B wins
3
10 — B wins
4
4 — 14 — draw
7 — draw
3 — 14 — draw
4
11 — draw
4 — 15 — A wins

0 — draw

3 — 13 — B wins
3
10 — B wins
4
7 — draw
4 — 14 — draw
3
3 — 14 — draw
4
11 — draw
4 — 15 — A wins
4 — draw
3 — 14 — draw
3
11 — draw
4
4 — 15 — A wins
8 — A wins
3 — 15 — A wins
4
12 — A wins
4 — 16 — A wins

FIGURE 7.4

Even if it is possible to tell which player has the theoretical win, it may not be easy to find the winning strategy. That is, it may be possible to prove that a winning strategy exists without actually being able to find such a strategy. As an example, consider the game of Hex (see [18], p. 73), which was invented in the 1940s by the Danish engineer, Piet Hein. The game is played on a diamond shaped board made up of hexagons: A, playing white, and B, playing black, take turns placing markers, one at each turn, on vacant hexagons. A's object is to complete a connected path of white markers that touches the top and the bottom of the board. B's objective is similar, except that her path of black markers must go from side to side. Whoever completes such a path first is the winner.

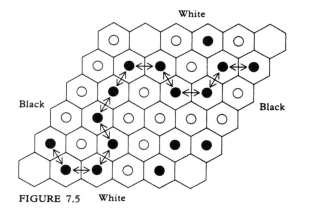

Figure 7.5 illustrates a completed game of Hex on a 6 × 6 board. B (black) has won; her winning path is indicated by the arrows.

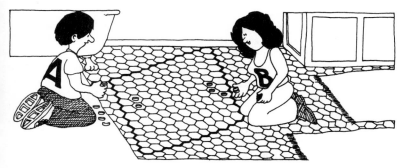

FIGURE 7.5   White

It is fairly easy to show that the first player has a winning strategy for any size Hex board (*n* units on a side). To see this, we must first see that the game cannot be drawn. Suppose a draw were possible, yielding a board that is completely filled, but no one has won. Imagine taking a pair of scissors and cutting out all cells on which a black marker is found. Since A has not won, the board must separate into pieces so that no one piece contains cells from both the bottom and top of the board (otherwise the piece would contain a winning path for A). Therefore, there must be a series of cuts that completely separates the top of the board from the bottom. These cuts must go from the left side of the board to the right side, and hence contain a winning path for B.

Since the game cannot end in a draw, either A or B must have a winning strategy. If B had the winning strategy, then A could make any move he likes and then, ignoring his first move, play as if he were the second player. The extra marker he has on the board can in no way hinder him; nor can it help B. And if at some stage in the game the strategy calls for A to place a marker in the cell he originally marked, since he already has a marker there, he can again make any move he likes. Thus, A will win—yielding a contradiction to the assumption that B has a winning strategy. Hence, A, the first player, must be the one who has the winning strategy.

This same argument works for any game that cannot be drawn and in which a player's earlier moves do not limit his or her possible later moves and cannot help the opponent.

Even though it is known that the first player in a game of Hex has a winning strategy, it is still not known, for large boards, what that strategy is.

PRACTICE PROBLEMS 7.B

1. [A] In the game of Tic Tac Toe, a player's first move cannot hurt him or her later on. Why, then, doesn't the argument used above for the game of Hex show that A has a winning strategy for Tic Tac Toe?

2. [A] Show that B cannot have a winning strategy in the game Qubic (a three-dimensional, $4 \times 4 \times 4$ version of Tic Tac Toe) where the object for A is to get four X's in a row and the object for B is to get four O's in a row.)

3. Consider the following game played on a $4 \times 4 \times 4$ Tic Tac Toe board. Each player at his or her turn places an X on any cell of the board. The first player to complete four X's in a row is the winner. Does an argument similar to the one used for Hex—to show that the first player has a winning strategy—apply to this game? Explain.

4. In the game of Bridg-it (see page 262), prove that the first player must have a winning strategy

POSITION—STATE OF THE GAME

Before we begin to analyze games, that is, determine if winning strategies exist, it is convenient to be able to refer to the "state of the game" or "position" at any point during play. When we do so, we may include all of the following factors:

1. The physical placement of the pieces on the board (if it is a board game) or some analogous notion (if it is not a board game).

2. Which player is to make the next move.

3. What moves are legal for that player to make at the time in question.

4. What moves have been made up to that time.

We may think of a strategy as a chart listing all possible positions that could arise during the play of the game together with a description of what action should be taken (what move should be made) in each position.

The position that exists just prior to the start of the game will be referred to as the **initial** or **starting position** of the game. A position in which the game is over will be called a **terminal position**. Each nonterminal position that could conceivably arise in the course of play may be regarded as the starting position of a different, but related, game. Hence, by Theorem 7.1, from each particular possible position in the game, either the player who is about to move can force a win, or the opponent can force a win, or both players can force a draw.

We adopt the viewpoint of the player, X, who has just moved, thereby "leaving the game in a particular position." If that position is one from which X can force a win, then the position is obviously a favorable or **winning position** (for X). Similarly, if X has left a position from which his or her opponent can force a win, then that position is an unfavorable or **losing position**. A position from which both players can force a draw is called a **drawing position**. We can thus classify each position that could arise in a game as a winning, a losing, or a drawing position—for the player who has just moved. (Obviously, there can be no drawing positions in games that cannot end in a draw.)

A terminal position in which the player who has just moved has won the game is clearly a winning position. Similarly, a terminal position in which the game is drawn is a drawing position, and a terminal position in which the player who has just moved has lost is a losing position.

If X leaves a nonterminal winning position (from which he or she can force a win), then X's opponent, Y, on the next turn, will be unable to leave a position from which Y too could force a win, or even a draw; otherwise the statement that X could force a win would not be true. In other words, Y's move will have to result in a losing position (for Y). Thus, any move made from a winning position must result in a losing position.

Similarly, if X leaves a nonterminal losing position, then it must be possible for Y to force a win by choosing moves properly. In other words, from any losing position, there must be at least one move possible that results in a winning position.

Finally, if X leaves a nonterminal drawing position, then it must be impossible for Y to force a win but it must be possible for Y to force a draw. Thus, any move made from a drawing position must either leave a drawing position or a losing position, and at least one possible move will actually leave a drawing position.

In summary: The possible positions in a game in which draws are possible may be divided into three lists—I, winning positions; II, losing positions; III, drawing positions—such that the following conditions are satisfied:

1. Terminal positions, if any, in which the player who just moved has won are on list I; terminal positions, if any, in which the player who just moved has lost are on list II; terminal positions, if any, in which the game is drawn are on list III.

2. Any move made from a position on list I results in a position on list II.

3. From any nonterminal position on list II, there is at least one move possible that results in a position on list I.

4. Any move made from a position on list III results in a position in list II or III.

5. From any nonterminal position in list III, there is at least one move possible that results in another position in list III.

If draws are not possible, we get only two lists that satisfy conditions 1, 2, and 3.

Once we know the winning, losing, and drawing positions for a game, the game has been analyzed and it is easy to tell which player (if either) can force a win and to describe the winning or drawing strategy.

If the starting position is a losing position, then A can win as follows: His first move must be chosen so that he leaves a winning position. This is possible by condition 3 above. Thereafter, every move B makes will leave a losing position (condition 2), and A will always be able to choose his move so that he leaves a winning position (condition 3). Since we are dealing with a finite game, a terminal position must eventually be reached. Since B always leaves losing positions and A always leaves winning positions, A will win regardless of who reaches the terminal position: If B reaches it, it must be a terminal position in which the player who just moved lost; whereas, if A reaches it, it must be one in which the player who just moved won.

Similarly, if the starting position is a winning position, then B has a winning strategy. Each time A moves, he will leave a losing position (condition 2), and each time B moves she will be able to leave a winning position (condition 3).

Finally, if the game starts at a drawing position, then both players have drawing strategies. A must play so that he leaves a drawing position (this is possible by condition 5); then, so must B; and so on. Any deviation from this strategy will result in a player leaving a losing position (condition 4), from which it will be possible for his or her opponent to win.

Thus, we can analyze a game by determining whether each possible position that could arise is a winning, losing, or drawing position.

Actually, though, it is not always necessary to investigate all possible positions. As long as we can find three lists of positions that satisfy the five conditions above (or two lists that satisfy conditions 1, 2, and 3) such that the starting position is on one of the lists, the same reasoning used above enables us to determine which player wins and what strategy he or she should follow. We illustrate with the following example.

An Example—Another Matchstick Game

From a pile of 11 matchsticks, two players take turns removing either 1 or 4 sticks. The player who removes the last stick is the winner. In this game, a position may be represented by a number—the number of matchsticks remaining. (Note that no draws are possible in this game and that the only terminal position, 0, is a winning position.)

Ignoring for the moment the question of how the lists in Figure 7.6 were obtained, it is easily verified that they do satisfy conditions 1, 2, and 3 above, and hence they give the winning and losing positions for the game.

From any position in the winning column, subtracting 1 or 4 gives a position in the losing column. Similarly, from each position in the losing column, it is possible to obtain a position in the winning column either by subtracting 1 or by subtracting 4, as the case may be.

Since the starting position, 11, is a losing position, A wins this game (if he plays properly). He may select either 1 or 4 sticks and thus place himself in a winning position. Whatever B then does will leave a losing position, from which A will be able to return to a winning position, etc. To fully describe A's strategy, we must say what he should do in each possible case. This is obvious from the chart.

The chart in Figure 7.6 contains all possible positions that could arise in the game. Actually, however, shorter lists of winning and losing positions are possible. For example, the lists in Figure 7.7 also satisfy conditions 1, 2, and 3; and A's strategy using these lists is

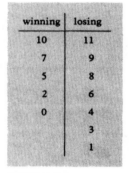

winning	losing
10	11
7	9
5	8
2	6
0	4
	3
	1

FIGURE 7.6

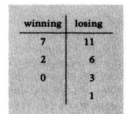

winning	losing
7	11
2	6
0	3
	1

FIGURE 7.7

easily stated: On his first turn, he should leave 7 sticks; on his second turn, he should leave 2 sticks (this will be possible no matter what B's first move is); on his third turn, he wins. The reason that we were able to omit some possible positions from the chart is that A was able to adopt a winning strategy in which the omitted positions could not occur.

Note that the choice of which winning positions we include on our list must relate to which losing positions are included, and vice versa; conditions 1, 2, and 3 must be satisfied. Hence, if we include 10 in our list of winning positions, then 9 must be included in the list of losing positions, so that 5 would have to be on the list of winning positions (8 cannot be since it is a losing position), and so on.

It is important to observe that although the game in question is a winning game for A, he can lose if he plays poorly. If he begins by taking 1 stick and then again takes 1 stick after B takes 1 stick, then he leaves 8 sticks—a losing position, and B can win with proper play.

As a final comment in this section, we note that it would not be correct for A to reason as follows: "Since 2 is a winning position, I can make random moves until there are a little more than 2 sticks left, and then I will be sure to leave 2 sticks."

B may be able to prevent him from leaving 2 sticks. In other words, in a game of the type we are considering in this chapter, *a strategy must start from the very first move;* it cannot begin in the middle of the game. In the game at hand, A must plan his strategy so that he is sure of being able to leave 2 sticks.

PRACTICE PROBLEMS 7.C

1. Analyze the matchstick game above by making a tree diagram and labeling the nodes as in Figure 7.4. Who will win? Find the winning strategy. Compare your answer with the charts in Figures 7.6 and 7.7.

THE STATE DIAGRAM OF A GAME

In addition to the tree diagram, there is another graph-theoretical approach to analyzing games. It is the **state diagram** of a game. In this diagram, each node represents a position or state of the game. A directed edge is drawn from node P to node Q if it is possible to move from position P to position Q in one move. For example, the state diagram of the game discussed in the preceding section is shown in Figure 7.8. Note that from the starting position, 11, we can go to either 10 or 7; from 10, we can go to 9 or 6; etc.

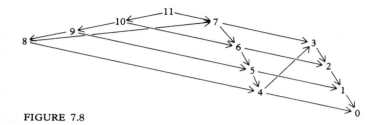

FIGURE 7.8

In the approach, sometimes referred to as the Sprague-Grundy method, the nodes of the diagram are labeled as winning, losing, or drawing, by starting at the terminal nodes and working backward.

In this example, 0 is a winning position, so we label it W. We now work backward, using our definition of winning and losing positions. (This game cannot end in a draw, so drawing positions need not be considered.) Since 0 can be reached by a move from 4 or from 1, these last two positions must be losing; so we label them L (Figure 7.9).

We continue, following the definitions of winning and losing positions: That is, a node from which at least one edge leads to a position labeled W must be labeled L; if all edges from a node lead to other nodes labeled L, then the node in question is labeled W. (In a game that can end in a draw, if some edge from a node leads to another node labeled D and no edges lead to nodes labeled W, then the node in question should be labeled D.)

For the game in question, we obtain Figure 7.10.

FIGURE 7.9

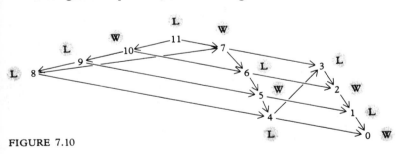

FIGURE 7.10

---

1. Use the Sprague-Grundy method to analyze the game for which the tree diagram is shown in Figure 7.1 (page 219).

2. Use the Sprague-Grundy method to analyze the matchstick game in Sample Problem 1.5 (page 3).

PRACTICE PROBLEMS 7.D

HOW DO WE FIND A WINNING STRATEGY?

Theoretically, the tree diagram approach used to analyze the game earlier in the chapter could be used to find a winning (or drawing) strategy for any game of the type we consider. For many games, the state diagram could also be used. However, with most games, there are just too many possible positions to make it practicable to draw a tree or state diagram. We therefore now investigate some other approaches that are often helpful in analyzing a game as well as some techniques for limiting the number of cases that need be considered.

FINDING A WINNING STRATEGY BY WORKING BACKWARD

In the tree diagram approach illustrated earlier, we were able, by working backward, to assign a value (A wins, B wins, or draw) to each node or position that could arise in the game. Similarly, in the state diagram approach, we again worked backward to assign a value W, L, or D to each position. Using this information, we were then able to find each player's proper strategy.

In general, even if we have not drawn a tree or state diagram of the game, we may still be able to work backward to determine that certain positions are winning or losing positions.

For example, we can analyze the game in Problem 7.1 by working backward, without drawing a tree or state diagram of the game.

If you have not already solved Problem 7.1 (page 214), try it again now.

Solution of Problem 7.1

If, in playing this game, 0 sticks remain after a player's turn, that player has won; so 0 is a winning position. (Note that, in this game, a position may be represented by the number of sticks remaining.)

If 1 or 2 sticks are left after a player's turn, that player will lose, as the other player may take the remaining stick(s). Thus, 1 and 2 are losing positions.

Continuing in this manner, we obtain Figure 7.11 for part (a) of the problem, in which each player may remove only 1 or 2 sticks at a turn.

winning	losing	reason
0		goal has been achieved
	1, 2	opponent can take last stick(s)
3		opponent must leave 1 or 2 (losing)
	4, 5	opponent may leave 3 (winning)
6		opponent must leave 4 or 5 (losing)
	7, 8	opponent may leave 6 (winning)

FIGURE 7.11

We can also represent this as in Figure 7.12.

number remaining	winning or losing
0	winning
1	losing
2	losing
3	winning
4	losing
5	losing
6	winning
7	losing
8	losing

FIGURE 7.12

If necessary, we could continue all the way up to 28; but you should notice a pattern developing. The numbers that are divisible by 3 seem to be winning positions, and all other numbers seem to be losing positions. To verify that this is in fact the case, we must show that conditions 1, 2, and 3 for a game (page 226) are satisfied by the two lists:

List I: All numbers divisible by 3 (of the form $3k$).

List II: All numbers not divisible by 3 (of the form $3k + 1$ or $3k + 2$, depending on the remainder upon division by 3).

Note that the only terminal position in this game, 0, is a winning position.

Since $0 = 3 \cdot 0$ is divisible by 3, condition 1 is satisfied.

If 1 or 2 is subtracted from a number divisible by 3, the result will not be divisible by 3 $[3k - 1 = 3(k - 1) + 2$ and $3k - 2 = 3(k - 1) + 1]$. Hence, condition 2 is satisfied.

Taking 1 from $3k + 1$ or 2 from $3k + 2$ leaves $3k$, a number divisible by 3; hence, condition 3 is satisfied.

Thus, in Problem 7.1(a), Thinker's strategy is clear. He should go first, taking 1, leaving 27 sticks. Thereafter, he should always leave a multiple of 3.

Note that our discussion above has more than answered Problem 7.1(a). It has also solved Problem 7.1(c) in the case that $m = 2$.

Problem 7.1(b) may be solved similarly. Working backward, we obtain the lists of winning and losing positions shown in Figure 7.13.

Again we are led to a conjecture: The winning positions are the multiples of 4 (numbers of the form $4k$); all other positions (numbers of the form $4k + 1$, $4k + 2$, or $4k + 3$) are losing positions. [Can you prove this? Try it using the method used in part (a).]

Thus, in Problem 7.1(b), Thinker should choose to go second, and should always leave a multiple of 4 sticks.

Can you now generalize the method to find a solution of Problem 7.1(c)? Problem 7.1(a) and 7.1(b) should suggest a conjecture for the

winning	losing
0	
	1, 2, 3
4	
	5, 6, 7
8	
	9, 10, 11
etc.	

FIGURE 7.13

winning positions in the general game. Or, if not, you should be able to make one by examining the lists of winning and losing positions in Figure 7.14.

winning	losing	reason
0		goal has been achieved
	$1, 2, \ldots, m$	opponent may take last stick
$m + 1$		opponent must leave $1, 2, \ldots,$ or m
	$m + 2, \ldots, 2m + 1$	opponent may leave $m + 1$
$2(m + 1)$		opponent must leave $m + 2, \ldots,$ or $2m + 1$
	etc.	

FIGURE 7.14

Conjecture For the game in Problem 7.1(c), the winning positions are all multiples of $m + 1$ [numbers of the form $(m + 1)k$] and the losing positions are all other numbers [numbers of the form $(m + 1)k + r$, with $1 \leq r \leq m$, where r is the remainder when the number is divided by $m + 1$].

We prove the conjecture by verifying conditions 1, 2, and 3 (note that, again, 0 is the only terminal position):

1. $0 = (m + 1) \cdot 0$

2. If we take t sticks, $1 \leq t \leq m$, from a multiple of $m + 1$, we do not get a multiple of $m + 1$. That is, for $1 \leq t \leq m$,

$$(m + 1)k - t = (m + 1)(k - 1) + (m + 1 - t),$$

where $1 \leq m + 1 - t \leq m$.

3. We can take r sticks from $(m + 1)k + r$ to leave $(m + 1)k$.

Thus, Thinker should always leave multiples of $m + 1$ sticks. Whether he should choose to go first or second depends on n. If n is a multiple of $m + 1$, Thinker should go second; otherwise, he should go first.

PRACTICE
PROBLEMS
7.E

1. Verify the conjecture for Sample Problem 7.1(b).

FINDING WINNING STRATEGIES BY SIMPLIFYING A GAME

In the above solution of Problem 7.1, although we essentially used the technique of working backward, we also used another technique—that of simplifying the problem. In working backward, we replaced the original problem by a sequence of simpler problems:

A game with 0 sticks;
A game with 1 stick;
A game with 2 sticks;
Etc.

Each of these was easily solved from the previous ones.

In addition, we conjectured a solution of the general problem [Sample Problem 7.1(c)] by first solving the special cases in 7.1(a) and 7.1(b). In fact, if part (c) of this problem were presented without parts (a) and (b), a good approach would be to consider special cases such as (a) and (b), and then to generalize in order to find a solution of part (c).

This technique of simplifying a problem was already encountered in Chapter 1. In general, we can try to simplify a game by reducing its size in one way or another.

For example, in Chapter 1, we solved the matchstick game that started with 6 or 7 sticks. We found that the second player would win in the 6 stick version and the first player would win when the game began with 7 sticks. Thus, in the 28 stick game—or any version of the game starting with at least 7 sticks—6 is a winning position and 7 is a losing position. Since at most 2 sticks may be taken at a time, any play of the 28 stick game must reach a point at which either 6 or 7 sticks remain (it is impossible to bypass both of these positions). When this point is reached, we need not continue playing, for we know who will win. We have thus shortened Problem 7.1(a) to the following: Starting with 28 sticks, play to leave exactly 6 sticks (and not 7). A little thought shows that this is essentially the same as the problem, "Starting with 22 sticks, play to take the last one." This in turn can be reduced to, "Starting with 16 sticks, play to take the last one." And so on until only the 4 stick game remains to be analyzed.

Thus, we can solve the 28 stick game by being able to solve the 6 stick game and the 4 stick game.

FINDING WINNING STRATEGIES WITH A FRONTAL ASSAULT

Although the procedure of working backward to reduce a game is an extremely important one, there are times when it is not much help. For example, if there are a large number of possible penultimate positions (positions from which the game can end in one move), then there is little to be gained by working backward rather than making a frontal assault.

We should therefore consider all possible opening moves that the first player could make, all possible responses by the second player, and so on. The Multiplication Principle makes it clear that the number of possible cases grows very rapidly as the game progresses. For example, if A has five possible choices for his first move and B may respond to each of these in any of four ways, then we already have twenty possibilities to examine. If A may make his next move in either of two ways, then there are forty branches in a tree diagram. And so on. Fortunately, there are several factors that may limit the number of possible cases which must be investigated.

HOW MANY POSSIBILITIES NEED BE CONSIDERED?

The number of possibilities we need investigate is limited by our objective—to find a winning (or drawing) strategy. If we find that a particular move by a particular player leads to a forced win for that player, then we need no longer consider his or her alternative moves. (We may wish to do so out of curiosity, but it is not essential.) On the other hand, we cannot conclude that the player will lose until we have investigated every possible move he or she could make.

SYMMETRY AS A LIMITING FACTOR

Another factor that can help limit the number of possibilities which need be considered is symmetry.

Imagine the following situation: Mel and Ira are playing a game of Tic Tac Toe (also known as Tit Tat Toe or Noughts and Crosses) at a blackboard. Their younger brothers, Lem and Ari, cannot see the blackboard directly but can follow the progress of the game through a mirror placed on a neighboring wall. Lem and Ari decide that they too will play Tic Tac Toe. In fact, Lem decides that he will always make the move that Mel appears to be making and Ari decides that he will always make the move that Ira appears to be making.

If the squares of Mel and Ira's board were numbered as in Figure 7.15a,	then it would appear to Lem and Ari that they were numbered as in Figure 7.15b.

1	2	3
4	5	6
7	8	9

(a)

3	2	1
6	5	4
9	8	7

(b)

FIGURE 7.15

Thus, if Mel's first move is as in Figure 7.16a,	then Lem's first move will be as in Figure 7.16b.

          (a)   (b)

FIGURE 7.16

It is clear that Lem will win his game if and only if Mel wins his, and that Ari will win if and only if Ira does. Furthermore, information about the appearance of either board during the game conveys information about the appearance of the other board. For example,

| if Mel and Ira's board appears as in Figure 7.17a, | then Lem and Ari's board must appear as in Figure 7.17b. |

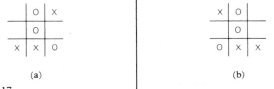

          (a)   (b)

FIGURE 7.17

Specifically, Lem and Ari's board may be obtained by flipping Mel and Ira's board about a vertical axis.

Clearly, if at some time during the game Mel can force a win, then, at that same time, Lem must also be able to force a win. A similar statement holds for Ira and Ari's winning positions. Thus, the strategic value of any position of Mel and Ira's board is the same as the strategic value of the corresponding position on Lem and Ari's board. In other words, if in analyzing the game we discover that a particular position is a winning (losing, drawing) position, then the corresponding reflected position is also a winning (losing, drawing respectively) position, and we need not analyze it further. We simply say that the two positions are equivalent "by symmetry."

Sometimes symmetries are easy to see. For example, in Tic Tac Toe, we could show that the game has seven symmetries. Figure 7.18 shows one position and its seven equivalent positions.

We say that the positions in parts a and b of the figure are equivalent by **vertical symmetry** (turned on the vertical axis); those in parts a and c are equivalent by **horizontal symmetry**; etc. Symmetry obtained by using 180° rotation, as in part g of Figure 7.18, is often referred to as **central symmetry**.

If we determine the strategic value of any of the eight positions in Figure 7.18, then the strategic value of each of the others is also known. It follows that, in analyzing a game such as Tic Tac Toe, we do not have to consider every possible sequence of moves. Symmetry enables

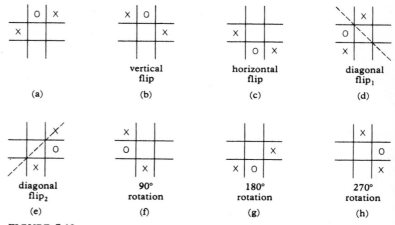

FIGURE 7.18

us to limit the number of positions that must be considered in analyzing the game. For example, as far as the opening move of the game is concerned, we have to consider only the three possibilities shown in Figure 7.19. Each other possible opening is equivalent to one of these three by symmetry.

FIGURE 7.19

1. **A** In Tic Tac Toe, which of the positions in Figure 7.20 are equivalent by symmetry?

(a)   (b)   (c)

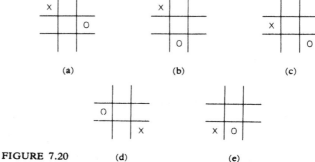

FIGURE 7.20   (d)   (e)

2. In Tic Tac Toe, which of the positions in Figure 7.21 are equivalent by symmetry?

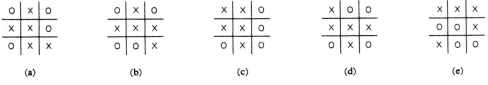

(a)   (b)   (c)   (d)   (e)

FIGURE 7.21

3. [A] Given the equivalent positions shown in Figure 7.22, what move in part b of this figure corresponds to the move of placing an X in the upper lefthand corner in part a?

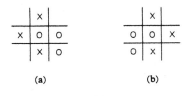

(a)   (b)

FIGURE 7.22

4. Given the equivalent positions shown in Figure 7.23, what move in part b corresponds to the move of placing an X in the upper lefthand corner in part a?

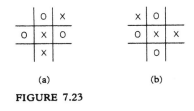

(a)   (b)

FIGURE 7.23

Although symmetry is an intuitively easy concept in Tic Tac Toe, in some games the various symmetries are not so obvious. In order to elaborate on this point, we should be a little more precise in explaining symmetry. Unfortunately, it is very difficult to give a precise definition without using concepts that go beyond the scope of this book. Therefore, we will rely on your intuition and simply say that two positions are equivalent by symmetry if to each move from one of the positions there is a corresponding move from the other so that the game proceeds in essentially the same way from both positions. If there is a sequence of moves by which a particular player can force a win from one position, then, from any equivalent positions, there will be a corresponding sequence of moves by which that player can force a win. In other words, two equivalent positions have the same strategic values—they are both winning positions, both losing positions, or both drawing positions.

Symmetry Applied to The Game of Problem 7.2

Whether or not two positions are equivalent depends basically on the rules of the game. For example, consider the game in Sample Problem 7.2 (page 214). This game is a variation of Tic Tac Toe. It is not hard to see that the seven symmetries of ordinary Tic Tac Toe are also present in this game. However, the nature of this new game gives it additional symmetries that are not symmetries of Tic Tac Toe.

Observe that the only consideration governing the legality of a move in the game of Problem 7.2 is whether or not certain squares are in the same row or the same column. Clearly then, changing the order of the rows or of the columns does not significantly alter the game. For example, cutting the righthand column off the board and pasting it onto the left of the board rearranges the board as indicated in Figure 7.24. For any move on the first board, there is a corresponding move on the second board, which leaves the resulting positions equivalent. Thus, for example, the positions in Figure 7.25 are essentially the same.

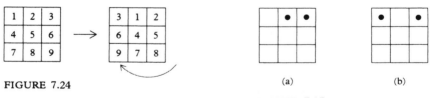

FIGURE 7.24

(a)   (b)

FIGURE 7.25

The sequence of moves shown in Figure 7.26a corresponds to the sequence in Figure 7.26b.

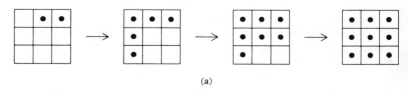

(a)

FIGURE 7.26

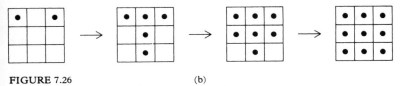

FIGURE 7.26   (b)

Thus, reordering the rows or columns of the board gives a symmetry of the game in Problem 7.2.

Does this also give a symmetry of ordinary Tic Tac Toe? No! For example, the position shown in Figure 7.27a (in which X wins) is clearly not equivalent to the position in Figure 7.27b (in which O wins), which was obtained by moving the rightmost column to the left.

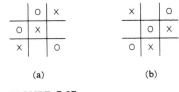

(a)   (b)

FIGURE 7.27

The reason that this type of symmetry does not apply to ordinary Tic Tac Toe is that reordering the rows (or columns) changes the diagonals, and the diagonals are important in Tic Tac Toe. On the other hand, they are not important in the game of Problem 7.2.

We can conclude that different games may have different symmetries even if they are played on the same board. In fact, even games that are not board games may have symmetries (for example, see the analysis of the game of Nim—Problem 7.3—below or the game of Sprouts in the exercises, page 265).

In general, the greater the number of symmetries that may be applied to a game, the more we can limit the number of positions we need consider in analyzing the game. For example, although there are many—thirty-three—possible opening moves that the first player could make in the game of Problem 7.2, owing to the large number of sym-metries of this game, only three of them are actually different (see Figure 7.28). All legal opening moves placing three checkers are

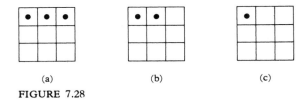

(a)   (b)   (c)

FIGURE 7.28

equivalent; all legal opening moves placing two checkers are equivalent; and all opening moves placing one checker are equivalent. For example, the position in Figure 7.29a may be seen to be equivalent to the position in Figure 7.29d.

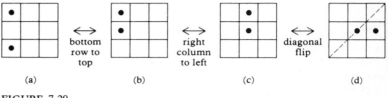

(a)   (b)   (c)   (d)

FIGURE 7.29

PRACTICE PROBLEMS 7.G

1. [A] In the game of Problem 7.2, which of the pairs of positions in Figure 7.30 are equivalent?

(a)   (b)

(c)

FIGURE 7.30

2. In the game of Problem 7.2, which of the pairs of positions in Figure 7.31 are equivalent?

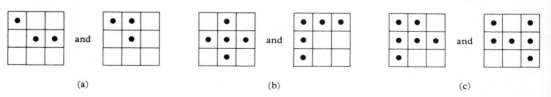

(a)   (b)   (c)

FIGURE 7.31

DÉJÀ VU—WE'VE SEEN IT BEFORE

For each of A's possible moves, B may choose a number of possible responses. This brings us to a third factor that limits the possibilities we need consider—déjà vu (this is pronounced day-zha-voo, and is French for "already seen"). If the same position results from two different sequences of moves, it need be considered only once when analyzing a game. Thus, for example, if we find that the case in which A moves as in Figure 7.32a and B responds as in Figure 7.32b leads to a win for B, then the case in which A moves as in Figure 7.32c and B responds as in Figure 7.32d also leads to a win for B. In fact, this is true not only for positions that are exactly the same but also of positions that are equivalent.

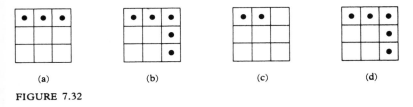

| (a) | (b) | (c) | (d) |

FIGURE 7.32

We are now ready to tackle Problem 7.2 (page 214). Try it again if you haven't already solved it. Use the various techniques we have just discussed in order to limit the number of cases you must consider.

Solution of Problem 7.2

We begin by simplifying the problem and considering the 2 × 2 version of the game. If A takes two boxes, then B can win by taking the remaining two boxes; if A takes one box, then B can win by taking the diagonally opposite box—see Figure 7.33. (A can only take one of the remaining two boxes, and so B gets the last one.) Thus, no matter what A does, B can force a win of the 2 × 2 game.

FIGURE 7.33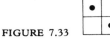

Therefore, the position shown in Figure 7.34a is a winning position in the 3 × 3 game, as is any position equivalent to it under the symmetries of the game (that is, any position in which four boxes remain, all of which lie in only two rows and two columns). For example, the positions in 7.34b and 7.34c are winning positions.

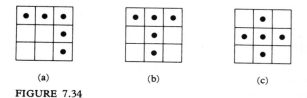

(a)   (b)   (c)

FIGURE 7.34

Furthermore, the same reasoning used in the 2 × 2 case shows that the positions shown in Figure 7.35 are also winning positions.

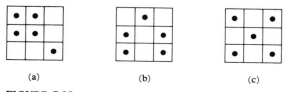

(a)   (b)   (c)

FIGURE 7.35

On the other hand, the position in Figure 7.36a and its equivalents are losing positions, because the next player can leave Figure 7.36b and win.

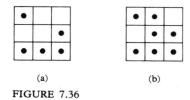

(a)   (b)

FIGURE 7.36

We saw above that there are, in essence, only three different possible first moves for A. We will consider them now.

Case 1 If A covers three boxes as in Figure 7.28a, then B can leave Figure 7.34a, a winning position.

Case 2 If A starts by covering two boxes, Figure 7.28b, then B can again leave the winning position in Figure 7.34a.

Since these two openings clearly lead to a win for B, we need consider them no further.

Case 3 A's only reasonable hope of winning is to start with one box as in Figure 7.28c.

B now has a number of alternatives.

If she places 1 or 2 checkers, all in the same row or the same column that A did, then we know, by virtue of déjà vu, that B will lose. If B leaves the position in Figures 7.28a or b or an equivalent, it would be the same as if B were the first player and she started by placing 2 or 3 checkers—a losing move.

Similarly, if B places 3 checkers anywhere, then A will be able to create an equivalent of one of the positions in Figure 7.34, and thus win.

Therefore, B's only reasonable moves are:

1. Place only 1 checker, in a row and column different from the one in which A placed his checker.

2. Place two checkers, neither of which is in the row or column in which A placed his checker.

3. Place two checkers, exactly one of which is in the row or column in which A placed his checker.

Because of symmetry, we need consider only the three positions in Figure 7.37.

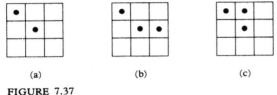

(a)   (b)   (c)

FIGURE 7.37

Since the positions in Figure 7.37b and c can be obtained in one move from the position in a, we consider b and c first. If either of these positions leads to a win for B, then Figure 7.37a would be a losing position since A could leave b or c. On the other hand, if b and c both are losing positions, then we would still have to investigate position a.

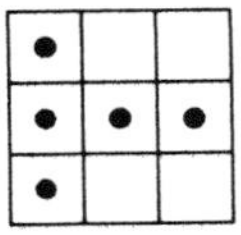

Position b may be disposed of quickly, because A can win by filling in the first column leaving Figure 7.38, an equivalent of the positions in Figure 7.34.

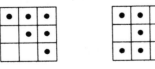

FIGURE 7.38

We now consider position c. For each possible move that A can now make, B has a countering move that leaves a winning position.

If A covers three squares, leaving one of the positions in Figure 7.39,

FIGURE 7.39   (a)   (b)

then B can win by leaving the corresponding position in Figure 7.40.

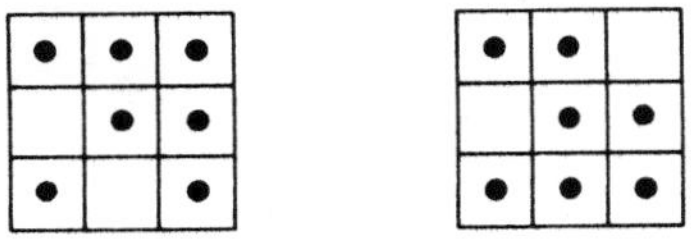

FIGURE 7.40   (a)   (b)

If A covers two squares, leaving one of the positions in Figure 7.41,

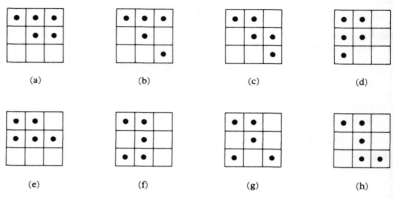

FIGURE 7.41

then B can leave the corresponding position in Figure 7.42.

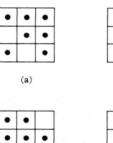

FIGURE 7.42

Finally, if A covers one square, leaving one of the positions in Figure 7.43,

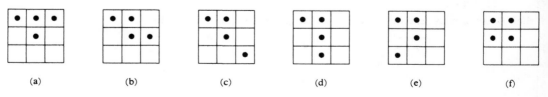

FIGURE 7.43

then B can leave the corresponding position in Figure 7.44.

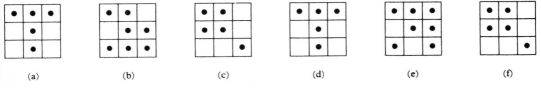

(a)   (b)   (c)   (d)   (e)   (f)

FIGURE 7.44

In each case, B leaves a winning position.

Since position c (of Figure 7.37) is a winning position, it is no longer necessary to consider position a.

This completes the analysis of the game. Recapping, B has a winning strategy: If A covers two or three squares, then B covers three or two respectively, in such a way as to leave a position that is equivalent to the 2 × 2 game. If A covers one square, then B covers two squares, one of which is in the same row or column as the square covered by A, and the other is not. After A's next move, B will be able either to leave a position equivalent to the 2 × 2 game, or to leave only two squares uncovered, which are neither in the same row nor the same column.

1. **A** In each of the positions in Figure 7.45, it is your move. Do you have a winning move? If so, what is it?

(a)   (b)   (c)   (d)

FIGURE 7.45

2. In each of the positions in Figure 7.46, it is your move. Do you have a winning move? If so, what is it?

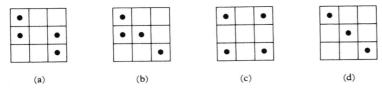

(a)   (b)   (c)   (d)

FIGURE 7.46

THE GAME OF NIM

While the techniques we have discussed so far in this chapter are often helpful in analyzing games, they do not always enable us to reduce the necessary case analysis to manageable proportions. Many games (such as chess and checkers) are still not analyzed for this reason. Sometimes, though, seemingly unrelated mathematical considerations lead us through the labyrinth. One such example is an application of the binary system (see Chapter 5) to the game of Nim, presented in Problem 7.3.

Before we present this application, we try to solve the problem with the techniques we have discussed so far in this chapter.

If you haven't already solved Problem 7.3 (page 214), try it again now.

Solution of Problem 7.3(a)

As with the one pile matchstick game of Problem 7.1, we might hope that consideration of some special cases would lead to a solution of the general problem. We therefore begin by considering the version of the game presented in Problem 7.3(a), in which the piles contain 3, 5, and 7 matchsticks at the beginning of the game.

Before attempting to find winning and losing positions for this version, we investigate symmetries of the game. Since it obviously makes no difference which pile contains 3 sticks, which contains 5, and which contains 7, switching the order of the piles is clearly a symmetry. Hence, there is no loss in generality in considering the pile with the fewest sticks first and the pile with the most sticks last. Thus, we may represent the number of sticks remaining in the three piles by an ordered triple (a, b, c), $0 \leq a \leq b \leq c$.

Now we may work backward to find winning and losing positions. (Note that there are no drawing positions, since a draw is not possible.) These are shown in Figure 7.47. Thus, $(3, 5, 7)$ is a losing position. It is therefore best, in the version of Problem 7.3(a), to go first and to leave either $(3, 4, 7)$, $(3, 5, 6)$, or $(2, 5, 7)$.

Solution of Problem 7.3(b)

In the general game, we are still in the dark. The winning positions above are still winning positions and the losing positions are still losing positions, but it is not at all clear as to how we may generalize to other positions. In all the winning positions above, except $(3, 5, 6)$, the sum of the first two numbers is the third number. We might suspect that this relationship holds in general. That this is not the case

winning	losing	reason
(0, 0, 0)		objective of the game
	(0, 0, a), a > 0	opponent may leave (0, 0, 0)
(0, 1, 1)		opponent must leave (0, 0, 1)
	(0, 1, a), a > 1 (1, 1, a), a ≥ 1	opponent may leave (0, 1, 1)
(0, 2, 2)		opponent must leave (0, 1, 2) or (0, 0, 2)
	(1, 2, 2) (2, 2, a), a > 1 (0, 2, a), a > 2	opponent may leave (0, 2, 2)
(0, a, a)		opponent must leave (0, c, a), c < a
	(0, a, b), 0 < a < b (b, a, a), 0 < b < a (a, a, b), 0 < a ≤ b	opponent may leave (0, a, a)
(1, 2, 3)		opponent must leave (0, a, b), a < b; (1, 1, a), a > 1, or (1, 2, 2)
	(1, 2, a), a > 3 (1, 3, a), a > 3 (2, 3, a), a > 3	opponent may leave (1, 2, 3)
(1, 4, 5)		opponent must leave (0, a, b), a < b; (1, 2, a), a > 3; (1, 3, a), a > 3; or (1, 4, 4)
	(1, 4, a), a > 5 (1, 5, a), a > 5 (a, 4, 5), 1 ≤ a ≤ 3	opponent may leave (1, 4, 5)
(2, 4, 6)		opponent must leave (0, a, b), a < b; (1, 2, a), a > 3; (1, 4, a), a > 5; (2, 2, a), a > 1; (2, 3, a), a > 3; (2, 4, 4); or (2, 4, 5)
	(2, 4, 7) (2, 5, 6) (3, 4, 6)	opponent may leave (2, 4, 6)
(2, 5, 7)		opponent must leave (0, a, b), a < b; (1, 2, a), a > 3; (1, 5, 7); (2, 2, a), a > 1; (2, 3, a), a > 3; (2, 4, 5); (2, 4, 7); (2, 5, 5); or (2, 5, 6)
	(3, 5, 7)	opponent may leave (2, 5, 7)
(3, 4, 7) (3, 5, 6)		opponent must leave one of the losing positions listed above.

generalizes
previous
steps

FIGURE 7.47

may be seen from the fact that (1, 3, 4), (1, 5, 6), and (2, 3, 5) are losing positions.

Actually, the secret to generalization lies in writing all numbers in the binary system—see Chapter 5—a point first observed by C. L. Bouton in 1902. Consider the winning positions above written in the binary system, as in the examples in Figure 7.48. The number (written

(3, 4, 7)	(2, 5, 7)	(2, 4, 6)	(3, 5, 6)	(1, 4, 5)
11	10	10	11	1
100	101	100	101	100
<u>111</u>	<u>111</u>	<u>110</u>	<u>110</u>	<u>101</u>
222	222	220	222	202

FIGURE 7.48

in base ten) of 1's in each column is shown below the line. In all cases, this number is even in all columns. On the other hand, in every losing position, at least one column sum is odd, as in Figure 7.49.

(1, 3, 4)	(1, 5, 6)	(2, 3, 5)	(3, 5, 7)
1	1	10	11
11	101	11	101
<u>100</u>	<u>110</u>	<u>101</u>	<u>111</u>
112	212	122	223

FIGURE 7.49

We define a position to be even if the number of 1's in every column is even (when the numbers are written in the binary system), and to be odd otherwise. We then claim that the set of winning positions is exactly the set of even positions, and the losing positions are the odd positions.

To see this, we must show that conditions 1, 2, and 3 of winning and losing positions (page 226) are satisfied.

The terminal position (0, 0, 0) is an even position, so condition 1 is satisfied.

Condition 2 says that any move made from an even position must result in an odd position. To see that this is the case, observe that the binary representations of two different numbers must differ in at least one digit—one has a 1 in a position in which the other has a 0. When a move is made from an even position, two of the binary numbers are left unaltered, and the third is changed—at least one 0 is changed to a 1 or one 1 is changed to a 0. But then, the sum of any column in which such a change takes place is either increased by 1 or decreased by 1. In either case, it is changed from an even number to an odd number, and hence the resulting position is odd.

111	111
1010	1010
<u>1101</u>	<u>1011</u>
2222	2132
(a)	(b)

FIGURE 7.50

For example, consider the even position (7, 10, 13) in Figure 7.50a. If two sticks are taken from 13 to leave $11 = (1011)_{two}$, the second 1 from the left in the binary representation of 13 is changed to a 0, and the 0 is changed to 1. Hence, the second and third column sums are changed from even to odd (Figure 7.50b).

As for condition 3, which says that from any odd position it must be possible to reach an even position in one move, we proceed as follows: Determine the leftmost column containing an odd number of

1's. At least one of the entries in that column must be a 1 (otherwise the number of 1's in that column would be zero, an even number). Choose any pile having a number of sticks whose binary representation has a 1 in the column in question. Remove as many sticks from that pile as is necessary to obtain a number whose binary representation results in all columns having an even number of 1's. For example, from the situation in Figure 7.51a we must select the second pile, since it is the only pile having a 1 in the leftmost odd column (the third column from the left). The remaining two piles sum to 1201211 (Figure 7.51b), so, to get an even number of 1's in each column, the new number in the second pile must have 1001011 as its representation (that is, it must be 75). This number will automatically be smaller than the number that was originally in the second pile, because the leftmost place in which the two numbers differ contained a 1 in the original number and a 0 in the new number. (Note that

$$2^k > 2^{k-1} + 2^{k-2} + \cdots + 2 + 1$$

$$\geq a_{k-1}2^{k-1} + a_{k-2}2^{k-2} + \cdots + a_1 \cdot 2 + a_0,$$

$0 \leq a_i \leq 1$ for all i, regardless of which of the a_i are 0 and which are 1. See Chapter 5, page 152.)

This same consideration guarantees that it will always be possible to replace a number having a 1 in the leftmost odd column by a smaller number, so that all the columns contain an even number of 1's.

As another example, from the position in Figure 7.52, any pile may be chosen to work with, because all three binary representations contain 1's in the leftmost odd column. If we work with the first pile, we must replace it by 101 to obtain an even number of 1's in each column; with the second pile, 10100 should be left; with the third pile, leave 10001.

Thus, playing Nim with three piles containing p, q, and r sticks respectively, we should choose to go first if the position is odd, second if it is even. (Obviously, whether the position is odd or even will depend on p, q, and r.)

It should be clear that these same considerations apply even if there are more than three piles.

Now that we have this simple method of determining strategy, the Figure 7.47 chart is no longer necessary.

```
101101
1010101          101101
1100110          1100110
2211312          1201211

  (a)              (b)
```

FIGURE 7.51

```
1011
11010
11111
23132
```

FIGURE 7.52

---

1. [A] In the game of Nim, are the following winning or losing positions?
   (a) (3, 6, 10)   (b) (3, 4, 11, 12).

2. In the game of Nim, are the following winning or losing positions?
   (a) (4, 7, 11)   (b) (2, 8, 10, 14)   (c) (3, 8, 12, 16).

3. $\boxed{A}$ In each of the following positions in the game of Nim, it is
   your move. You are playing against an expert. What should you do?
   (a) (6, 7, 8)   (b) (11, 12, 14)   (c) (6, 10, 12)   (d) (2, 4, 7, 11).

4. In each of the following positions in the game of Nim, it is your
   move. What should you do?
   (a) (3, 8, 12, 16)   (b) (7, 8, 9)   (c) (7, 11, 13).

PAIRING STRATEGIES

Our discussion of the game of Nim illustrated how theoretical considerations can sometimes help in finding the winning strategy. A
particular theoretical consideration that is often helpful is the concept
of a **pairing strategy**. Sometimes the moves of a game or the cells of the
game board can be paired in such a way that one of the players can
always force a win by making the move that pairs with the opponent's
previous move.

For example, in the game of Problem 7.1(c)(page 214), we may think
of the moves as being paired as follows: 1 is paired with m; 2 is paired
with $m - 1$; 3 is paired with $m - 2$; and so on. (If m is odd, $(m + 1)/2$
will be paired with itself.) Then, once a winning position is reached,
the player who reached it can win by thenceforth making the move that
is paired with the move his or her opponent makes.

As another example of a pairing strategy, consider Problem 7.4.

**If you haven't already solved it, try Problem 7.4 (page 215)
again now.**

Solution of Problem 7.4

A may view the dots as being paired, as in Figure 7.53.

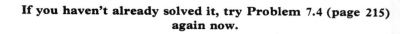

FIGURE 7.53

If he places his first X through the middle dot, then he will be
able to force a win by always marking the dot paired with the one his

opponent just marked, unless it is possible to win directly on the move. Thus, a typical game might proceed as in Figure 7.54.

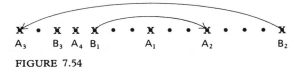

FIGURE 7.54

After A's first move, he plays the move paired with B's move, until B, at her third move, leaves A the opportunity to win.

By following the above strategy, A will never leave an opportunity for B to win. For, if A cannot win himself, then, after his turn, the left half of the board will be symmetrical with the right half; and hence, if no winning opportunity existed in one half, none will exist in the other half either.

_____

Note that the pairing in the game above is based on symmetry. This is true of many pairing strategies. However, even in games with symmetrical boards, it is not always true that one player can force a win by playing symmetrically to his or her opponent. Sometimes, a symmetry strategy will result in a quick loss.

VARIATIONS OF A GAME

Once a game is completely analyzed and a best strategy is known for one player, X, the game becomes boring; and, of course, a knowledgeable opponent will not want to play. (However, if X can find a non-thinking opponent, N, he or she can easily pick up a few dollars by baiting N—letting N win a few times and thus setting N up for the kill.)

Sometimes, though, if a game is too difficult or too easy to analyze, we may make some changes in the hope that a more interesting game will result. Many different kinds of alterations are possible.

We can vary the size or shape of the "board." For example, instead of playing the game with 28 matchsticks as described in Problem 7.1(a), we may play the same game but start with 36 sticks. Or, instead of having one pile of sticks, we may have several. Instead of playing Tic Tac Toe on a 3 × 3 board, we may play it on a 4 × 4 board, or a 3 × 3 × 3 board. An interesting variation of chess is played on a cylinder—the vertical edges of the board are bent back until they touch each other.

In addition to varying the board, we can also vary the rules of movement. Instead of being allowed to take only 1, 2, or 3 matchsticks, we may allow 1, 2, 3, or 4 to be taken; or we may take as many as we like, as long as they are all in the same pile. Or we may allow a player to make two consecutive moves, or forbid a player to make a certain move, and so on. In a game such as chess, we may introduce new pieces: a grasshopper (which moves along the same lines as a queen but which, in order to move, must hop over a piece and land on the next square) or a reflecting bishop (which may bounce off the edges of the board as if it were a ball on a billiard table).

Still another way of varying a game is to alter the winning objective of the game. For example, instead of trying to take the last matchstick in Nim, you may try to force your opponent to take the last one. Or, instead of trying to get three in a row in Tic Tac Toe, you may try to force your opponent to get three in a row. Playing a game with the objective of trying to force your opponent to achieve what would normally be considered a win is usually referred to as the **misère form** of the game (pronounced miz air).

The game in Problem 7.2 is a variation of Tic Tac Toe in which both the rules for moving and the objective of the game have been changed.

Some other variations of Tic Tac Toe will be considered in the exercises at the end of the chapter, as will many other games and some of their variations.

THE CHAPTER IN RETROSPECT

In this chapter you have been introduced to games in which definite strategies are involved. These games are not only fun to play, but analyzing games of this type further improves the ability to think critically and solve problems.

We have also discussed some techniques for finding strategies. Some, such as simplification, we have encountered before. Most of them you have probably used in some form or other in everyday problem solving. The "déjà-vu" technique means, practically speaking, rely on your previous experience in similar situations. And the symmetry technique, again in practical terms, will remind you to recognize when two situations are essentially equivalent.

Game analysis is closely allied to problem solving. Note again how mathematics creeps in where you least expect it—congruences in the matchstick game, the binary system in the game of Nim, and graph theory in some of the exercises.

In the next chapter, we will see some one-person games in which mathematics plays a part.

Exercises

The exercises below are of two basic types. In the first group (Exercises 7.1 through 7.72), a game is described. Your task is to analyze that game. Assume that two experts, A and B, are playing the game in question; if A makes the first move, will one player win (which one?) or will the game end in a draw? If one player can force a win, describe that player's winning strategy; otherwise, describe both players' drawing strategies. Depending on the specific game being considered, such a description might be a statement of some general principle that the player should follow in selecting his or her moves; or it might be a list of all possible positions with which the player could be faced, combined with a designation of what move the player should make from each possible position; or perhaps just a list of winning and losing positions is sufficient. In any case, it should be made clear that the strategy will work no matter what moves the opponent makes.

In many of these exercises, we examine only some of the smaller cases of the game. A more general discussion of most of these games may be found in Martin Gardner's books ([18], [19], [20], [21], [22], [23], [24], [25]) and in his articles in Scientific American.

Boards on which many of the games discussed can be played are illustrated on the insides of the front and back covers of this book. For counters, use coins or anything else your ingenuity suggests.

Be warned that we, the authors, are not positive of the answers to all these problems. Those about which we are unsure have been indicated by ★★★. This does not necessarily mean that the problem is very difficult, but rather that we haven't had an opportunity to do a complete case analysis.

The remaining problems in the chapter (Exercises 7.73 to 7.83) consist of end position problems. That is, a particular position of the game is presented, and the challenge is to determine what move the player who is about to move should make to ensure a win. Or, in the event that this player has no winning move, can he or she assure at least a draw? In addition, you must explain how you know that your answer is correct.

Matchstick Games—One Pile

Each of the following games begins with one pile of matchsticks—or toothpicks, if you like. The number of sticks in the pile is given in the column headed n. If the number appearing in this column is n itself, you should analyze the game for all values of n in general.

The players, A and B, alternate turns. The number of matchsticks that each player may take at a turn is indicated in the column headed **The rule of play**.

The definition of a win for the game is indicated in the final column, headed **Object**

of the game. In most cases, it is either that the player who takes the last stick wins, which we refer to as the **regular form** of the game, or it is that the player who takes the last stick loses, which we refer to as the **misère form** of the game. In the misère form, the object is to force your opponent to take the last stick. Note that, in problems having more than one part (a, b, etc.) and more than one objective (regular, misère), each objective should be considered for each part.

Exercise	n	The rule of play	Object of the game
7.1.	(a) 32 [A] (b) n	2, 3, or 4. If only 1 stick remains, the game is drawn.	(i) Regular (ii) Misère
7.2.	n	3, 4, 5, or 6. If 1 or 2 sticks remain, the game is drawn.	(i) Regular (ii) Misère
★ **7.3.** [A]	n	$j, j + 1, \ldots, m$. If only 1, 2, …, $j - 1$ sticks remain, the game is drawn.	(i) Regular (ii) Misère
7.4.	(a) 28 (b) n	1, 3, or 4.	(i) Regular [A] (ii) Misère
★ **7.5.** [H]	28	1, 2, 3, 4, or 5; but a player may not use the same number his or her opponent just used. A player who cannot move loses.	Regular
7.6.	60	Any perfect square (1, 4, 9, 16, etc.).	Regular
★ **7.7.** [H]	Any odd n	1, 2, or 3.	[A] (i) Regular: the winner possesses an odd number of sticks when all of the sticks have been taken. (ii) Misère: the loser possesses an odd number of sticks.

Matchstick Games—More Than One Pile

Each of the following games begins with more than one pile of matchsticks or toothpicks. The number of piles and the number of sticks in each pile is given in the column headed $k; (a_1, \ldots, a_k)$. For example, $3; (3, 4, 5)$ means that there are three piles containing 3, 4, and 5 sticks respectively.

As in the one pile games, the rule of play column indicates how many sticks each player may take at a turn, and the object of the game column defines what is considered to be a win (regular—last stick wins; misère—last stick loses).

Exercise	$k; (a_1, \ldots, a_k)$	The rule of play	Object of the game
★ **7.8.** [A]	$2; (m, n), m < n$	1, 2, or 3 from the same pile.	(i) Regular (ii) Misère
7.9. [H] (Wythoff's Nim)	[A] (a) $2; (11, 15)$ (b) $2; (12, 20)$	As many as desired from one pile, or the same number from both piles.	(i) Regular (ii) Misère
7.10. [H]	$3; (3, 5, 7)$	As many as desired from one pile, or the same number from two piles.	(i) Regular (ii) Misère
7.11. [H] [A]	$3; (3, 7, 9)$	As many as desired from one pile, or the same number from all nonempty piles.	Regular
7.12.	(a) $3; (3, 5, 7)$ (b) $3; (m, n, p)$ $m < n < p$	As many as desired from up to two piles.	(i) Regular [A] (ii) Misère
★ **7.13.** [H] [A]	(a) $4; (3, 5, 7, 9)$ (b) $4; (m, n, p, q)$ $m < n < p < q$	Same as Exercise 7.12.	Regular
★ **7.14.** [H] [A] (Moore's Nim)	$k; (a_1, \ldots, a_k)$	As many as desired from up to m piles (m given).	Regular
7.15.	(a) $3; (3, 5, 7)$ (b) $3; (m, n, p)$	Take 1 or an entire pile.	[A] (i) Regular (ii) Misère

Other Matchstick Games

7.16. [A] (Modified Kayles) Beginning with 13 matchsticks in a row as in Figure 7.55,

FIGURE 7.55

each player, at his or her turn, may take either 1 or 2 adjacent sticks. (Once a stick is removed, the sticks to the left and to the right of the removed sticks are not considered to be adjacent.)

(a) [H] Analyze the regular form of this game.

(b) Analyze the misère form of this game.

(c) [H] Beginning with n matchsticks in a row, each player, at his or her turn, may take either one or two adjacent sticks. Analyze the regular form of this game.

7.17. [H] (a) Beginning with 13 matchsticks in a circle as in Figure 7.56, each player, at his or her turn, may take either 1 or 2 adjacent sticks.

FIGURE 7.56

(i) Analyze the regular form of this game.

(ii) Analyze the misère form of this game.

(b) Beginning with n matchsticks arranged in a circle, each player, at his or her turn, may take either one or two adjacent sticks. Analyze the regular form of this game.

7.18. [H] Beginning with one pile containing n matchsticks, each player must, at his or her turn, separate any existing pile into two smaller subpiles. The first player who cannot move is the loser. For example, if the original pile contains 10 sticks, A might divide it into two piles containing 5 and 5 sticks respectively, or 3 and 7 sticks, etc. In the latter case, B might then subdivide the 3 pile into 2 and 1, leaving piles of 1, 2, and 7. If A then divides the 2 pile, leaving 1, 1, 1, and 7, then B would have to work with the pile containing 7 sticks, and so on.

Analyze this game.

★ ★ **7.19.** (Grundy's Game) Beginning with one pile containing 14 matchsticks, each player, at his or her turn, must separate any existing pile into two *unequal* subpiles. (Thus a pile of 4 can only be separated into piles of 1 and 3; and a pile of 2 cannot be subdivided.) The first player who cannot move is the loser.

(a) [S] [A] Analyze the game.

(b) [A] Analyze the game if the pile has 15 matchsticks.

★ **7.20.** [A] Beginning with one pile containing n matchsticks, A must subdivide the pile into two smaller piles. B must take one of these piles (the other pile is discarded) and subdivide it. A then selects one of these piles (discarding the other) and subdivides it. Etc. A player who cannot move loses.

(a) Analyze the game if there is no restriction as to whether or not the two subpiles are of equal size.

(b) Analyze the game if each subdivision must be into two unequal piles.

Related Games

★ ★ **7.21.** [S] [A] Four aces, four 2's, four 3's, and four 4's are removed from a deck of cards and placed face up on the table. A selects one of the cards, turns it face down, and calls out the face value of the card (aces count as 1).

B then selects one of the remaining face up cards, adds its face value to A's total, and turns the card over. The face value of A's next card is added to the previous total, and so on. The player who reaches 22 or forces his or her opponent to exceed 22 is the winner.

Analyze the game. ([11], Problem 476)

★ **7.22.** [H] [A] On a normal die, the number 1 is opposite 6, 2 is opposite 5, and 3 is opposite 4. A places such a die on the table and calls out the number showing on its top face. B must then give the die a quarter turn, so that the number previously on top is now on one of the sides. The new top number is added to the total obtained by A, who then gives the die a quarter turn and adds the new top number to the previous total; and so on. The player who reaches 26 or forces his or her opponent to exceed 26 is the winner.

If A can originally place the die on the table in any manner that he chooses, analyze the game.

(Note that if, for example, a 1 is on top of the die, then the next player may turn the die in such a way that either 2, 3, 4, or 5 becomes the top number, but may not turn it so that 1 remains on top or 6 becomes the top number.) ([45], Problem 180)

7.23. Consider the following modification of the game in Exercise 7.22. Instead of trying to reach 26, the winning number is determined by B at the start of the game. B may choose any number over 20. After B chooses the number, A makes the first move, as in Exercise 7.22.

Is there any number B can select so that she can be sure of winning the game?

7.24. [A] Consider the following variation of the games in Exercises 7.22 and 7.23. A is allowed to select the winning number but, after he does so, B is allowed to select the starting position of the die. A then turns the die as before and the counting starts with the number that is then on top.

If A is allowed to select any number over 20, analyze this game.

7.25. [A] A is allowed to select any date of the year other than December 31. B may then select any date later in the same month or the same day of any later month. For example, if A selects June 16, then B may choose any later date in June or the 16th of any month from July through December. Using the same rule with regard to the date that B has selected, A must select a new date, and so on. The winner is the player who arrives at December 31.

(a) What date should A select to begin with in order to ensure the fastest possible win?

(b) What if A must start with a date in January? ([27], May 1971)

7.26. [A] Two rooks are placed on diagonally opposite corners of an 8×8 chessboard. A moves the white rook, which is located in the upper lefthand corner of the board, and B moves the black rook, which begins in the lower righthand corner. All cells that are in the same row or column as a rook are under attack by it. A player may move her or his rook to any cell it attacked prior to the move, provided that it neither passes over nor lands upon a cell that is under attack by the opponent's rook. A player who cannot make a legal move is the loser.

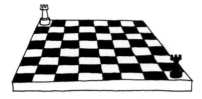

Analyze the game. ([13], Problem 393)

★ 7.27. [A] Consider the following variation of the game in Exercise 7.26. A and B command queens rather than rooks. A queen attacks all cells that are in the row, column, or either of the diagonal lines on which the queen is located. A's queen starts one cell to

the right of the upper lefthand corner of the board and B's queen begins one cell to the left of the lower righthand corner.

Analyze the game.

★ 7.28. [H] [A] (Northcott's Nim) (a) Each player has three counters placed on a 3 × 8 checkerboard as in Figure 7.57. At a turn, each player may move any one of his or her counters as far as desired to the right or left, as long as it does not leave the same row it started in and does not pass over or land on the piece the opponent has in that row. The first player who cannot move is the loser.

Ⓐ							Ⓑ
Ⓐ							Ⓑ
Ⓐ							Ⓑ

FIGURE 7.57

Analyze this game.

(b) If the game is played on an $m \times n$ checkerboard (m rows and n columns) with each player having m counters, and if A starts at the left end of each row and B starts at the right end of each row, analyze the game.

Tic Tac Toe Variations

Each of the following games is a variation of Tic Tac Toe. For each, the playing field, the rules of movement, and the object of the game are given in the chart below. As before, we refer to the first player to move as A and to the second player as B.

The phrase "in a row" means in cells that are in a straight line (horizontal, vertical, or diagonal) and contiguous. Contiguous means that the faces, edges, or the corners of consecutive cells are touching (see Figure 7.58 for some possible three-in-a-row positions).

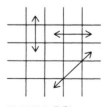

FIGURE 7.58

Exercise	Size of board	The rules of play	Object of the game
7.29. [S] [A]	3 × 3	Each player may place either an X or an O at each turn.	(a) The player who completes three X's or three O's in a row is the winner. (b) Misère—the player described above loses.
7.30. [S] [A]	3 × 3	One player plays X's and the other player plays O's. No Player may play in the middle or top rows unless the X or O placed there is supported by either an X or O directly below it.	(a) The player to complete three in a row wins. (b) Misère.
7.31. [A]	3 × 3	Each player may place either an X or an O, but, to place a piece in the middle or top rows, it must be supported, as in Exercise 7.30.	The player to complete three X's or three O's in a row is the winner.
7.32. [A]	4 × 4	Same as Exercise 7.30.	The player to complete three in a row wins.
7.33. [A]	3 × 3	One player has three X's, the other has three O's. Play begins as in regular Tic Tac Toe. However, when all six pieces are placed, each player may, at his or her turn, move one of his or her pieces one space horizontally or vertically but not diagonally.	The player to complete three in a row is the winner. However, if a position is repeated three times, then the game is considered as a draw.
7.34.	4 × 4	Play as in regular Tic Tac Toe (A places an X at each turn; B places an O).	[A] (a) The player to complete three in a row is the winner. ★ ★ ★ (b) Misère.

Exercise	Size of board	The rules of play	Object of the game
7.35. [A]	7 × 7	Same as Exercise 7.34.	The player to complete four in a row is the winner.
7.36. [A]	5 × 5	A places two X's, then B places two O's, etc.	Same as Exercise 7.35.
★ **7.37.** [H] [A]	4 × 4	At each turn, each player places both an X and an O in two horizontally or vertically adjacent cells.	If either player completes three X's in a row, then A wins. If three O's in a row are completed, then B wins. If three X's and three O's are completed simultaneously, then the player who made the move completing both is the winner.
7.38. [A]	3 × 3	Each player at a turn may place as many X's as he or she desires, provided that they are all in one horizontal (row) or one vertical (column) direction. The X's placed need not be in contiguous cells.	The player who fills in the last box loses.
7.39. [H] [A] (P. Hein's Tac Tix)	3 × 3	As in Exercise 7.38, but the X's must all be in contiguous boxes in the same horizontal or vertical direction.	(a) The player who fills in the last box wins. (b) Misère.
7.40. [A]	3 × 3 × 3 (three dimensional)	Same as Exercise 7.34.	(a) The first player to complete three in a row is the winner. ★ ★ (b) [S] Misère.
7.41. [A]	3 × 3 × 3	Same as Exercise 7.40, except that no piece may be placed at an upper	(a) The first player to complete three in a row is the winner.

Exercise	Size of board	The rules of play	Object of the game
		level unless it is supported by a piece below it.	(b) [H] Misère.
7.42. [A]	$3 \times 3 \times 3$	Same as Exercise 7.37.	Same as Exercise 7.37.
7.43.	$3 \times 3 \times 3$	Same as Exercise 7.37, except that each double piece placed at an upper level must *completely* rest on pieces below it.	The player who completes three of a kind (regardless of whether three X's or three O's) in a row is the winner.
7.44.	$1 \times n$, n odd, $n \geq 3$	Each player places one X at each turn.	The first player to complete three or more in a row is the winner.
7.45. [A]	(a) 1×10 (b) 1×12	Same as Exercise 7.44.	Same as Exercise 7.44.
★ 7.46. [A]	(a) 1×10 (b) 1×9 (c) 1×11	Same as Exercise 7.44.	The first player to complete three or more in a row is the loser.
7.47. [A]	1×16	Each player places two X's at each turn.	(a) Same as Exercise 7.44. (b) Same as Exercise 7.46.
7.48.	(a) 7×7 (b) [A] $5 \times 5 \times 5$	Same as Exercise 7.44.	The first player to complete three or more in a horizontal, vertical, or diagonal row is the winner, as in ordinary Tic Tac Toe.

Dots

The game of Dots is played as follows: Given a rectangular array ($m \times n$) of dots, the two players alternate turns drawing straight lines that connect two dots which are adjacent in a horizontal or vertical direction. Whenever a player draws the line segment completing the fourth side of a unit box (: :), that person writes her or his initial in the box and *must* go again, continuing the turn as long as she or he continues to complete boxes. Note that even if it is possible for a player to complete a box, the player is not obliged to do so. The winner is the player who completes the most boxes.

7.49. Analyze the game of Dots on each of the following boards:

(a) $\boxed{\text{A}}$ 2×2 (: :)   (b) $\boxed{\text{A}}$ 2×3 (: : :)

(c) $\boxed{\text{A}}$ 2×4 ★ ★ ★ (d) 3×3

★ ★ ★ (e) 3×4

Hex (P. Hein)

The game of Hex was described in the text (pages 222–223).

7.50. (a) $\boxed{\text{A}}$ $\boxed{\text{H}}$ Analyze the game on the following boards.

(i) 3×3 (ii) 4×4 (iii) 5×5

(b) Analyze the misère version (the first player to complete a path connecting his or her two sides of the board is the loser) on the following boards:

★ (i) $\boxed{\text{S}}$ 3×3 ★ (ii) $\boxed{\text{S}}$ 4×4

★ ★ ★ (iii) 5×5 ★ ★ ★ (iv) 6×6

Bridg-it (D. Gale)

The game of Gale or ·Bridg-it (as it became known when it was sold commercially) is played on a board consisting of two different colored $n \times (n + 1)$ rectangles intermeshed with each other. For example, to construct what we refer to as the 3×4 board, we use the first color (say black) to make a 3×4 rectangular array of dots, as in Figure 7.59a, and then use the second color (say, gray) to form

a 4×3 array (Figure 7.59b) which is centered in the first array to obtain Figure 7.59c.

The players alternate turns drawing line segments between two horizontally or vertically (but not diagonally) adjacent dots. A uses a black pencil to connect two black dots; then B uses a gray pencil to connect two gray dots; and so on. A's objective is complete a path connecting the two black sides of the board, and B tries to complete a path between the gray sides. One player may not draw a line segment that crosses a segment drawn by his or her opponent.

7.51. $\boxed{\text{A}}$ (a) Analyze the game of Bridg-it on the following boards:

(i) 2×3 (ii) 3×4 ★ ★ (iii) 4×5

(b) Analyze the misère form (the player who completes a path loses) on the following boards:

(i) 2×3 ★ ★ (ii) 3×4

Slither (D. L. Silverman)

The game of Slither is played on a board consisting of a rectangular array of dots. As in the game of Dots (Exercise 7.49), players alternate turns drawing line segments between horizontally or vertically adjacent dots. However, in Slither, a player's moves are restricted by the following rule: At any time during the game, the completed line segments must form a continuous path that does not intersect itself. A player who cannot make a legal move is the loser (the winner in the misère form). For example, in Figure 7.60a, the player who is about to move may make any of the moves indicated by the dotted lines.

In Figure 7.60b, the player who must move loses because he or she has no legal move left.

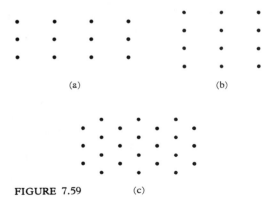

FIGURE 7.59   (c)

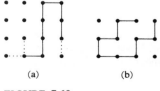

(a)   (b)

FIGURE 7.60

7.52. [H] [A] (a) Analyze the game of Slither on the following boards:
(i) 3 × 3 (ii) 3 × 4 (iii) 4 × 4
(iv) 4 × 5 ★★★ (v) 5 × 5

(b) Analyze the misère form on the following boards:
(i) 3 × 3 (ii) 3 × 4 (iii) 4 × 4
★★★ (iv) 4 × 5 ★★★ (v) 5 × 5

Cram

The game of Cram is one in which two players take turns placing dominoes on a checkerboard. It is assumed that a domino is the correct size to exactly cover two adjacent cells of the board. Each of the following exercises deals with a variation of this game. Note that when we refer to an $m \times n$ board, we mean a board having m rows and n columns.

Exercise	Size of board	The rules of play	Object of the game
7.53. [A]	2 × n	Each player places one domino at each turn, covering two adjacent cells of the board.	(i) The player who cannot move loses. (ii) Misère form (the player who cannot move wins).
7.54. [H] [A]	3 × 4	Same as Exercise 7.53.	Same as Exercise 7.53.
7.55. [A]	(a) $m \times n$, m odd, n even (b) $m \times n$, m even, n even	Same as Exercise 7.53.	Same as Exercise 7.53(i).
★ **7.56.** [H] [A] (Crosscram; G. Andersson)	2 × n	Same as Exercise 7.53, except that A can place dominoes only horizontally and B can place them only vertically.	Same as Exercise 7.53.
7.57. [A]	(a) 3 × 3 (b) 3 × 4 (c) [S] 4 × 4 ★ (d) 4 × 5	Same as Exercise 7.56.	Same as Exercise 7.53.
7.58. [A]	3 × 4	Same as Exercise 7.53, except that A can place dominoes only vertically and B can place them only horizontally.	Same as Exercise 7.53.

Exercise	Size of board	The rules of play	Object of the game
7.59. [A]	(a) 3 × 3 (b) 3 × 4 (c) 4 × 4	Same as Exercise 7.56.	(i) The player who places a domino completing a row or column of the board is the winner. (ii) Misère form (the player who completes a row or column is the loser).
7.60. [A]	5 × 5	Same as Exercise 7.56.	Same as Exercise 7.59(i).

Hexapawn (M. Gardner)

The game of Hexapawn is played on an $m \times n$ checkerboard. The two players sit opposite each other and each controls n pieces, called pawns. (Assume that A controls white and B controls black.) At the beginning of the game, each player's pawns are placed in the n cells of the row closest to that player. Each player at his or her turn may move one pawn in one of the following ways:

1. A pawn may be moved one space vertically forward (toward the opponent's side of the board) provided that the cell on which the pawn will land is vacant.

2. A pawn may be moved one space diagonally forward provided that the cell on which the pawn will land is occupied by an opponent's pawn. In this case, the opponent's pawn is captured and removed from the board.

If the cell vertically forward of the cell on which a pawn is located is occupied and if neither of the diagonally forward cells are occupied by the opponent's pawns, then that pawn may not move.

Note that even if a pawn is able to capture, it is not obliged to do so.

The game may be won in two ways: either one of your pawns reaches the opponent's side of the board, or your opponent is unable to move when it is his or her turn.

7.61. [A] Analyze the game on the following boards:
  (a) 3 × 3   (b) [H] 4 × 4
  (c) 3 × n, n ≤ 10   ★ ★ (d) 5 × 5

7.62. If the rules of Hexapawn above are changed so that pawns must capture when able to do so, analyze both the regular and misère forms (in the latter, the player who would win the regular form is the loser) of the game on the following boards:
  (a) 3 × 3   (b) 4 × 4   (c) 3 × n, n ≤ 10

Hip (M. Gardner)

The game of Hip is also played on a checkerboard. Players alternate turns placing one checker on the board. (Assume that A places white checkers and B places black checkers.) The game continues until checkers of one color complete the vertices of a square. The square need not have horizontal and vertical sides; the sides may be oblique, as in Figure 7.61. In the regular version of the game, the player to complete a square is the loser. In the misère form, he or she is the winner.

FIGURE 7.61

★ **7.63.** [A] (i) Analyze the regular form on the following boards:

(a) 3 × 3   ★★★   (b) 4 × 4

(ii) Analyze the misère form on the following boards:

(a) 3 × 3   (b) 5 × 5   (c) $n \times n$ for $n > 5$

7.64. [A] If the game of Hip is played with both players using checkers of the same color, analyze the misère form of the game on each of the following boards:

(a) 3 × 3   (b) 4 × 4   (c) $n \times n$ for n odd
(d) $n \times n$ for n even

Other Checker Games

7.65. [A] Consider the following game which is played on an $m \times n$ checkerboard (m rows, n columns). The game begins when A places a checker anywhere on the bottom row of the board. B may then move it one space to the right, one space to the left, or one space up the board. Then A does the same. Etc. No square of the board may be occupied by the checker more than once during the game. The game ends when the checker reaches the top row of the board. In the regular version of the game, the player who reaches the top row is the winner; in the misère version, he or she is the loser.

Analyze both (i) the regular and (ii) the misère versions of this game on the following boards:

(a) 5 × 5   (b) 6 × 6   (c) $m \times n$

7.66. [H] [A] Two players alternate turns placing checkers on a square tabletop. At each turn, a player places one checker so that it does not touch any of the checkers already on the table. The player unable to place a checker as above is the loser.

Can either player adopt a winning strategy?

Sprouts (J. H. Conway and M. S. Paterson)

The game of Sprouts starts with n dots on a piece of paper. Players alternate turns drawing lines connecting two of the dots or one dot to itself. New lines may not intersect previously drawn lines, nor may they pass through any dot other than the dot(s) that the line connects. After the line is drawn, a new dot is created somewhere on the line. A dot may have no more than three line segments sprouting from it (we say that a dot has three lives). If a player cannot move when it is his or her turn, then he or she loses.

For example, if $n = 1$ (there is one dot), A has no choice but to connect it to itself. A new dot, 2, is created (see Figure 7.62a).

Since both dots now have two segments sprouting from them, B cannot connect either of them to itself (this would result in four segments sprouting from the dot). Hence, B must connect 1 to 2. This may be done in two ways, inside or outside (Figures 7.62b and c); but either way, after the new point 3 is created, 1 and 2 are dead (can no longer be used) and 3 has only one life remaining (it already has two sprouts), and so A cannot move. B is thus the winner.

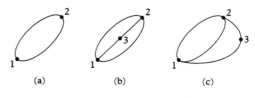

(a)   (b)   (c)

FIGURE 7.62

7.67. [H] [A] (a) Analyze the game of Sprouts for the case $n = 2$.

(b) Analyze the game of Sprouts for the case $n = 3$.

Sim (G. J. Simmons)

The game of Sim begins with n given points, no three of which are colinear (on the same straight line). Players alternate turns connecting two points by a line segment, A using a black pencil and B using a gray pencil. Two points may be connected, even if the line segment between them crosses a previously drawn line segment (but no new vertex is created). The game ends when there is a monochromatic triangle formed (that is, there are three of the given points such that all three line segments between them are of the same color). In the regular version of the game, the player who completes the triangle is the loser; in the misère form, he or she is the winner.

7.68. $\boxed{\text{H}}$ $\boxed{\text{A}}$ (a) Analyze the game of Sim for the following cases:
  ★ ★ (i) The regular form, for $n = 5$.
  (ii) The misère form, for $n = 5$.
  (iii) The misère form, for $n > 5$.
  (b) Can the regular form, for $n = 6$, end in a draw?

7.69. As a variation of Sim, consider the game in which both players use the same color. The game ends when a triangle is formed (that is, all three line segments connecting three of the given points have been drawn).
  Analyze the regular and misère forms of this game in the following cases:
  (a) $n = 4$ (b) $n = 5$ ★ (c) $n = 6$

Games of Entrapment

In most of the games appearing so far in this chapter, both players have the same or similar objectives. In some games, however, this is not the case. One class of games in which the players have different objectives are games of entrapment—one player tries to trap or capture the other, while the other player tries to escape.

One of the better known games of this type is Fox and Hounds, also called Sheep and Wolves. There are several variations of this game, played on either an 8×8 or a 10×10 checkerboard. On the smaller board, the game requires four checkers of one color to represent the hounds, and one checker of another color—the fox. (On the larger board, there are five hounds.) The hounds begin the game on the bottom row of the board and can move only as checkers do (they use only the black cells of the board and move forward one space diagonally at each move). The fox starts the game on a black cell of the top row and moves as a king moves in checkers (it too stays on the black cells of the board, but it moves one space diagonally, forward or backward). There is no jumping in this version of the game, and so the fox can be blocked if a hound appears in its path. The object of the hound is to entrap the fox, so that it cannot move at all. The fox tries to escape. It escapes if it passes the line of hounds, because the hounds cannot move backward.

In another version of the game, there are twice as many hounds, and the fox can jump over a single hound but cannot jump over two hounds in a row.

In all versions of this game, the hounds can win with careful play. We will not ask you to prove this here, but we do consider a Fox and Hounds end position problem in Exercise 7.82.

The following exercises are some related games.

Fox and Geese

The game of Fox and Geese is closely related to the second version of Fox and Hounds, but it is played on a different board. The board for Fox and Geese consists of 33 cells in the shape of a cross, as indicated in Figure 7.63.

The X's represent geese and the O represents the fox. The geese move one cell at a time in a horizontal or downward vertical direction. (They may not move vertically up-

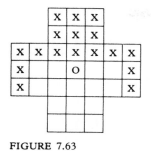

FIGURE 7.63

ward.) The fox moves one cell horizontally or vertically (including upward). In addition, the fox can jump over (and remove) a goose located on a square that the fox could move to, provided that the following square is vacant. The fox tries to consume the geese and to avoid entrapment. The geese try to entrap the fox, so that it cannot move. The geese move first.

7.70. ⊞ Analyze the game of Fox and Geese.

Dwarfs and Giant

Dwarfs and Giant is another game of entrapment, in which three dwarfs try to trap one giant. There are several different boards on which this game may be played. Two of these are shown in Figure 7.64.

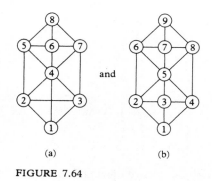

(a)   (b)

FIGURE 7.64

On board *a* in the figure, the dwarfs start on cells 1, 2, and 3, and the giant may start on

any cell that is numbered 5 or greater. On board *b*, the dwarfs start on 1, 2, and 4, and the giant must start on 3. At his or her turn, the player moving the dwarfs moves one dwarf forward or sideways from the cell in which it is located to a vacant cell to which it is directly connected by a line segment. Thus, for example, a dwarf on cell 2 in board *a* may move to cell 3, 4, or 5 if it is vacant; and a dwarf on cell 2 in board *b* may move to 3, 5, or 6. The moves of the giant are similar, except that the giant may also move backwards. The dwarfs try to entrap the giant at the top of the board (cell 8 in *a* and cell 9 in *b*), occupying cells 5, 6, and 7 in *a* and cells 6, 7, and 8 in *b*. The giant tries to avoid entrapment, either by breaking past the dwarfs, or by causing the exact same position to be repeated three times (in which case the giant is awarded the win). The dwarfs move first.

7.71. ⊞ Ⓐ (a) Analyze the game on board *a* of Figure 7.64.

(b) Analyze the game on board *b*.

(c) What happens if the giant starts on 5, 6, 7, 8, or 9 in *b*?

7.72. ⊞ Ⓐ The streets of Orderly, Maryland were planned very carefully, as the map in Figure 7.65 shows.

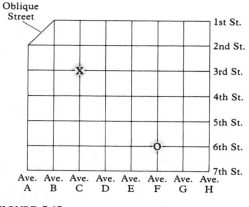

FIGURE 7.65

Slippery Sam once held up the First National Bank of Orderly, located at the intersection of 3rd Street and Avenue C. The police quickly set up roadblocks at all roads out of town. But, as they were shorthanded that week, this left only Fred Fleetfoot, the flatfoot, to track Sam down. Fred knew that he must capture Sam before nightfall; otherwise Sam would probably be able to slip though the roadblocks. Fred succeeded in his task.

To commemorate the occasion, the town council manufactured the following game: Using the diagram in Figure 7.65 as a playing board, one player (X) plays Sam and the other player (O) plays Fred. They start at the locations indicated. (The Orderly police station is located at the intersection of 6th Street and Avenue F.) Each player, in his or her turn, moves to a neighboring intersection. Fred's objective is to capture (land on) Sam. Sam's objective is to elude Fred long enough for night to fall: If Sam has not been caught after Fred's twenty-first move, then Sam wins.

Analyze this game if

(a) Sam goes first.

(b) Fred goes first.

End Position Problems

7.73. You are playing a game of Nim, which began with seven piles of matchsticks. At the moment, only five nonempty piles remain, containing 13, 15, 18, 41, and 45 sticks respectively.

It is your move. What should you do?

7.74. (a) You are playing the following version of Tic Tac Toe on a 4 × 4 board. (The 3 × 3 version of the game was discussed in the chapter.) Each player, at his or her turn, is allowed to place as many X's as desired as long as they are in the same row or column. The player who places the final X is the winner.

At the moment, the board looks like Figure 7.66. (The letters a, b, c, and d, and the numbers 1, 2, 3, and 4 have been added only to facilitate reference to the individual squares.)

It is your move. What move should you make?

	1	2	3	4
a		X		X
b	X	X		X
c	X	X		X
d				X

FIGURE 7.66

(b) The board is as above, but you are playing Tac Tix (see Exercise 7.39). That is, each player may place as many X's as he or she chooses, as long as they are in the same row or column and they occupy contiguous cells.

What should your move be now?

7.75. $\boxed{A}$ Starting with a 6 × 6 array of dots, you and your opponent take turns placing X's through the dots (see Exercise 7.48). Each places one X per turn. The first to complete three or more adjacent X's in a horizontal, vertical, or diagonal row as in Tic Tac Toe is the winner.

It is your turn to move from the position in Figure 7.67.

Where should you place your next X?

	1	2	3	4	5	6
a	•	•	•	•	•	•
b	•	X	•	•	•	•
c	•	•	•	•	X	•
d	•	•	•	•	•	•
e	•	•	•	•	•	•
f	•	•	•	•	•	•

FIGURE 7.67

7.76. $\boxed{H}$ You are playing Dots (see Exercise 7.49) on a 4 × 4 board. The position is as shown in Figure 7.68 and it is your move.

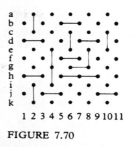

FIGURE 7.68

What should you do to ensure a win? (Recall that a player is not obligated to complete a box if it is possible to do so, but that, if a player does complete a box, then he or she must make another move.)

7.77. In a game of Hex on a 6 × 6 board, you are White. The position is as shown in Figure 7.69.

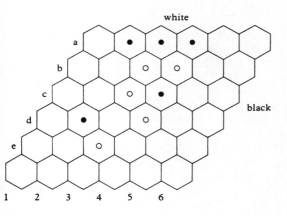

FIGURE 7.69

It is your move. What should you do? And how should the play continue?

7.78. In a game of Bridg-it (see Exercise 7.51) on a 5 × 6 board, you are gray (vertical). The position is as shown in Figure 7.70 and it is your move. What should you do, and how will the game proceed from there?

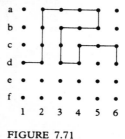

FIGURE 7.70

7.79. [A] In a game of Slither (see Exercise 7.52) on a 6 × 6 board, the position is as shown in Figure 7.71.

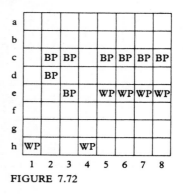

FIGURE 7.71

It is your move. What move should you make? How should you continue to play thereafter?

7.80. You are White in a game of Hexapawn (see Exercise 7.61) on an 8 × 8 board. The position is as shown in Figure 7.72.

	1	2	3	4	5	6	7	8
a								
b								
c		BP	BP		BP	BP	BP	BP
d		BP						
e			BP		WP	WP	WP	WP
f								
g								
h	WP			WP				

FIGURE 7.72

The outcome appears bleak. But it is your move and, by making the correct move, you can actually win. What move should you make?

7.81. In a rematch of Hexapawn, on a 9 × 9 board, you reach the position shown in Figure 7.73.

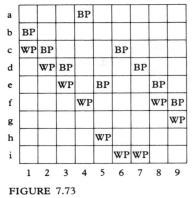

FIGURE 7.73

You are White and it is your move. What should you do?

7.82. [A] You are playing Fox and Hounds (see Games of Entrapment, page 266) on an 8 × 8 board. It is your move. The position is as shown in Figure 7.74.

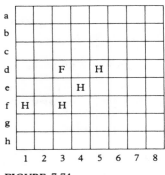

FIGURE 7.74

(a) What should you do if you are the fox?
(b) What should you do if you are the hounds?

7.83. You are playing a game of three dimensional Tic Tac Toe on a 4 × 4 × 4 board (see Figure 7.75).

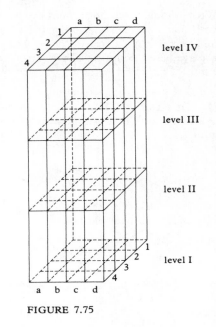

FIGURE 7.75

To facilitate playing the game, the board may be visualized as in Figure 7.76.

Your objective is to complete four O's in a straight line. The line may lie in a horizontal, vertical, or any of several diagonal directions. Your opponent's objective is to complete four X's in a straight line.

In visualizing the board as in Figure 7.76, you must be careful to note whether or not a set of X's or O's constitutes a win. For example, the four X's in Figure 7.77 are actually in a straight line as are the four O's (see Figure 7.78).

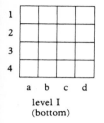

level I
(bottom)

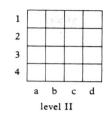

level II

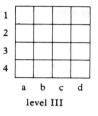

level III

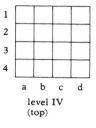

level IV
(top)

FIGURE 7.76

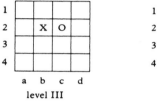

level I (bottom) · level II · level III

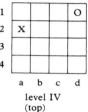

level IV (top)

FIGURE 7.77

In the game you are currently playing, you are O, and the position shown in Figure 7.79 (or, equivalently, Figure 7.80) has been reached. It is your move.

How can you be sure of winning the game?

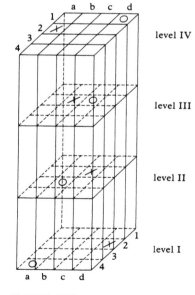

FIGURE 7.78

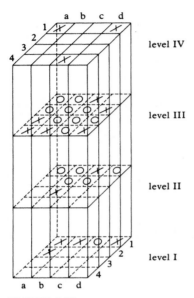

FIGURE 7.79

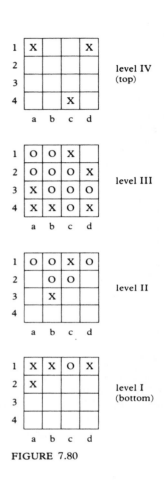

FIGURE 7.80

8

Solitaire Games and Puzzles

In Chapter 7, we considered two person games—games in which each player competes against an opponent. In this chapter, we consider "solitaire" or one person games, many of which are well known and of ancient origin. In a sense, all recreational problems are one person games; however, here we are primarily interested in games where counters or pieces are moved or manipulated.

With most puzzles of this type, finding the solution appears at first to be just a matter of trial and error; but, oftentimes, mathematics plays an important role in the ultimate solution.

As usual, the chapter begins with a number of sample problems. Some of these may be familiar to you because versions of them have been sold commercially. Try them, by trial and error if necessary; models for them can easily be made out of paper. If you are unable to obtain a systematic solution, do not be discouraged. You will get a feeling for the game just by playing it several times.

In many cases, the mathematical ideas necessary for the solutions will seem at first to bear no relationship to the problem; you may feel as if the ideas have been pulled out of a hat. In a sense, this is the reason why these particular problems were chosen. It's exciting to see how mathematics can be applied in unlikely looking situations. The usefulness of seemingly unrelated ideas is an important facet of problem solving in general, in all fields of endeavor. Many of the best problem solvers are people who are able to go beyond the apparent confines of a problem and draw on knowledge and ideas from other areas.

Problem 8.1

A popular puzzle sold in many stores consists of a wooden board supporting three posts, on one of which is found a set of seven rings of diminishing diameters, with the largest ring on the bottom and the smallest on the top. The object of the puzzle is to transfer all seven rings from the post they are on to a different post, subject to the following rules:

1. Only one ring may be moved at a time; thus, a move consists of taking the uppermost ring from one post and placing it on another post.

2. At no time may a larger ring be placed on top of a smaller ring.

(a) What is the smallest number of moves in which the required transfer can be effected?

(b) If, instead of seven rings, there are n rings, what is the smallest possible number of moves (expressed as a function of n) in which the transfer may be carried out?

Problem 8.2

(a) Assuming that a domino is exactly big enough to cover two neighboring cells of a checkerboard, can an ordinary 8×8 checkerboard from which two diagonally opposite corner cells have been removed be covered by a set of 31 dominoes?

(b) Is it possible to cover all but one cell of an 8×8 checkerboard by using 21 straight trominoes? A straight tromino looks like ☐☐☐, and is assumed to be just the right size to cover three cells of the checkerboard. ([26], pp. 20–21)

FIGURE 8.1

Problem 8.3

Consider a board containing 33 holes, as shown in Figure 8.1. There is a peg in each hole except the center hole.

Whenever it occurs that two adjacent (horizontally or vertically) holes are occupied and that the next hole in that same line is empty, the two pegs from the occupied holes may be removed and one of them then placed in the third hole. (In other words, one peg jumps over the other, landing in the vacant hole; the peg that was jumped is removed.)

The object of the game is to leave only one remaining peg. When this is accomplished, in which holes could the final peg possibly lie?

Problem 8.4

Consider a flat box containing the numbers 1 through 15 arranged as in Figure 8.2.

1	2	3	4
5	6	7	8
9	10	11	12
13	14	15	

FIGURE 8.2

A move consists of sliding any number adjacent to the empty space (in the diagram above, 12 or 15) into that space, thereby creating a new arrangement of the numbers.

Find a sequence of moves—if it is possible to do so—starting with the position shown in Figure 8.2 and resulting in

(a) the arrangement shown in Figure 8.3a.
(b) the arrangement shown in Figure 8.3b.

Problem 8.5

Given four cubes with faces colored as in Figure 8.4, arrange them in a column so that each side (front, back, left, and right) of the stack shows each of the four colors (red, white, blue, and green) exactly once.

15	14	13	12
11	10	9	8
7	6	5	4
3	2	1	

(a)

	15	14	13
12	11	10	9
8	7	6	5
4	3	2	1

(b)

FIGURE 8.3

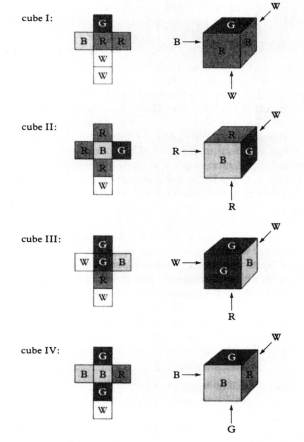

FIGURE 8.4

The problems considered in this chapter are of two basic types. We are asked either to perform some particular manipulative task or to determine exactly which manipulative tasks are possible to perform. In the latter case, we frequently use some form of mathematical reasoning to limit possibilities—to show that certain tasks are impossible. Thus, we know what we cannot do, but not what we can do. We are then left with the problem of showing that the tasks which remain are actually possible. This frequently reduces to a problem of the first type—we can show that a task is possible by actually carrying it out.

A frequent question associated with problems of the first type is how to perform the required manipulation in the minimum number of moves. This question is often very difficult to answer; but, in some puzzles of this type, mathematics can again come to our aid.

Problem 8.1 is such a puzzle, and there is a story* that goes with it.

THE TOWER OF BRAHMA

> In the great temple at Benares, beneath the dome which marks the center of the world, rests a brass plate in which are fixed three diamond needles, each a cubit high and as thick as the body of a bee. On one of these needles, at the creation, God placed sixty-four discs of pure gold, the largest disc resting on the brass plate, and the others getting smaller and smaller up to the top one. This is the tower of Brahma. Day and night unceasingly the priests transfer the discs from one diamond needle to another according to the fixed and immutable laws of Brahma, which require that the priest on duty must not move more than one disc at a time and that he must place this disc on a needle so that there is no smaller disc below it. When the sixty-four discs shall have been thus transferred from the needle on which at the creation God placed them to one of the other needles, tower, temple, and Brahmans alike will crumble into dust, and with a thunderclap the world will vanish.

As a result of this story, the game described in Problem 8.1 has become known as the Tower of Brahma. It is also known as the Tower of Hanoi, although we are not sure why.

Before reading the solution of Problem 8.1 (page 274), you should try the problem again if you haven't already solved it.

---

* The story is taken from Ball [3], p. 304. He attributes the story to De Parville at about the time that the puzzle was brought out by E. Lucas (c. 1883).

We now try to solve Problem 8.1(a).

After some trials, one can see that, before the seventh ring can be moved, the top six rings must first be moved and must be in the position shown in Figure 8.5. (If any other ring were still on A, the

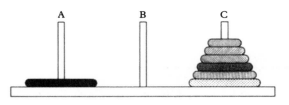

FIGURE 8.5

seventh ring could not be moved; and if there were rings on both B and C, then the seventh—the largest—ring would have no place to go.) At this point, the seventh ring is moved to peg B; and then the six rings, step by step, must be moved so that they end up on top of ring 7 on peg B.

Thus, if

$$m_7 = \text{the least number of moves required} \\ \text{to move the pile of seven rings,}$$

and

$$m_6 = \text{the least number of moves required} \\ \text{to move a pile of six rings,}$$

then

$$m_7 = \underbrace{m_6}_{\substack{\text{to move 6 rings} \\ \text{to peg C}}} + \underbrace{1}_{\substack{\text{to move ring 7} \\ \text{to peg B}}} + \underbrace{m_6}_{\substack{\text{to move 6 rings} \\ \text{back to peg B}}}$$

$$= 2m_6 + 1.$$

The problem now reduces to a similar one for fewer rings. If

$$m_5 = \text{the least number of moves required} \\ \text{to move a pile of five rings,}$$

then

$$m_6 = 2m_5 + 1$$

and so

$$m_7 = 2(2m_5 + 1) + 1 = 2^2 m_5 + 2^1 + 1.$$

Continuing this way,

$$m_7 = 2^2(2m_4 + 1) + 2^1 + 1 =$$
$$= 2^3m_4 + 2^2 + 2^1 + 1 = \cdots$$
$$= 2^6m_1 + 2^5 + 2^4 + 2^3 + 2^2 + 2^1 + 1$$
$$= 2^6 + 2^5 + 2^4 + 2^3 + 2^2 + 2^1 + 1 \ (m_1 = 1)$$
$$= 2^7 - 1 \ (\text{as we saw in Chapter 5, page 152})$$
$$= 127.$$

Solution of Problem 8.1(b)

Similarly, it can be verified, for $n = 1, 2, 3, 4, \ldots, 7$, that

n	m_n
1	$1 = 2^1 - 1$
2	$3 = 2^2 - 1$
3	$7 = 2^3 - 1$
4	$15 = 2^4 - 1$
⋮	⋮ ⋮
7	$127 = 2^7 - 1$

Based on the pattern in the above chart, our guess is that, for Problem 8.1(b), the least number of moves required to move a pile of n rings is $m_n = 2^n - 1$. This is only a conjecture, however, and needs proof. We use the method of mathematical induction.*

Let S_n be the statement that the least number of moves in which a pile of n rings can be moved from one peg to another is $m_n = 2^n - 1$.
S_1 is true, as one ring may be moved in $2^1 - 1 = 1$ move.
Assume that S_n is true. That is, assume $m_n = 2^n - 1$.
To move $n + 1$ rings, we must first move n rings from A to C, then move the $(n + 1)$st ring from A to B, and finally move the n rings from C to A. Thus

$$m_{n+1} = 2m_n + 1$$
$$= 2(2^n - 1) + 1$$
$$= 2^{n+1} - 1.$$

But this is statement S_{n+1}. Thus, if S_n is true, so is S_{n+1}.

---

* This topic is discussed in Appendix B. Mathematical induction has been mentioned before in this book and we hope you have been curious enough to read this appendix. In any case, we finally carry out an inductive proof in detail.

Since S_1 is true, the inductive argument implies that S_2 is true; therefore, the inductive argument implies that S_3 is true; etc. Thus, S_n is true for every positive integer n. This completes the proof.

By mathematical induction, then, the minimum number of moves required to move a pile of n rings is $2^n - 1$.

When, then, will the world come to end? Assuming that the Brahmans move one ring per second and that they never rest, the world will come to an end $2^{64} - 1$ seconds after it was created. But you need not rush to get your affairs in order; $2^{64} - 1$ seconds is over 500 billion years.

DISSECTION PROBLEMS

Another type of solitaire puzzle of ancient origin involves geometrical dissection and reassembly. In this type, you cut up a given geometrical figure and reassemble the pieces to form some other specified geometrical figure. One of the best known and most fundamental theorems of mathematics, the Pythagorean Theorem, was probably first proved c. 550 BCE using a dissection proof—although it was known by the ancient Chinese (c. 1100 BCE). The theorem says that the sum of the squares of the two legs of a right triangle is the square of the hypotenuse—that is, if a and b are the lengths of the legs of a right triangle and if c is the length of the hypotenuse, then $c^2 = a^2 + b^2$.

The proof in question proceeds by dissecting the square of side $a + b$ in two different ways. In the first way, the square is broken into four right triangles, congruent to the given one, and two squares of sides a and b respectively. The second dissection breaks the square of side $a + b$ into four right triangles, congruent to the given one, and one square of side c (see Figure 8.6). Since the four triangles in Figure 8.6a are all congruent to, and hence cover the same total area as the four triangles in Figure 8.6b, the remaining two squares in Figure 8.6a must cover the same area as the remaining square in Figure 8.6b— $a^2 + b^2 = c^2$, as required.

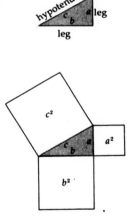

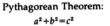

Pythagorean Theorem:
$a^2 + b^2 = c^2$

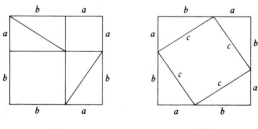

FIGURE 8.6 (a) (b)

There are many different proofs of the Pythagorean Theorem, including a number of other proofs by dissection [see Exercise 8.5(c)].

A much deeper theorem of geometry says that if A and B are any two planar polygonal figures (figures with straight line segments as edges) having the same areas, then it is possible to dissect A into a finite number of pieces that can then be reassembled to form B. For example, a square may be cut into four pieces that can be reassembled to form the letter T (see Figure 8.7).

Most recreational problems of this type ask not only that we dissect and reassemble, but that we use the fewest cuts possible in doing so. Many such problems are still unsolved in the sense that, although a dissection and reassembly have been found that appear to make use of a minimal number of cuts, no one has actually been able to prove that fewer cuts will not suffice. A description of some of the techniques used in solving dissection problems may be found in [37]. In general, the solution of these problems seems to require considerable trial and error and a great deal of patience.

In some sense, the discussion up to this point does not seem to belong in this chapter, for the problems discussed so far are not really manipulative—at least, not until the dissection part of the problem has been completed. In many problems, however, we are given the pieces that we are to use, and the problem is to assemble them.

Probably the best known recreations of this type, purported by some to date back over 4000 years, are Tangrams. Traditionally, a square is cut up into pieces as indicated in Figure 8.8 (other ways of cutting

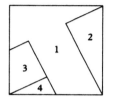

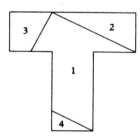

FIGURE 8.7

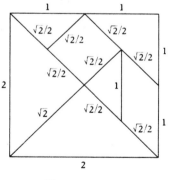

Tangram square

FIGURE 8.8

a square or even a rectangle have also been used occasionally), and the object is to combine all of the pieces to make various pictures. For example, a shark may be made as in Figure 8.9.

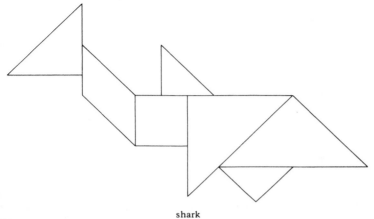

shark

FIGURE 8.9*

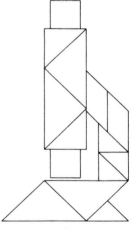

FIGURE 8.10*

Sometimes more than one set of Tangrams are combined to make more intricate pictures. For example, a microscope can be made using two Tangram sets, as Figure 8.10 shows (at a decreased scale).

The solutions of problems such as these require measurement of line segments and angles, but the solutions also seem to depend, to a large extent, on trial and error. For this reason, we do not dwell on problems of this type, but instead turn our attention to related problems whose solutions depend more heavily on reasoning.

POLYOMINOES

In what follows, we are primarily interested in dissecting a checkerboard or modified checkerboard into pieces that are made up of a fixed number of cells (squares). (Much of the theory in this area has been developed by S. Golomb and may be found in his book [26].) Pieces that contain two cells look like ☐☐ and are called dominoes, because of their resemblance to ordinary dominoes. Pieces with three cells are called trominoes; pieces with four cells are called tetrominoes; pieces with five cells are called pentominoes; and so on. Although all dominoes look alike (☐☐ or ☐, which are essentially the same), there are actually two distinct kinds of trominoes—the straight tromino and the right tromino, shown in Figure 8.11.

FIGURE 8.11

Similarly, there are five or seven tetrominoes—see Figure 8.12— depending on whether d and f (and similarly, e and g) of the figure are

* Based on [53], Problems 84 and 322.

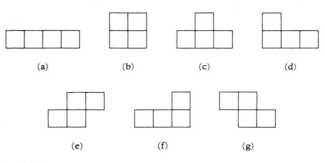

FIGURE 8.12

considered as the same or different. The only way that f can be obtained from d is by flipping the piece over so that the face that was originally facing up ends up facing down. If the front and back of a tetromino are indistinguishable from each other, then this flipping is permissible and the two pieces would be considered as being the same; however, if the front and back differ (for example, if they are different colors), then the two pieces must be considered to be different. Unless otherwise indicated, tetrominoes and other polyominoes here are assumed to be reversible, and so we consider only the five tetrominoes shown in Figure 8.12a, b, c, d, and e.

It is easily seen that a set of 32 dominoes can cover a standard 8×8 checkerboard; or, equivalently, that a standard checkerboard can be dissected into a set of 32 dominoes. Similarly, an $m \times n$ checkerboard can be dissected into a set of $mn/2$ dominoes, if either m or n is even. If both m and n are odd, then obviously an $m \times n$ board cannot be covered by dominoes, since each domino covers an even number of cells (2) and the board contains an odd number of cells. However, the remaining board can be covered with dominoes, if an appropriate cell is removed.

We are now ready to solve Problem 8.2(a), page 274.
Try it again if you haven't already solved it.

Solution of Problem 8.2(a)

In this problem, two cells have been removed from an ordinary 8×8 board. Because an even number (62) of cells remain, our first impulse is to think that the required covering can be done. However, after trying for a while to do it, we begin to suspect that maybe it cannot be done after all. The key to the solution of the problem is to realize that both cells that have been removed from the board are of the same color. Thus, 32 black but only 30 white cells remain. A domino must necessarily cover one cell of each color. Thus, 31 dominoes must cover 31 black cells and 31 white cells. Therefore, 31 dominoes cannot cover the given board.

On the other hand, if any two cells of opposite color are removed, then the remaining board can be covered [Exercise 8.9(a)].

The idea of coloring plays a role in many other problems, especially to show that certain positions cannot be obtained. As an example, we consider Problem 8.2(b).

We again ask you to try Problem 8.2(b), page 274, if you haven't already solved it.

This time, since we are interested in trominoes, we recolor the board in three colors, so that any straight tromino placed on the board must necessarily cover one cell of each color (see Figure 8.13).

Solution of Problem 8.2(b)

FIGURE 8.13

A set of 21 trominoes must therefore cover 21 cells of each color. Because there are 22 black cells and only 21 of each of the other colors, the cell that will not be covered must be one of the black ones. In other words, all of the cells colored gray or white in the Figure 8.13 board must be covered. Note, however, that this coloring is somewhat arbitrary. Any other coloring scheme in which every tromino must cover one cell of each color could also be used. Applying any symmetry of this board (see Chapter 7, page 235) produces another coloring having 22 black cells and 21 cells of each of the other colors. In particular, using vertical reflection, we get Figure 8.14. Again, the uncovered cell must be one of the black ones. The only cells that are black in both colorings are those shown in Figure 8.15. These

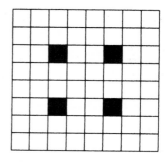

FIGURE 8.14   FIGURE 8.15

are equivalent to each other under all symmetries of the board, and so we may select any one of them to be the cell that remains uncovered.

Now that we know which cell to leave uncovered, we must find a covering that works. One such covering is shown in Figure 8.16.

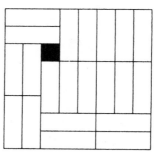

FIGURE 8.16

Many other problems of polyominoes have been investigated (see [26], for example). In particular, pentominoes have generated much interest. Some questions relating to these and the other polyominoes may be found in Exercise 8.9 at the end of the chapter.

SOMA

It is natural to attempt to generalize the ideas presented so far to higher dimensions. Instead of considering planar regions made up of a number of connected squares, we might investigate three dimensional figures made up of a number of connected cubes.

As with squares, there is only one way of sticking two cubes together, and there are two ways of attaching three cubes. However, the extra dimension enables us to increase the number of ways in which four cubes can be connected. Considering two shapes obtained by connecting the cubes to be the same if and only if one can be turned in 3-space to obtain the other, we obtain the eight different possibilities shown in Figure 8.17.

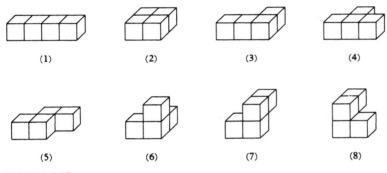

FIGURE 8.17

Note that, although (7) and (8) are very similar, they are different because, no matter how (7) is turned, it cannot be made to look exactly like (8).

Piet Hein, whom we already met in connection with the game of Hex, observed that the last six of the solids pictured in the preceding figure may be combined with the tricube in Figure 8.18 to form a $3 \times 3 \times 3$ cube. In fact, this may be done in many different ways. This observation led Hein to invent the puzzle Soma, the object of which is to use the seven pieces—(3), (4), (5), (6), (7), (8), and (9)—to form various three dimensional shapes. Coloring can sometimes be useful here too to show that certain configurations are not possible. (For example, see Exercise 8.10.)

(9)

FIGURE 8.18

In the solution of the tromino problem—Problem 8.2(b)—we colored a grid with three colors to determine which square could be removed from a standard checkerboard so that the remaining board could be covered with straight trominoes. The same basic idea is also useful in the solution of Problem 8.3.

PEG SOLITAIRE

The game of Peg Solitaire dates back at least as far as the early 1700s and is probably much older. Through the ages, it has been played on boards of many different sizes and shapes, but there are only two basic variations of the game.

In the first version, a move consists of one peg jumping over an adjacent peg and landing in a vacant hole on the other side of the jumped peg. The jumped peg is then removed from the board. The object of the game is to leave one peg (or a specified number of pegs) remaining.

In the second version, the object is to move a set of pieces from the given starting position to a given terminal position, or to interchange the positions of two sets of pieces of different colors. A piece may be moved from the square (hole) it occupies to any adjacent vacant square, or a piece may jump, as in the first version, except that the jumped piece is not removed from the board. (See Exercise 8.12 for problems of this type.)

The game in Problem 8.3 obviously falls into the first category, and is played on a cross-shaped board. This version of the game has been sold commercially under the name Hi Q.

**If you haven't already solved Problem 8.3 (page 274),
try it again now.**

Solution of Problem 8.3

We can think of the peg board as having squares rather than holes. If we then color the board as in the tromino problem, we obtain Figure 8.19.

FIGURE 8.19

The key to the solution of Problem 8.3 is the observation that any move in the game involves one cell of each color. Two of the three cells involved were originally occupied but end up vacant, and the third cell was originally vacant but ends up occupied. Thus, if G, B, and W respectively denote the number of gray, black, and white cells that are occupied by pegs, then any move changes G, B, and W by 1 (two of them decrease by 1 and the third increases by 1).

But this means that the relative parities of G, B, and W are not changed by any move. That is, if G and B were originally of the same parity—both were odd or both were even—then they are still of the same parity after any move. (The parity of each is changed, and therefore they remain of the same parity as each other.) Similarly, if G and B were originally of opposite parities—one odd and one even— then they remain of opposite parity after any move. A similar discussion applies to G and W, and to B and W.

In the starting position of Problem 8.3, using the coloring indicated above, we obtain Figure 8.20. Note that G and W are of the same parity and B is of the opposite parity.

In a final position, with only one peg remaining, we must have one of the following:

$$G = 1, B = W = 0; \qquad B = 1, G = W = 0; \qquad W = 1, G = B = 0.$$

But G and W must be of the same parity after each move, since they are of the same parity to begin with. Thus, the only possibility is $B = 1$, $G = W = 0$; and so the one remaining peg must end up on a black square.

Applying this same reasoning to the symmetrical pattern (Figure 8.21), we find that the last peg must again be on a black square. The only squares that are black in both coloring schemes are shown in Figure 8.22.

G = 11
B = 10
W = 11

FIGURE 8.20

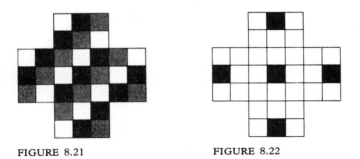

FIGURE 8.21   FIGURE 8.22

To see that it is possible to have the final peg remain in any of these five squares, it is only necessary to find a sequence of jumps that accomplishes the desired task. We know of no algorithm or theoretical approach for finding such a sequence; rather, it seems to be a matter of reasoned trial and error, and of patience. To present one sequence that works, we number the cells of the board as in Figure 8.23. Note that the first digit in each box represents the row number (from top to bottom) and the second digit represents the column number (from left to right).

		13	14	15		
		23	24	25		
31	32	33	34	35	36	37
41	42	43	44	45	46	47
51	52	53	54	55	56	57
		63	64	65		
		73	74	75		

FIGURE 8.23

A jump will be denoted by the number of the cell in which the jumping piece starts followed by the number of the cell where it lands. For example, 46–44 means that the piece in 46 jumps over the piece in 45 and lands in 44.

The following sequence of moves leaves one piece in 44 : 46–44, 65–45, 57–55, 54–56, 52–54, 73–53, 43–63, 75–73, 73–53, 35–55, 15–35, 23–43, 43–63, 63–65, 65–45, 45–25, 37–57, 57–55, 55–53, 31–33, 34–32, 51–31, 31–33, 13–15, 15–35, 36–34, 34–32, 32–52, 52–54, 54–34, 24–44.

Just prior to the last move in the sequence above, there are two pegs left—one in 24 and one in 34. If the last move had been 34–14 instead of 24–44, then a lone peg would be left in 14.

To obtain a single peg in 47, 74, or 41, we need only apply rotational symmetry to the sequence that leaves a single peg in 14.

Note that, since we start with 32 pegs and end with 1 peg, a total of 31 jumps must be made. However, if we count successive jumps by the same peg as part of the same move, it becomes meaningful to ask "What is the smallest number of moves in which one peg can be left?" For several years, it was thought that the answer was 19, until E. Bergholt (in 1912) presented the 18 move solution given above, and claimed that no one would ever beat it. It was not until recently, though, that it has actually been proven that no solution with fewer moves is possible. (If a different space is left vacant in the starting position, then it is possible to leave a lone peg in fewer than 18 moves, but the remaining peg will not end up in 44.) In general, questions of minimality are very difficult, because theoretical considerations seem to be of little help.

THE FIFTEEN PUZZLE

The ideas of coloring a grid or checkerboard and of parity can also be used in the solution of yet another puzzle, which was first introduced in the 1870s by Sam Loyd, one of the greatest American problematists. The puzzle was originally entitled the Boss Puzzle, but has become known as the Fifteen Puzzle for reasons that will soon be obvious.

The game consists of a flat box containing 15 movable pieces, numbered 1 through 15. To begin with, the pieces are arranged with a blank space in the lower righthand corner, as in Figure 8.24. The object is to arrive at different specified arrangements of the digits by sliding the pieces about. Any piece adjacent to the blank space may be slid into the blank space, thereby creating a new blank space and a new arrangement. No piece may be removed from the box at any time.

1	2	3	4
5	6	7	8
9	10	11	12
13	14	15	

FIGURE 8.24

Since there is only one blank space, a move may be described by indicating the number of the piece being moved. Thus, 12, 11, 7 indicates the sequence of moves in Figure 8.25.

1	2	3	4
5	6	7	8
9	10	11	12
13	14	15	

$\xrightarrow{12}$

1	2	3	4
5	6	7	8
9	10	11	
13	14	15	12

$\xrightarrow{11}$

1	2	3	4
5	6	7	8
9	10		11
13	14	15	12

$\xrightarrow{7}$

1	2	3	4
5	6		8
9	10	7	11
13	14	15	12

FIGURE 8.25

As originally proposed by Loyd, the problem was to begin with the starting arrangement indicated in Figure 8.24 and to obtain the arrangement in Figure 8.26. He offered a $1000 prize to the first person submitting a correct solution. Although many people attempted this problem, Loyd knew that his money was safe—the problem is impossible to solve.

How does one show that the task in question is impossible? And how does one know what arrangements are possible? As with Hi Q, we will first use theoretical considerations to show that certain positions are impossible to obtain and then we will show how all others can be obtained. We present two approaches to the solution of this problem. The first makes use of parity arguments and the second combines parity with coloring.

Let us first, though, simplify the puzzle and consider a 2 × 2 version of the game. We start with Figure 8.27 and ask what arrangements of the pieces may be obtained by following the above rules.

The only moves we have available are to rotate the pieces clockwise or counterclockwise. There are only twelve arrangements possible, all obtainable by clockwise rotation. These are shown in Figure 8.28.

1	2	3	4
5	6	7	8
9	10	11	12
13	15	14	

FIGURE 8.26

1	2
3	

FIGURE 8.27

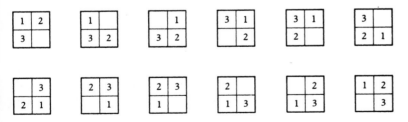

FIGURE 8.28

We are essentially dealing with arrangements of the integers 1, 2, and 3 within the 2 × 2 box. We can associate a triple (a, b, c) with each arrangement as follows: List the integers in the order given by the pattern in Figure 8.29, ignoring the blank. For example, the arrange-

FIGURE 8.29

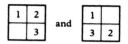

FIGURE 8.30

FIGURE 8.31

ments in Figure 8.30 both correspond to (1, 2, 3), and that in Figure 8.31 corresponds to (2, 3, 1), etc.

Any ordering of the numbers 1, 2, and 3 (or, in general, 1, 2, ..., n) is called a **permutation**. By the Multiplication Principle (see Chapter 1, page 19), there are $n(n - 1)(n - 2) \cdot \ldots \cdot 2 \cdot 1$ distinct permutations of the n integers 1, 2, ..., n. For $n = 3$, the $3 \cdot 2 \cdot 1 = 6$ permutations are: (1, 2, 3), (1, 3, 2), (2, 1, 3), (2, 3, 1), (3, 1, 2), and (3, 2, 1).

The permutations (3, 1, 2), (2, 3, 1), or (1, 2, 3) all correspond to arrangements that could be obtained (see Figure 8.28) in the 2 × 2 game starting with Figure 8.27.

There are three other possible permutations of the numbers 1, 2, and 3: (3, 2, 1), (2, 1, 3), and (1, 3, 2). These respectively correspond to the arrangements in Figure 8.32 (among others), which are not attainable from the starting position in Figure 8.27.

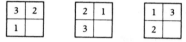

FIGURE 8.32

Note that the permutation we associate with a given arrangement depends on the pattern or listing order that we use. We have chosen this particular listing in the 2 × 2 case because the permutations that arise distinguish between obtainable arrangements and those that are not obtainable. If, instead of using the pattern in Figure 8.29, we list numbers in a different order—for example, upper left corner, upper right corner, lower left corner, lower right corner (see Figure 8.33), then we could get all six possible permutations. For example, using the new listing order, the attainable arrangement in Figure 8.34 would give rise to the permutation (1, 3, 2).

FIGURE 8.33

FIGURE 8.34

In this case, considering permutations would be of no value to us because the method would not eliminate any permutations and so we would not be able to say that only arrangements corresponding to certain permutations are attainable.

Fortunately, by choosing our listing order properly, we have been able to limit the number of attainable permutations to three. This enables us to draw conclusions about the attainability of an arrangement from knowledge of its associated permutation—and we can indeed then say that only arrangements corresponding to certain permutations are attainable.

This is the rationale behind the approach we will use to analyze the general game. By properly choosing a listing order, we will be able to show that only arrangements corresponding to certain permutations are attainable and that, in fact, all arrangements corresponding to the specified permutations actually may be attained. What is meant by "properly choosing a listing order" will become clear shortly.

Before we proceed to our analysis of the general game, let us next consider the 2 × 3 version. The starting position is shown in Figure 8.35. This time, there are 360 possible arrangements, too many for us to list them all. However, we can still associate a permutation with each arrangement. If we adopt the listing order upper left, upper middle, upper right, lower right, lower middle, lower left (see Figure 8.36), then the starting position gives rise to the permutation (1, 2, 3, 5, 4).

It turns out that, of the 120 possible permutations of the numbers 1, 2, 3, 4, 5, only 60 are attainable with this listing order.

FIGURE 8.35

FIGURE 8.36

EVEN AND ODD PERMUTATIONS

To see why this is the case, we introduce the concept of an inversion. In any permutation of the numbers 1, 2, ..., n, we say that an **inversion** occurs each time a larger number precedes a smaller number.

Thus, for example, the permutation (1, 2, 3, 5, 4) contains one inversion (5 precedes 4) and the permutation (1, 5, 2, 4, 3) contains four inversions (5 precedes 2, 5 precedes 4, 5 precedes 3, and 4 precedes 3).

Definition 8.1 A permutation is said to be **even** if it contains an even number of inversions, and is said to be **odd** otherwise.

Thus (1, 2, 3, 5, 4) is an odd permutation, (1, 5, 2, 4, 3) is an even permutation, and (1, 2, 3, 4, 5) is an even permutation (it contains zero inversions, and 0 is an even number).

---

1. **A** How many inversions are there in each of the following permutations? Is each even or odd?
   (a) (7, 1, 4, 6, 3, 2, 5)   (b) (5, 1, 6, 8, 2, 4, 7, 3)
   (c) (1, 2, 3, 4, 5, 6).

2. How many inversions are there in each of the following permutations? Is each even or odd?
   (a) (3, 8, 1, 4, 5, 6, 7, 2)   (b) (6, 5, 4, 3, 2, 1)
   (c) (3, 7, 4, 2, 8, 6, 1, 5).

PRACTICE PROBLEMS 8.A

When we make a move, thereby changing from one permutation to another, how is the number of inversions affected?

To answer this question, we first consider what happens when two adjacent numbers are interchanged in a permutation—that is, if

$(\ldots, a, b, \ldots)$ is changed to $(\ldots, b, a, \ldots)$. The position of the other numbers appearing in the permutation relative to a and b and relative to each other are unchanged; the only change that occurs in the number of inversions in the permutation does so by virtue of the change in the positions of a and b relative to each other. If $a < b$, then the new permutation $(\ldots, b, a, \ldots)$ contains one inversion (b before a) that the original permutation does not have; similarly, if $b < a$, then the original permutation contains one inversion (a before b) that the new permutation does not have. In either case, the number of inversions in the new permutation differs from the number of inversions in the original permutation by plus or minus 1. Thus, the two permutations have opposite parities (one is even and one is odd).

Now suppose we want to change the permutation $(\ldots, a, b, c, \ldots)$ to $(\ldots, b, c, a, \ldots)$, or vice versa. This can be done by first changing $(\ldots, a, b, c, \ldots)$ to $(\ldots, b, a, c, \ldots)$ and then changing $(\ldots, b, a, c, \ldots)$ to $(\ldots, b, c, a, \ldots)$. Since each of these changes is of the type discussed above, each induces a change in the parity of the permutation. Since two such changes occur, the new permutation $(\ldots, b, c, a, \ldots)$ has the same parity as the original permutation $(\ldots, a, b, c, \ldots)$.

As an example, to change from

$(6, 3, 1, 4, 5, 2)$ to

$(6, 1, 4, 3, 5, 2)$,

first change

$(6, 3, 1, 4, 5, 2)$ to $(6, 1, 3, 4, 5, 2)$

and then change

$(6, 1, 3, 4, 5, 2)$ to $(6, 1, 4, 3, 5, 2)$.

Note that each time we interchange two adjacent digits, the number of inversions changes by $+1$ or -1.

Since two such changes occur, $(6, 1, 4, 3, 5, 2)$ has the same parity as $(6, 3, 1, 4, 5, 2)$.

Generalizing, the permutation $(\ldots, a, b_1, b_2, \ldots, b_r, \ldots)$ may be changed to the permutation $(\ldots, b_1, b_2, \ldots, b_r, a, \ldots)$ in r steps: First change it to $(\ldots, b_1, a, b_2, \ldots, b_r, \ldots)$, then change this to $(\ldots, b_1, b_2, a, \ldots, b_r, \ldots)$, and so on until $(\ldots, b_1, b_2, \ldots, a, b_r, \ldots)$ is finally changed to $(\ldots, b_1, b_2, \ldots, b_r, a, \ldots)$. Since each step involves a change of parity, $(\ldots, a, b_1, b_2, \ldots, b_r, \ldots)$ and $(\ldots, b_1, b_2, \ldots, b_r, a, \ldots)$ are of the same parity if r is even, and are of opposite parity if r is odd.

We are now ready to see what happens when we make a move in the 2×3 game. Recall that we are using the pattern in Figure 8.36 to determine the permutation associated with each arrangement.

Clearly, a horizontal move (when a piece slides to the left or right) does not change the permutation at all. Neither does a vertical move (sliding a piece up or down) in the rightmost column, as in Figure 8.37.

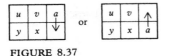

FIGURE 8.37

A vertical move in the middle column (Figure 8.38) changes the permutation (x, a, b, c, y) to (x, b, c, a, y) or vice versa. This is the type of change discussed above—a is shifted two places to the right or left. Since two is an even number, such a shift does not change the parity of the permutation.

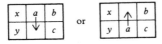

FIGURE 8.38

Similarly, a vertical move in the leftmost column (Figure 8.39) changes the permutation (a, b, c, d, e) to (b, c, d, e, a) or vice versa. This time, a is shifted four places; but, again, since four is even, no change occurs in the parity of the permutation.

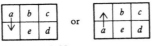

FIGURE 8.39

Thus, any move in the 2×3 version of the game does not change the parity of the associated permutations. In other words, an even permutation gets changed to an even permutation and an odd permutation gets changed to an odd permutation.

Therefore, if we start with an odd permutation $(1, 2, 3, 5, 4)$, we can obtain only odd permutations. Since it can be shown that the number of odd permutations is equal to the number of even permutations*, half the possible permutations are even. Thus, half the possible permutations cannot be obtained.

---

* It is not difficult to show that the number of even permutations is the same as the number of odd permutations and that this number is therefore half the total number of permutations. Interchanging the last two numbers of a permutation either introduces a new inversion or else removes one, and thereby changes the parity of the permutation. Therefore, pairing each permutation with the permutation obtained by interchanging the last two numbers of the original permutation gives a one-to-one correspondence between the odd permutations and the even ones:

even	odd
(1, 2, 3)	(1, 3, 2)
(2, 3, 1)	(2, 1, 3)
(3, 1, 2)	(3, 2, 1)

The fact that any move preserves the parity of the associated permutations depends heavily on the way in which the permutations are obtained—that is, on the listing order. In the 2×2 game, if we use the listing order upper left, upper right, lower left, lower right, then any vertical move changes the parity of the permutation, and all six possible permutations can occur. However, if we use the listing order upper left, upper right, lower right, lower left, then the parity of the permutations is preserved by any move. This explains what is meant by "properly choosing a listing order"—we choose a listing order so that each possible move does not change the parity of the associated permutations.

We are now ready to consider the 4×4 game.

If you haven't already solved Problem 8.4(a), page 274, try it again now.

Solution of Problem 8.4(a)

As in the 2×2 and 2×3 cases, we wish to consider for the 4×4 case each possible arrangement as a permutation by listing all the numbers in some specified order. In fact, we want to choose the listing order in such a way that any move preserves the parity of the associated permutations. One possible listing order that works is indicated by the path in Figure 8.40.*

With this listing order, the starting position (Figure 8.41) corresponds to the permutation (1, 2, 3, 4, 8, 7, 6, 5, 9, 10, 11, 12, 15, 14, 13). This permutation contains 9 inversions and hence is odd. If we can show that, using this listing order, the parity of permutations is preserved, then only odd permutations can arise.

FIGURE 8.40

1	2	3	4
5	6	7	8
9	10	11	12
13	14	15	

FIGURE 8.41

1	2	3	4
5	6	7	8
9	10	11	12
13	15	14	

FIGURE 8.42

Because the arrangement suggested by Loyd (Figure 8.42) gives rise to the permutation (1, 2, 3, 4, 8, 7, 6, 5, 9, 10, 11, 12, 14, 15, 13), which is even (8 inversions), and the arrangement of Problem 8.4(a)—see Figure 8.43—gives rise to the permutation (15, 14, 13, 12, 8, 9, 10, 11,

15	14	13	12
11	10	9	8
7	6	5	4
3	2	1	

FIGURE 8.43

---

* There are other patterns which give listing orders that work. This particular pattern was chosen because the proof that it works is more easily presented than are the proofs for most other patterns.

7, 6, 5, 4, 1, 2, 3), which is even (96 inversions), neither of these arrangements is possible to achieve from the starting position indicated.

It remains to show that the suggested listing order is "properly chosen," that is, using this order, any move preserves parity.

As before, any horizontal move does not change the permutation and hence certainly preserves parity.

Vertical moves may be subdivided into several cases:

First, there are those that do not change the permutation at all. These are shown in Figure 8.44.

FIGURE 8.44

Second, there are vertical moves which shift the number being moved, a, two places (Figure 8.45). Thus $(\ldots, a, b, c, \ldots)$ is changed to $(\ldots, b, c, a, \ldots)$ or vice versa, which does not change parity.

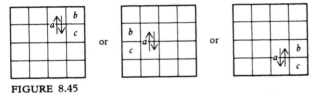

FIGURE 8.45

The third class of vertical moves are those in which the number being moved, a, is shifted four places (Figure 8.46). These moves change the permutation from $(\ldots, a, b, c, d, e, \ldots)$ to $(\ldots, b, c, d, e, a, \ldots)$ or vice versa, and do not change parity.

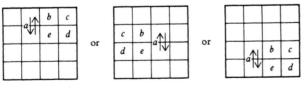

FIGURE 8.46

Finally, the remaining vertical moves shift the number being moved six places (Figure 8.47). Here $(\ldots, a, b, c, d, e, f, g, \ldots)$ is changed to $(\ldots, b, c, d, e, f, g, a, \ldots)$ or vice versa, and again parity is unchanged.

Thus, any vertical move preserves parity, and the argument is complete.

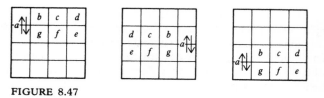

FIGURE 8.47

So far, we have shown that, starting with an arrangement giving rise to an odd (even) permutation, it is not possible to obtain one corresponding to an even (odd) permutation; but we have not yet shown that all arrangements corresponding to odd (even) permutations actually are obtainable. They are! A formal proof of this fact is straightforward but quite tedious. Rather than present such a proof, we leave it to you to convince yourself—play with the game and try to obtain various arrangements.

PRACTICE PROBLEMS 8.B

1. **A** Given the arrangement in Figure 8.48, is it possible to obtain each of the arrangements in Figure 8.49? In each case in which it is possible, give a suitable sequence of moves.

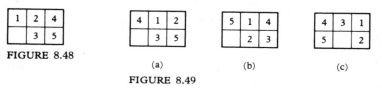

FIGURE 8.48

(a) (b) (c)

FIGURE 8.49

2. Given the arrangement in Figure 8.50, is it possible to obtain each of the arrangements in Figure 8.51? In each case in which it is possible, give a suitable sequence of moves.

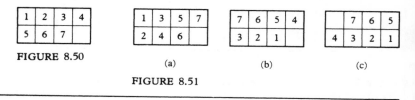

FIGURE 8.50

(a) (b) (c)

FIGURE 8.51

We are not yet finished with Problem 8.4; we must still show how to obtain the arrangement in Problem 8.4(b). (Note that the associated permutation (15, 14, 13, 9, 10, 11, 12, 8, 7, 6, 5, 1, 2, 3, 4) is odd—93 inversions—and therefore the arrangement is obtainable.)

Try Problem 8.4(b), page 274, if you haven't already solved it.

One way of obtaining the desired arrangement is as follows:

Solution of
Problem 8.4(b)

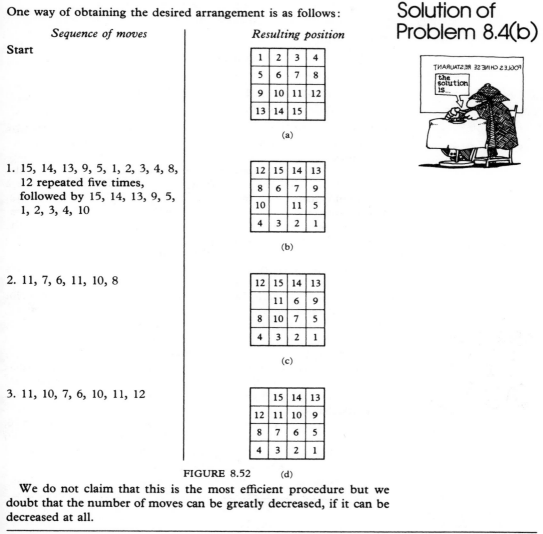

Sequence of moves   *Resulting position*

Start

1	2	3	4
5	6	7	8
9	10	11	12
13	14	15	

(a)

1. 15, 14, 13, 9, 5, 1, 2, 3, 4, 8, 12 repeated five times, followed by 15, 14, 13, 9, 5, 1, 2, 3, 4, 10

12	15	14	13
8	6	7	9
10		11	5
4	3	2	1

(b)

2. 11, 7, 6, 11, 10, 8

12	15	14	13
	11	6	9
8	10	7	5
4	3	2	1

(c)

3. 11, 10, 7, 6, 10, 11, 12

	15	14	13
12	11	10	9
8	7	6	5
4	3	2	1

FIGURE 8.52   (d)

We do not claim that this is the most efficient procedure but we doubt that the number of moves can be greatly decreased, if it can be decreased at all.

1. Starting with the arrangement in Figure 8.53, obtain, if possible, each of the arrangements in Figure 8.54.

PRACTICE
PROBLEMS
8.C

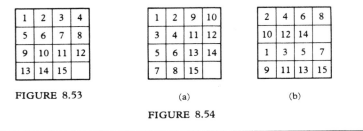

FIGURE 8.53

(a)   (b)

FIGURE 8.54

☆ COLORING AND THE FIFTEEN PUZZLE—
A SECOND APPROACH

Before leaving the Fifteen Puzzle, we return to a comment that we made when we first introduced it—namely, that coloring a checkerboard has something to do with this problem.

Suppose that the squares of the box were colored as a gray and white checkerboard. Then the color of the square on which the blank space falls would change with each move. Thus, if the blank space were on a white square to begin with, then it would be on a gray square after one move, on a white square after two moves, and so on. That is, the blank would be a white square after an even number of moves and a gray square after an odd number of moves.

Suppose, further, that a listing order is adopted such that each move results in a change of parity of the associated permutations. Then, with each move, the parity of the permutation changes, as does the color of the square on which the blank space falls. Thus, if the starting position is even and the blank is a white square, then, thereafter, whenever the permutation is even, the blank will be a white square and whenever the permutation is odd, the blank will be a gray square.

Actually, there is no nice pattern that gives such a listing order. However, if we count the blank space as if it were the number 16, and include 16 in each permutation, then any listing order will work. In particular, we can adopt the listing order given by a line scanner—the arrangement in Figure 8.55 gives the permutation (1, 2, 3, 4, 5, 6, 7, 8, 9, 10, 11, 12, 13, 14, 15, 16), which is even. We leave it to you to verify that, with this listing order and with the blank being counted as 16, then every move results in a change of parity.

FIGURE 8.55

PRACTICE PROBLEMS ♫♫ 8.D

1. Verify the comment above.

If the blank in Figure 8.55 starts on a white square, then an arrangement is obtainable from the figure if and only if

(a) it gives an even permutation and the blank is on white

or

(b) it gives an odd permutation and the blank is on gray.

FIGURE 8.56   (a)   (b)

Thus, Figure 8.56a gives	Figure 8.56b gives
(1, 2, 3, 4, 5, 6, 7, 8, 9, 10, 11, 12, 13, 15, 14, 16)	(15, 14, 13, 12, 11, 10, 9, 8, 7, 6, 5, 4, 3, 2, 1, 16)
which is odd (1 inversion); the blank is white, hence this is not obtainable.	which is odd (105 inversions); the blank is white, hence this is not obtainable.

(c)

Figure 8.56c gives (16, 15, 14, 13, 12, 11, 10, 9, 8, 7, 6, 5, 4, 3, 2, 1), which is even (120 inversions); the blank is white, hence this arrangement is obtainable.

This agrees with what we found above.

COLORED CUBES

Another puzzle in which color plays an important role is the colored cubes puzzle of Problem 8.5. This time, however, the coloring pattern has been imposed upon us by the statement of the problem, rather than our being able to introduce a coloring pattern that suits our needs.

The colored cubes puzzle has been marketed by several companies under several different titles. One marketed version was called "Instant Insanity"—an attempt to describe the effect it might have on a prospective solver. After all, there are 41,472 essentially different ways to arrange the cubes, of which only one actually yields a solution.

The order in which the cubes are stacked clearly is not important in the solution; so we consider only arrangements with cube I on the bottom, cube II next, and so on. There are six possible ways to choose the front face of cube I; once it is chosen, the back face is determined and there are four ways in which to select the top face. Hence, by the Multiplication Principle, there are 24 different ways in which cube I can be arranged. The same is true for each of the other cubes. Therefore, there are $24 \cdot 24 \cdot 24 \cdot 24 = 331{,}776$ possible arrangements with cube I on the bottom, cube II on top of cube I, etc. Because we need not consider as different any arrangement obtained by rotating the entire stack of four cubes (Figure 8.57), we must divide by 4, giving 82,944 arrangements. The arrangement obtained by keeping the front–back of each cube fixed and interchanging left and right (rotating 180° around the front–back axis) is also the same as the original. Therefore, we must divide by 2, getting only 41,472 arrangements.

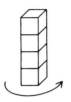

FIGURE 8.57

If you haven't yet solved Problem 8.5, page 275, try it again now.

Solution of Problem 8.5

At first, it seems as if we will have to resort to massive systematic trial and error; no other approach presents itself. However, there are at least two different ways of looking at the problem to treat it mathematically.

In both approaches, the key observation is that once the position of a particular face of the cube is determined, the position of the opposite face is also determined. Hence, the only significant information about each cube is which pairs of colored faces are opposite each other.

For cube I, we observe these three pairs: G–W, B–R, and R–W.

For cube II, we get: R–R, R–G, and B–W.

For cube III, we get: G–R, W–B, and G–W.

And for cube IV, we get: G–G, B–R, and B–W.

We are now ready to present the first of the two solutions.

Suppose we assign numbers to each color as follows: 1 to blue, 2 to red, 3 to green, and 5 to white.

We can then assign to each pair of opposite faces the product of the numbers assigned to the faces themselves. We thus obtain the chart in Figure 8.58.

When the cubes are stacked properly, there will be a $B = 1$, $R = 2$, $G = 3$, and $W = 5$ on each side of the stack, giving a product of $1 \cdot 2 \cdot 3 \cdot 5 = 30$. In particular, the front and back of the column must each give products of 30, yielding a total product of $30 \cdot 30 = 900$ for the front and back faces. But the front and back of the column are made up of pairs of opposite faces from each cube. Hence, we must

	opposite face	corresponding product
	G–W	$3 \cdot 5 = 15$
cube I	B–R	$1 \cdot 2 = 2$
	R–W	$2 \cdot 5 = 10$
	R–R	$2 \cdot 2 = 4$
cube II	R–G	$2 \cdot 3 = 6$
	B–W	$1 \cdot 5 = 5$
	G–R	$3 \cdot 2 = 6$
cube III	W–B	$5 \cdot 1 = 5$
	G–W	$3 \cdot 5 = 15$
	G–G	$3 \cdot 3 = 9$
cube IV	B–R	$1 \cdot 2 = 2$
	B–W	$1 \cdot 5 = 5$

FIGURE 8.58

choose one pair of opposite faces from each cube so that the product of the products corresponding to the chosen pairs is 900. The same is also true of the side faces of the stack.

Hence, the problem reduces to selecting one pair of opposite faces from each cube so that the associated product is equal to 900 (to represent front and back of the stack) and then selecting a different pair of opposite faces from each cube so that the associated product is again 900 (representing the sides of the stack).

We first consider all possible products of one pair of opposite faces from cube I and one pair from cube II. We obtain Figure 8.59.

We could now multiply each of these by each of the possible products for cube III and then multiply by the possibilities for cube IV, to see which products give 900. It is a little easier first to compute the possible products for cubes III and IV alone, and then to see which of these multiplied by which of the ones above yields 900. We get the table in Figure 8.60.

The cases in which the necessary multiplier is found in the list for

cube I · cube II

$15 \cdot 4 = 60$
$15 \cdot 6 = 90$
$15 \cdot 5 = 75$
$2 \cdot 4 = 8$
$2 \cdot 6 = 12$
$2 \cdot 5 = 10$
$10 \cdot 4 = 40$
$10 \cdot 6 = 60$
$10 \cdot 5 = 50$

FIGURE 8.59

cube III · cube IV	multiplier needed to give 900
$6 \cdot 9 = 54$	impossible
$6 \cdot 2 = 12$	75 ✓
$6 \cdot 5 = 30$	30 (not in the I-II list)
$5 \cdot 9 = 45$	20 (not in the I-II list)
$5 \cdot 2 = 10$	90 ✓
$5 \cdot 5 = 25$	36 (not in the I-II list)
$15 \cdot 9 = 135$	impossible
$15 \cdot 2 = 30$	30 (not in the I-II list)
$15 \cdot 5 = 75$	12 ✓

FIGURE 8.60

cubes I and II have been checked, indicating that there are three ways in which a pair of opposite faces from each cube can be chosen so that the product is 900. These are shown in Figure 8.61.

	cube I	cube II	cube III	cube IV
combination 1	15(G–W)	5(B–W)	6(G–R)	2(B–R)
combination 2	15(G–W)	6(R–G)	5(W–B)	2(B–R)
combination 3	2(B–R)	6(R–G)	15(G–W)	5(B–W)

FIGURE 8.61

We must select one of these combinations to represent the front–back of the stack and another combination to simultaneously represent the sides. But combinations 1 and 2 cannot be used simultaneously (they both make use of the G–W pairing on cube I) and combinations 2 and 3 cannot be used simultaneously (they both use the R–G pairing on cube II). Hence, we must use combinations 1 and 3.

Since we are considering rotation of the stack to be a symmetry— that is, not to yield a new solution—there is no loss of generality in assuming that combination 1 represents front–back and that the green face on cube I is in front. This forces the white face of cube II, the red face of cube III, and the blue face of cube IV to be in front.

Combination 3 now shows us how to arrange the side faces of the cubes, without affecting front and back. There are two solutions, depending on whether the blue face of cube I is on the right or left. (Actually, the solutions are equivalent, because one can be obtained from the other by turning each cube upside down without changing front or back.) The two solutions we get are shown in Figure 8.62.

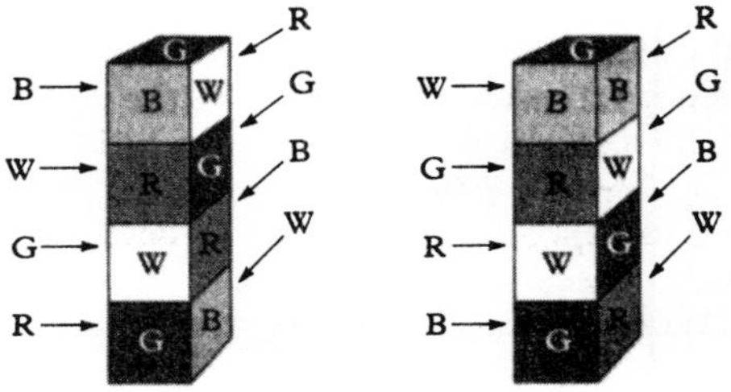

	front	back	solution 1 right side	solution 1 left side	solution 2 right side	solution 2 left side
cube I	G	W	B	R	R	B
cube II	W	B	R	G	G	R
cube III	R	G	G	W	W	G
cube IV	B	R	W	B	B	W

FIGURE 8.62

Before we leave this approach to the solution of Problem 8.5, we should comment on the reason why we chose the numbers 1, 2, 3, and 5 to represent the various colors. We want the numbers to be relatively small so that they are easy to work with; and we want them to be pairwise relatively prime (no two have a common factor other than 1) so that each possible color combination corresponds to a different product. For example, suppose we used 1, 2, 3, and 4 instead of 1, 2, 3, and 5. And suppose we found that we obtain the proper product (which in this case would be 576) when cube II has value 4 and when the other three cubes are properly chosen. We would now have to consider two possible arrangements of cube II, because we could get 4 in two ways—B–W (1 · 4) or R–R (2 · 2). By choosing numbers that are pairwise relatively prime, we help limit the possibilities.

✩ COLORED CUBES—A SECOND APPROACH

We now turn our attention to a second approach to the solution of Problem 8.5. This time the approach is graph-theoretical.

We associate a graph with each cube as follows: We first represent the four colors—blue, red, green, and white—as points (or nodes). We then draw lines (edges) connecting points that correspond to opposite faces of each cube in Figure 8.4. We obtain Figure 8.63.

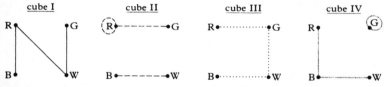

FIGURE 8.63

We now superimpose these graphs on one diagram (Figure 8.64).

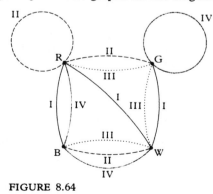

FIGURE 8.64

When the cubes are stacked properly, each color appears once in front and once in back. We therefore can select one opposite pair from each cube (one edge from the graph of each cube) such that each color (each node) is used exactly twice. In other words, we can select four lines (one corresponding to each cube) from the composite graph and assign an orientation to each (designate one end or direction to represent the front and the other end or direction to represent the back) so that each of the four points has one selected line entering it and one leaving it. It is not difficult to see from the composite diagram that this can be done only as shown in Figure 8.65.

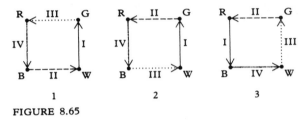

FIGURE 8.65

Again we must select two of these—one for the front–back of the stack and one for the sides. But, as before, combination 2 cannot be used with 1 (they both use the R–B pair from cube IV) or with 3 (they both use the R–G pair from cube II). Hence, we must use combinations 1 and 3.

Putting G in the front of cube I and B on the right, we obtain the situation shown in Figure 8.66.

	front	back	right side	left side
cube I	G	W	B	R
cube II	W	B	R	G
cube III	R	G	G	W
cube IV	B	R	W	B

FIGURE 8.66

This is the same as solution 1 that we found above.

Similarly, putting G in the front of cube I and B on the left, we obtain the other solution found above.

THE CHAPTER IN RETROSPECT

This chapter has presented some solitaire games that have mathematical solutions, and it has shown that mathematics pops up in unusual places.

In solving the problems of this chapter we have made use of

mathematical induction, geometry, coloring arguments and results from number theory, graph theory, and permutation theory.

In the case of some of the puzzles, you might ask why a particular approach is used. The answer in each case is that it works. How the original solvers of these problems were led to these ideas is sometimes a mystery. Often, insight and discovery come as a result of experimentation, or just playing around. There are no known bounds to human ingenuity.

Now try to apply what you have learned to the exercises that follow. And, no matter what your field of endeavor, be on the lookout to see if mathematics can be of help to you.

Exercises

8.1. In the Tower of Brahma puzzle, label the three posts A, B, and C. If we wish to transfer n rings from post A to post C in as few moves as possible, to which post should we make the first move? (Note: The answer will depend on n.)

8.2. 【H】 【A】 In order to prevent tampering by unauthorized individuals, a row of switches at a defense installation is wired so that, unless the following rules are followed in manipulating the switches, an alarm will be activated:

1. The switch on the right may be turned on or off at will.

2. Any other switch may be turned on or off only if the switch to its immediate right is on and all other switches to its right are off.

What is the smallest number of moves in which such a row of switches, which are all on, may be turned off without activating the alarm if:

(a) There are three switches in the row?
(b) There are four switches in the row?

(c) There are five switches in the row?
(d) There are six switches in the row?
★★ (e) There are n switches in the row (n odd)?
★★ (f) There are n switches in the row (n even)? (See [29].)

★ **8.3.** 【A】 Another row of switches at the defense installation is governed by the following rules of movement:

1. Any switch may be turned on at will;

however, doing so activates a relay that automatically turns off the switch to its immediate left (unless that switch is already off).

2. The switch on the right may be turned off at will; any other switch may be turned off only if the switch to its immediate right is on and all other switches to its right are off.

What is the smallest number of moves in which such a row of switches, which are all on, may be turned off if:

(a) There are three switches in the row?

(b) There are four switches in the row?

(c) There are five switches in the row?

(d) There are six switches in the row?

(e) Etc., up to ten?

Dissection Problems

8.4. Given that the area of a rectangle (Figure 8.67) with base of length b and

$$A = bh$$

FIGURE 8.67

height of length h is bh, use a dissection argument to show that:

(a) $\boxed{A}$ The area of a parallelogram (Figure 8.68) with base b and altitude h is bh.

$$A = bh$$

FIGURE 8.68

(b) The area of a right triangle (Figure 8.69) with legs of length b and h is $\frac{1}{2}bh$.

$$A = \frac{1}{2}bh$$

FIGURE 8.69

(c) The area of any triangle (Figure 8.70) is $\frac{1}{2}bh$, where b is the length of one side and h is the length of the altitude to that side.

$$A = \frac{1}{2}bh$$

FIGURE 8.70

★ (d) $\boxed{A}$ The area of a trapezoid (Figure 8.71) is $\frac{1}{2}h(b_1 + b_2)$ where b_1 and b_2 are the lengths of the parallel sides, and h is the length of the perpendicular distance between them.

$$A = \frac{1}{2}h(b_1 + b_2)$$

FIGURE 8.71

8.5. $\boxed{H}$ (a) Given Figure 8.72, make two cuts and use the pieces to form a square.

FIGURE 8.72

(b) $\boxed{A}$ Given Figure 8.73, cut it into the fewest number of pieces that can be reassembled to form a right triangle.

FIGURE 8.73

(c) Give proofs of the Pythagorean Theorem suggested by each of the diagrams in Figure 8.74. ([14], pp. 72–73)

8.6. $\boxed{H}$ $\boxed{A}$ Given a

(a) 2 × 2 checkerboard,

(b) 4 × 4 checkerboard,

(c) 3 × 3 checkerboard with central cell missing,

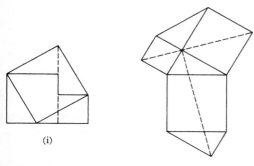

FIGURE 8.74 (ii)

(d) 5 × 5 checkerboard with central cell missing,

(e) 3 × 3 checkerboard with corner cell missing,

in how many ways can each of the above boards be cut along the lines of the checkerboard into two identical pieces if:

(i) coloring is counted and pieces cannot be turned over;

(ii) coloring is not counted but pieces cannot be turned over;

(iii) coloring is not counted and pieces may be turned over?

Whether or not two pieces are considered to be identical depends on whether (i), (ii), or (iii) applies. Similarly, two ways of cutting the board will be considered to be the same if the pieces obtained from cutting in one way are congruent to the pieces obtained by cutting in the other way subject to the conditions in (i), (ii), or (iii), as the case may be.

8.7. Given a

(a) 2 × 2 checkerboard,

(b) 4 × 4 checkerboard,

(c) 3 × 3 checkerboard with central cell missing,

★ (d) 5 × 5 checkerboard with central cell missing,

(e) 3 × 3 checkerboard with corner cell missing,

in how many ways can each of the above boards be cut along the lines of the checkerboard into 4 identical pieces if:

(i) [A] coloring is counted and pieces cannot be turned over;

(ii) coloring is not counted but pieces cannot be turned over;

(iii) [A] coloring is not counted and pieces may be turned over?

Two cuttings are considered to be the same as described in Exercise 8.6.

Tangrams

8.8. [A] Seven pieces make up the tangram square (Figure 8.75).

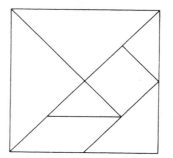

Trace this figure, cut out the pieces and use them for Exercise 8.8 and for other tangrams.

FIGURE 8.75

(a) Fit the above seven pieces to form the letter T in Figure 8.76. ([53], Problem 20)

(b) Fit the above seven pieces to form the boat in Figure 8.77. ([53], Problem 169)

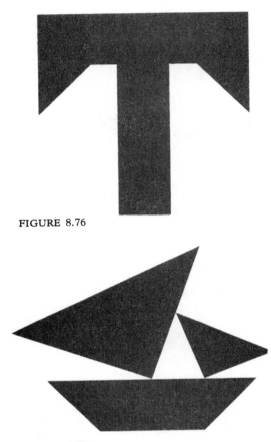

FIGURE 8.76

FIGURE 8.77

Polyominoes (S. Golomb)

8.9. (a) [H] [A] Can an 8×8 checkerboard with one red and one black cell missing be covered by 31 dominoes, regardless of which red and black cells are missing? Justify your answer.

(b) [H] [A] Can an $m \times n$ checkerboard (with m and n odd) that has one cell removed be covered with dominoes?

(c) [H] [A] Can the 5 tetrominoes (Figure 8.12, page 282) be used to form a rectangle?

(d) [S] Draw the 12 pentominoes (5-cell shapes).

★ (e) [A] Find the smallest region on the 8×8 checkerboard into which each of the 12 pentominoes, taken one at a time, will fit.

★ ★ (f) [H] [A] Arrange the 12 pentominoes to form two 5×6 rectangles.

(g) [H] Show that the region in Figure 8.78 cannot be covered by the 12 pentominoes.

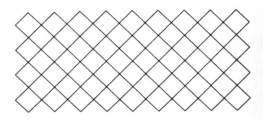

FIGURE 8.78

(h) [A] How many hexominoes (6 cells) are there?
(See [26], pp. 30, 33, 37)

Polycubes

8.10. (a) [A] Consider a $3 \times 3 \times 3$ cube made up of 27 unit cubes. Remove one of these unit cubes and cover the rest of the configuration with $2 \times 1 \times 1$ dicubes. Which unit cube could be removed in order to accomplish this?

(b) Can the $3 \times 3 \times 3$ cube be made up of 9 of the tricubes shown in Figure 8.18 page 285)?

(c) Show by a coloring argument that a $2 \times 3 \times 4$ solid cannot be made from the polycubes numbered (1), (2), (3), (4), (5), and (7) in Figure 8.17 (page 285).

Peg Solitaire

8.11. (a) [H] [A] In Hi Q (Problem 8.3, page 274), which of the original pegs could possibly be the one remaining in the center?

(b) **H** In the French version of Hi Q (see Figure 8.79), show that you cannot start with the center hole empty and end with one peg anywhere.

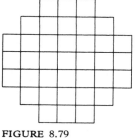

FIGURE 8.79

(c) **H** In the French version, show that no matter which hole starts empty, you cannot end with one peg in the center hole.

(d) **H** **A** For which holes other than the center do the statements in (b) and (c) hold true?

(e) **A** On the regular board, start with Figure 8.80 (the cell numbering is as in the text of this chapter, page 287) and get one in the center.

```
            24
        33  34  35
            44
            54
```
FIGURE 8.80

(f) Start with Figure 8.81 and get one in the center.

```
            24
            34
    42  43  44  45  46
            54
            64
```
FIGURE 8.81

(g) Start with Figure 8.82 and get one in the center.

```
            34
        43  44  45
    52  53  54  55  56
```
FIGURE 8.82

(h) **H** Given the board in Figure 8.83, pegs (indicated by ●) jump diagonally as in checkers, and the jumped peg is removed. Can you leave only one peg?

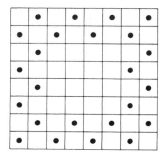

FIGURE 8.83

(i) **H** Given the board in Figure 8.84, pegs (indicated by ●) jump toward or away from the center or along the circle the peg is on. The jumped peg is removed. A piece cannot jump into the center except on the last move. Show that you cannot end with one peg in the center.

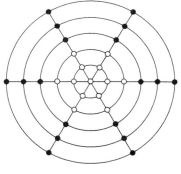

FIGURE 8.84

(j) **A** On a 5 × 5 board with the 9 center squares filled, you may jump horizontally, vertically, or diagonally (removing the jumped peg). Successive jumps by the same peg count as one move. What is the fewest number of moves that allows you to leave one peg in the center?

(k) **H** **A** Can (j) be done using only horizontal and vertical jumps?

(l) ⓗ The board is a 10-cell triangle (see Figure 8.85). Move as in Hi Q, with horizontal and diagonal jumps. Show that, if you begin with any corner cell empty and the remaining cells occupied, it is not possible to leave only one peg remaining.

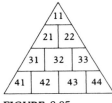

FIGURE 8.85

(m) ⓗ Ⓐ The board is a 15-cell triangle (see Figure 8.86). Leave the top cell vacant. Move as in Hi Q, with horizontal and diagonal jumps, so that only one peg remains. Which cell could contain the last peg?

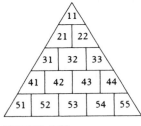

FIGURE 8.86

8.12. Ⓐ Consider the arrangement of black and white pegs in a row shown in Figure 8.87a, with the center cell vacant. White pegs move to the right; black, to the left. A peg may advance one space to a vacant hole, or may jump over another peg to an adjacent vacant hole. The jumped peg is not removed.

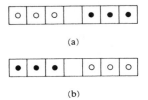

(a)

(b)

FIGURE 8.87

(a) What is the fewest number of moves required to reverse the black and white pegs; that is, to attain the position in Figure 8.87b?

(b) The same as (a) but with n black and n white pegs.

(c) Given the arrangement of black and white pegs shown in Figure 8.88, white pegs move to the left or up, black pegs move to the right or down. Otherwise the rules are the same as in (a). How many moves are required for the black and white pegs to interchange positions?

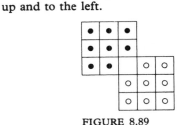

FIGURE 8.88

★ ★ (d) The same rules as (a), but with 8 black and 8 white pegs arranged on the board in Figure 8.89, the black pegs moving down and to the right and the white moving up and to the left.

FIGURE 8.89

8.13. Given $2n$ checkers in a straight line, the object of this puzzle is to obtain n stacks of two checkers each by making n moves of the following type: One checker may be moved (to the right or left) over two other checkers so that it lands on top of the next checker in line. (Note that stacks count as two checkers and empty spaces do not count as any checkers.)

Explain how this may be accomplished for each of the following values of n.

(a) $n = 4$

(b) $\boxed{\text{H}}$ $n = 5$

(c) $\boxed{\text{H}}$ General n.

Sliding Block Puzzles

8.14. $\boxed{\text{H}}$ Some variations of the Fifteen Puzzle use letters instead of numbers.

Consider the 3×3 box shown in Figure 8.90, with the center empty and with the word ROTATION spelled clockwise around the perimeter.

R	O	T
N		A
O	I	T

FIGURE 8.90

Move the pieces as in the Fifteen Puzzle so that the word ROTATION is again spelled in a clockwise direction around the perimeter, but the letter R is in the middle box of the top row.

8.15. $\boxed{\text{A}}$ When Foster Quarrels died, he left a will that provided that his estate should be divided equally among his five cousins, Ann, Dan, Jan, Nan, and Van. Unfortunately, the cousins did not get along well with each other. In fact, they detested each other so vehemently that it would have been unwise to allow any two of them to be in the same room.

It came time to read the will at the apartment of the late Mr. Quarrels, which is pictured in Figure 8.91. The lawyer placed Ann in the kitchen, Dan in the dining room,

dining room	den	bedroom
kitchen	foyer	bathroom

FIGURE 8.91

Jan in the den, Nan in the bedroom, and Van in the bathroom. However, Van insisted that he wait in the kitchen, and Ann wanted to go to the bathroom. How could the lawyer make this change in the fewest possible moves, without any two cousins meeting each other in the process? (The other cousins need not end up in the same rooms in which they were originally placed.)

Colored Cubes

8.16. (a) Suppose that we keep the first three cubes in Problem 8.5 (Figure 8.4, page 275), but that cube IV is replaced by the cube in Figure 8.92a. Show that it is not possible to stack the cubes as required in Problem 8.5.

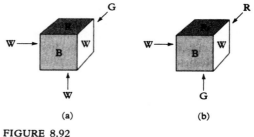

(a)   (b)

FIGURE 8.92

(b) Suppose that we keep the first three cubes in Problem 8.5, but that cube IV is replaced by the cube in Figure 8.92b. Find all possible ways of stacking the cubes as required in Problem 8.5.

8.17. $\boxed{\text{H}}$ $\boxed{\text{A}}$ (P. A. MacMahon) Given 6 colors, there are 30 different ways to color a cube so no two faces of the cube receive the same color.

Select any one of the 30 cubes—say, the

cube that is blue on top, yellow on the bottom, red in front, white in back, green on the left, and orange on the right. Use 8 of the remaining 29 possible cubes to form a $2 \times 2 \times 2$ model of the chosen cube. In constructing such a model, any faces of the unit cubes that touch each other must be of the same color.

Other Solitaires

8.18. $\boxed{\text{H}}$ Use a coloring argument to show that a reentrant knight's tour (see Chapter 6) is not possible on a board having an odd number of cells.

8.19. $\boxed{\text{A}}$ Several modern games (Oware or Kalah, and Ruma are examples) are variants of the game Mancala, whose origin can be traced back more than 3000 years. The following is an interesting solitaire version of this game.

One bowl and n cups are placed in a circle, and k marbles are placed in each cup (but none are put in the bowl). Taking the marbles from any cup, "sow" them, one at a time, in a clockwise direction around the circle, starting with the cup (or bowl) after the one from which the marbles were taken and ending when the marbles are all used up. Note that, if the

number of marbles is sufficiently large, you may proceed around the circle more than once, the bowl and original cup receiving one (or more) marble(s).

If the last marble lands in an empty cup, you lose. If the last marble lands in a nonempty cup, you must use the marbles from that cup for sowing next. If the last marble lands in the bowl, you may select any cup to sow from.

Following these rules, the object of the game is to transfer all the marbles to the bowl.

Can this be accomplished if:

(a) $n = 3, k = 1$?   (b) $n = 3, k = 2$?
(c) $n = 3, k = 3$?   (d) $n = 4, k = 1$?
(e) $n = 4, k = 2$?   (f) $n = 4, k = 3$?
(g) $n = 4, k = 4$?   (h) $n = 5, k = 2$?
(i) $n = 6, k = 2$?

(Adapted from [10], Problem 185)

9

Potpourri

In previous chapters, we considered a wide variety of recreational problems, with each chapter devoted to problems relating to a particular area of mathematics. But we have only scratched the surface of the many topics usually considered to be part of recreational mathematics. It is not possible to include all of this material in any one book; but we would like to present here, in our final chapter, a sampling of some well-known recreational problems that do not require any specific mathematical tools. All that is needed to solve them is some careful thought.

There is no text for this chapter, other than the brief remarks that introduce some of the problems. The chapter is what its name says—a potpourri.

Decimation

The term "decimation" stems from an ancient custom of executing one-tenth of a mutinous crew by lining them up and selecting every tenth man for the gallows. However, the prototype of this kind of problem, often referred to as the Josephus Problem, is based on an event that occurred in the year 67. After the fall of Jotapata, during the rebellion against Rome, Josephus and forty other Jews fled and took refuge in a cave. It was decided that they would kill themselves rather than allow themselves to fall into Roman hands. Josephus, who was not ready to die, devised the following plan. He arranged the forty-one people (including himself) in a circle and suggested that, counting clockwise, every third person should be killed, until only one person remained. That person would then commit suicide. Josephus placed himself and a friend so that they would be the last two survivors.

9.1. In what positions relative to the one from which the counting started did Josephus place himself and his friend?

9.2. Nine children are seated in a circle playing the game of Buzz. The game starts with one player saying "one"; the person on his or her right says "two"; the individual on that person's right says "three"; and so on, continuing around and around the circle. Whenever anyone is about to say a number that contains a seven as one of its digits or that is divisible by seven, the person says "buzz" instead of the number. Failure to do so eliminates that person from the game and removes him or her from the circle. The next player then resumes the count with the next number.

The children have no trouble recognizing numbers that contain the digit seven, but they do not yet know the multiples of seven. Assume that everyone who comes to a multiple of seven that does not contain the digit seven is eliminated.

Where is the eventual winner of the game seated with reference to the player who starts by saying "one"?

9.3. [A] The magician removed ten cards, with face values ace through ten, from a deck of playing cards. He arranged them in some order and held them in a pile in his hand, face down. He then spelled aloud the word "ace"—A C E—removing one card from the top of the deck and placing it on the bottom (often referred to as "ducking a card") for each letter. He then turned the next card face up and placed it on the table. It was the ace. Using the nine cards remaining in his hand,

he spelled T W O, again ducking one card with each letter. The next card was then placed face up on the table. It was the two. He continued in this manner, ducking a card with each letter in the face value of the next card to be exposed, until only the ten remained in his hand.

How had he originally arranged the cards?

9.4. At Helen's birthday party, she and eleven friends sat in a circle playing Hot Potato. The rules of the game call for passing a potato in a clockwise direction while music plays. When the music stops, the person holding the potato is eliminated and must leave the circle. Unfortunately, the record player at Helen's house was broken, so Helen's mother was asked to sing. Not wishing to do so, she mentioned a number, n, greater than ten, and suggested that every nth player to handle the potato should be out. The counting started with Helen and moved clockwise.

After six players had been eliminated, Jeffrey realized that he was going to be the next to go. Quick as a wink, he suggested that passing in a clockwise direction is unfair to lefties and they should therefore pass counterclockwise. Everyone agreed, and Jeffrey eventually was the winner.

What was the smallest number, n, greater than ten, that Helen's mother could have suggested; and where was Jeffrey sitting in relation to Helen?

9.5. Ten teenagers got together to play basketball. Lauren and Lisa were appointed team captains and had to choose their teams. The eight remaining players happened to be standing in a circle.

"How should we choose?" asked Lauren.

Lisa suggested the following scheme: "Starting with Burt as number one, I'll take the fourth person, counting around the circle in a clockwise direction. You'll take the fourth person after that. I'll take the fourth after

that, and so on. Each person will leave the circle as soon as he or she is selected."

"That isn't fair," objected Lauren. "You get all the good players that way. Let me get the first player that's counted out, you get the next player, and so on."

"OK," replied Lisa. "But let me pick a number other than four."

Lauren agreed.

What number should Lisa pick so that she still gets the good players?

Coin Weighing

9.6. [A] Clu D. Tector was lecturing on methods for detecting counterfeit coins. To illustrate a point, he removed nine coins from his pocket and said that, although the coins appeared to be identical, one was counterfeit and lighter than the rest. He produced a balance scale and challenged the audience to suggest a plan for determining, in the fewest possible weighings, which of the coins was the counterfeit.

When no one found the optimal scheme, Clu mentioned that two weighings should suffice.

Can you figure out how to do it?

9.7. [S] After Clu (see Exercise 9.6) demonstrated how two weighings would suffice, one of the members of the audience posed a similar problem: Removing four coins from her purse, she said that one of them was counterfeit and did not weigh the same as the other coins, but she would not say whether it was heavier or lighter. She then produced a fifth coin and said it was a good coin. The problem she posed was to devise a plan for finding the counterfeit coin in no more than two weighings.

Clu did it. Can you?

★ **9.8.** [H] Referring to the problem posed in Exercise 9.7, Clu said, "Let's consider a more difficult variation. If you have twelve coins, one of which is either heavier or lighter than the rest, can you find the counterfeit coin in no more than three weighings and determine whether it is heavier or lighter?"

★★ **9.9.** [H] Clu placed five coins on the table. They appeared to be identical, but Clu remarked that they were all made of different alloys and that no two weighed the same.

In no more than seven weighings on a balance scale, can you determine which coin is lightest, which is next lightest, and so on?

★★ **9.10.** [H] For his next weighing demonstration, Clu told his audience that the five coins (see Exercise 9.9) weighed 10 grams, 20 grams, 30 grams, 40 grams, and 50 grams. He then proceeded to demonstrate how one can determine, in no more than five weighings on a balance scale, which coin is which.

How did he do it?

9.11. As his final topic for the evening, Clu pointed out that a regular scale, as distinct from a balance scale, can often be even more helpful. He displayed five sacks of coins and said that four of the sacks contained ordinary coins, weighing 1 ounce each, but that the remaining sack contained counterfeit coins, each weighing 2 ounces.

The problem Clu posed is to determine in one weighing which is the sack of counterfeit coins.

How can this be done if:

(a) [A] The scale you use shows the weight correct to the nearest ounce?

(b) [H] The scale indicates only pounds?

Shunting

In Chapter 1, we considered some shunting problems in which the object was to switch the position of a locomotive and railroad cars on a track. We now present several other variations of this type of problem.

9.12. An accident on the main track between Eastville and Westville forced the Eastville–Westville local to take a seldom used side track. Unfortunately, three freight cars had been left standing on this track. Fortunately, there was a triangular arrangement of tracks in the vicinity (Figure 9.1), but only one car or engine could fit on any of the four parts

(1, 2, 3, 4) of the arrangement at one time. The engine D of the local can go in either direction, pulling or pushing cars in the process. Every time the engine reverses its direction of movement, fuel is wasted; hence, the engineer wants to pass the stalled cars with as few engine reversals as possible.

(a) If the engineer, Connie Duktor, wishes to leave the three freight cars exactly as she found them, and she also wants to arrive at Westville with her train exactly as it was when she started (with cars in the original order and facing the original direction), how can she pass the freight cars with no more than 22 engine reversals?

(b) [S] If she does not care in what order or direction she leaves the stalled cars on the main track but does want to arrive in Westville with her train exactly as it was when she left Eastville, how can she pass the freight cars with no more than 16 engine reversals?

9.13. [A] After passing the freight cars (see Exercise 9.12), the Eastville–Westville local continued on its way. Unfortunately, the Westville–Eastville local was using the same track. Fortunately, there was a siding (Figure 9.2). Unfortunately, the siding had room enough for only one car or engine at a time.

How can the trains pass with the fewest total engine reversals?

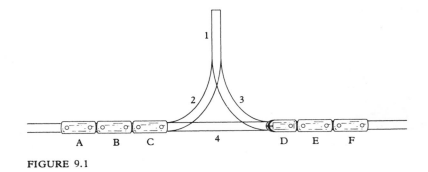

FIGURE 9.1

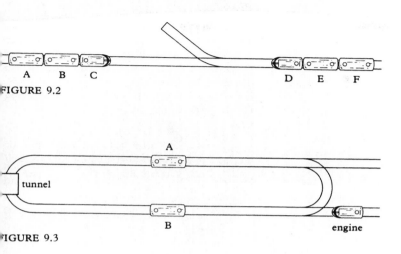

FIGURE 9.2

FIGURE 9.3

.14. ☐H☐ On the track in Figure 9.3, inter-
hange the position of cars A and B and
eturn the engine to its starting position. By
he way, only the engine can pass through the
unnel; the railcars are too big.

yllogisms

.15. All bangs are bengs.
     All bengs are bings.
     All bings are bongs.
Can we conclude that all bangs are bongs?

.16. ☐S☐ All gangs are gengs.
          Some gengs are gings.
          No gengs are gongs.
Can we conclude that
(a) No gangs are gongs?
(b) No gings are gongs?
(c) Some gings are gongs?
(d) Some gangs are gings?

17. All Cyclopses have only one eye.
People cannot get a driver's license unless
ey have depth perception.
No one eyed creatures have depth percep-
 on.
Polyphemus is a Cyclops.

Is it possible that Polyphemus can get a
driver's license?

9.18. ☐A☐ According to Lewis Carroll in *The
Hunting of the Snark* (1876):
   All Boojums are snarks.
   Every Bandersnatch is a frumious animal.
   Only animals which frequently breakfast at
five o'clock tea can be snarks.
   No frumious animals breakfast at five
o'clock tea.
   Are any Bandersnatches Boojums?

9.19. ☐S☐ ☐A☐ It is true that:
   1. All colonial hymenopters are capable
of reproducing by parthenogenesis.
   2. All bees live in hives.
   3. Every insect that helps pollinate fruit
orchards is beneficial to the farmer.

4. Only colonial hymenopters live in hives.

5. All bees help pollinate fruit orchards.

Can we conclude that all insects that are beneficial to the farmer are capable of reproducing by parthenogenesis?

9.20. [A] Nobody who would walk in a cemetery after midnight is superstitious.

All people who believe that knowledge is obtainable through reason are rationalists.

Everyone who walks under a ladder is taking a chance that a bucket of paint might fall on his or her head.

Only people who believe in the supernatural are afraid of ghosts.

Only people who are superstitious do not walk under ladders.

No one who is a rationalist believes in the supernatural.

Only people who are afraid of ghosts would not walk in a cemetery after midnight.

Is there any relationship between people who would take a chance that a bucket of paint might fall on their heads and people who believe that knowledge is obtainable through reason?

9.21. Everyone who believes that mermaids exist has visited Atlantis.

Only people who have found favor with the gods are protected by Neptune.

No one who has not tasted real ambrosia has partaken of food with the gods.

People who do not believe that mermaids exist have never seen a mermaid.

Everyone who has tasted real ambrosia has a hangover.

Only people who have partaken of food with the gods have ever been invited to Mount Olympus.

No one who has not survived a shipwreck has visited Atlantis.

Everyone who has found favor with the gods has been invited to Mount Olympus.

No one who is not protected by Neptune has survived a shipwreck.

What, if anything, can we conclude about the connection between seeing a mermaid and having a hangover?

Grab Bag

9.22. [A] A jigsaw puzzle contains 500 pieces. A "section" of the puzzle is a set of one or more pieces that have been connected to each other. (Note that a single piece by itself is also called a section.) A "move" consists of connecting two sections.

What is the smallest number of moves in which the puzzle can be completed? ([41], Vol. 26, p. 169)

9.23. A bacterium in a petri dish filled with agar-agar will divide in two every 20 minutes, forming two bacteria. Starting with one bacterium, it takes 24 hours to completely fill the petri dish with bacteria.

How long does it take until the petri dish is half full?

9.24. [H] [A] Fi Bonacci breeds a strain of self-fertilizing worms that reproduce in the following manner. Each worm produces its first offspring at the age of two weeks and continues to produce exactly one offspring each week thereafter.

If Mr. Bonacci started breeding these worms just over twelve weeks ago, and if he had exactly one newborn worm at that time, how many worms does Fi have now, assuming that none has died?

9.25. [A] A phonograph record of radius 15 centimeters is playing on a turntable. The grooves on the record are $\frac{1}{2}$ millimeter wide. The grooves begin 5 millimeters from the edge of the record and end 5 centimeters from the center.

How far does the needle travel during the time that the record is playing?

9.26. [A] A tired old inchworm fell into a gopher hole that is 10 inches deep. The worm started climbing up the side of the hole, progressing at the rate of 1 inch every 5 minutes. At the end of each 15 minutes of climbing, the worm had to pause to rest for 1 minute. During each of these rest periods, the worm slipped back 1 inch.

How many minutes does it take the worm to climb out of the hole?

9.27. [A] A man puts all his socks in a drawer, without pairing them. He owns two kinds of socks, black ones and brown ones. All the black ones match each other, and all the brown ones match each other. One day, the man is in a hurry and, without looking to see what color they are, he pulls some socks from the drawer. If the drawer contains 8 black socks and 12 brown ones:

(a) What is the smallest number of socks he must pull in order to be sure that he has a pair of black socks?

(b) What is the smallest number of socks he must pull in order to be sure that he has at least one pair of socks of either color?

9.28. Paula Peedle's poodle Puddles is the best dressed dog in town. Not only does he have fancy outfits for all occasions, but he even has knitted socks to match. In all, Puddles has 12 red socks, 24 green ones, and 40 white ones, which Paula keeps in a large laundry bag. Whenever Paula dresses Puddles, she first draws the socks, one at a time, from the bag, until she has 4 socks of the same color. She then chooses the rest of Puddles' outfit to match the socks.

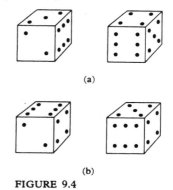

What is the greatest number of socks Paula ever has to draw from the bag?

9.29. [A] Hardluck Harry Harris was shooting craps with Shifty Steve Sutter. Steve had the dice. On his first pass, he rolled a 7 (see Figure 9.4a). On his next pass, Steve threw an 11. (Figure 9.4b).

(a)

(b)

FIGURE 9.4

"You cheat," shouted Harry. "Give me back my money."

Harry had no reason to think that all dice are made the same way.

How then did he know that Steve was cheating?

9.30. [A] Find the largest set of integers less than 100 such that it is not possible to choose a subset that sums to 100. ([15], Problem 36)

9.31. The ends of a piece of string are to be tied together so that the string will form a circle with circumference of 12 inches. Before the ends are tied, three knots are to be placed

in the string at such locations that, after the string is tied (thereby creating a fourth knot), every integral number of inches is the distance along the circumference of the circle between two of the knots.

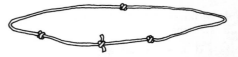

(a) [A] Where should the three knots be tied?

(b) What if the circumference is to be 19 inches and we can make four knots before the ends are tied?

THE BOOK IN RETROSPECT

In this book we have presented some of the wide variety of problems that are usually dealt with in books on mathematical recreations. We have also presented some related mathematical topics, to illustrate how mathematics can be applied in solving problems and also how mathematical theory can be motivated and developed. In the course of working through this book, your abilities in general problem solving have been exercised and therefore strengthened and you will be able to carry them over to problem solving in other areas.

We hope that you will return to this book from time to time, to try problems you have not yet attempted or to find better solutions for those that you have already solved.

Although this book contains over 450 problems, there are still areas of mathematical recreations that have not been discussed. In deciding which topics to include, we were guided partly by our desire to discuss certain mathematical concepts and applications, partly by which kinds of problems are classic or standard, and partly by our own preferences. Many other types of problems may be found in the literature.

For additional problems, we recommend the books appearing in the bibliography. These and other books of recreational problems may be found in libraries. (Check the card catalog under Mathematical Recreations.) To keep in contact with recent developments, we suggest Martin Gardner's "Mathematical Games" column in current issues of *Scientific American* magazine. We also recommend the *Journal of Recreational Mathematics*, although some of the articles require more mathematical background than does this book.

Appendix A

Some Basic Algebraic Techniques

The ability to solve algebraic equations is one of the most important of mathematical skills. In this appendix, we review some of the standard basic techniques:

A. How to solve a linear equation in one unknown.

B. How to solve a linear inequality in one unknown.

C. How to use the quadratic formula to solve a quadratic equation in one unknown.

D. How to use factoring to solve a quadratic equation in one unknown.

E. How to use the substitution method to solve a simultaneous system of two linear equations in two unknowns.

F. How to use the method of elimination to solve a simultaneous system of two linear equations in two unknowns.

G. How to solve a simultaneous system of three linear equations in three unknowns.

H. How to solve a system of two equations in two unknowns, where one equation is linear and the other is quadratic.

In our discussion of each of these methods, we give the general explanation on the left and an example on the right.

A. How to solve a linear equation in one unknown.

Method	*Example*
The object is to simplify the equation to the form $x = a$.	$3(x + 2) + \frac{5}{4}x = 7(\frac{2}{5} - 2x)$
1. Multiply out to remove parentheses (if there are any).	$3x + 6 + \frac{5}{4}x = \frac{14}{5} - 14x$
2. If there are fractions, multiply both sides of the equation by the least common denominator. (If you can't find the least common denominator, you may multiply by the product of all the denominators.)	$20(3x + 6 + \frac{5}{4}x = \frac{14}{5} - 14x)$ $60x + 120 + 25x = 56 - 280x$
3. Transpose all x terms to one side of the equation, and all constant terms to the other side.	Add $280x$ to both sides and subtract 120 from both sides: $60x + 120 + 25x + 280x - 120$ $= 56 - 280x + 280x - 120$
4. Combine the x terms, and combine the constants.	$365x = -64$
5. Divide both sides by the coefficient of x.*	$\frac{365}{365}x = \frac{-64}{365}$ $x = -\frac{64}{365}.$ (Note that there is exactly one solution.)
6. Check your answer in the original equation.	Check: $3(\frac{-64}{365} + 2) + \frac{5}{4}(\frac{-64}{365}) \overset{?}{=} 7[\frac{2}{5} - 2(\frac{-64}{365})]$ $\frac{-192}{365} + 6 - \frac{80}{365} \overset{?}{=} \frac{14}{5} + \frac{896}{365}$ $\frac{-272}{365} + \frac{2190}{365} \overset{?}{=} \frac{1022}{365} + \frac{896}{365}$ $\frac{1918}{365} \overset{\checkmark}{=} \frac{1918}{365}.$

---

* If the coefficient of x should be zero, then either there are no solutions (if the constant term is not zero) or else every value of x gives a solution (if the equation becomes $0x = 0$). Otherwise, the equation has exactly one solution.

B. How to solve a linear inequality in one unknown.

Method	*Example*
The object is to transform the inequality into the form	$3x - \frac{1}{5} < -4(7 - 2x)$

$$x > a \quad \text{or} \quad x < a.$$

1. Solve as if it were an equation. That is, repeat steps 1 to 5 above. The only difference is that if you multiply or divide both sides of an inequality by a negative number, then the sense of the inequality is reversed.*

$$3x - \tfrac{1}{5} < -28 + 8x$$
$$5(3x - \tfrac{1}{5} < -28 + 8x)$$
$$15x - 1 < -140 + 40x$$
$$15x - 1 - 40x + 1 <$$
$$\qquad -140 + 40x - 40x + 1*$$
$$-25x < -139$$
$$\tfrac{-25}{-25}x > \tfrac{-139}{-25}$$

(since -25 is negative)

$$x > \tfrac{139}{25}$$

The answer is all values of x that are greater than $\frac{139}{25}$.

Note that, in general, there are an infinite number of solutions, although, as in solving a linear equation—method **A** above—there will be no solutions if the inequality reduces to $0x > a$ where a is positive or to $0x < a$ where a is negative.

6. As a check, you can try some values in the indicated range to see that they do work, and try some values not in the indicated range to see that they don't work.

$6 > \frac{139}{25}$, therefore 6 should work:

$$3(6) - \tfrac{1}{5} = 17\tfrac{4}{5}$$
$$-4[7 - 2(6)] = 20$$
$$17\tfrac{4}{5} < 20.$$

On the other hand, $5 < \frac{139}{25}$, so 5 should not work:

$$3(5) - \tfrac{1}{5} = 14\tfrac{4}{5}$$
$$-4[7 - 2(5)] = 12$$
$$14\tfrac{4}{5} > 12.$$

---

* By properly choosing the side of the inequality to which x is transposed, we can avoid having to multiply or divide by a negative number.

* $15x - 1 \; < -140 + 40x$
$15x - 1 + 140 - 15x$
$\qquad < -140 + 40x + 140 - 15x$
$139 < 25x$
$\tfrac{139}{25} < x.$

C. How to use the quadratic formula to solve a quadratic equation in one unknown.

Method	*Example*

The object is to first transform the equation to the form $ax^2 + bx + c = 0$, and then to apply the quadratic formula, which is given below.

$$-2x(x - 4) = 2x^2 + 3$$

1. Multiply out to remove parentheses (if there are any).

$$-2x^2 + 8x = 2x^2 + 3$$

2. Transpose all terms to one side of the equation.

$$0 = 2x^2 + 3 + 2x^2 - 8x$$

3. Collect terms and write in order of descending powers of x.

$$0 = 4x^2 - 8x + 3$$

4. Use the quadratic formula to solve for x. The formula says that the solutions of $ax^2 + bx + c = 0$ are

$a = 4$, the coefficient of x^2
$b = -8$, the coefficient of x
$c = 3$, the constant term.

$$x = \frac{-b + \sqrt{b^2 - 4ac}}{2a}$$

and

$$x = \frac{-b - \sqrt{b^2 - 4ac}}{2a}.$$

$$x = \frac{-(-8) \pm \sqrt{(-8)^2 - 4(4)(3)}}{2(4)}$$

$$= \frac{8 \pm \sqrt{64 - 48}}{8}$$

$$= \frac{8 \pm \sqrt{16}}{8}$$

$$= \frac{8 \pm 4}{8}$$

(If $b^2 - 4ac$ is negative, then there are no solutions. If $b^2 - 4ac = 0$, then there is one solution: $x = \dfrac{-b}{2a}$. Otherwise, there are exactly two solutions.)

There are exactly two solutions:

$$x = \frac{8 + 4}{8} = \frac{12}{8} = \frac{3}{2}$$

and

$$x = \frac{8 - 4}{8} = \frac{4}{8} = \frac{1}{2}.$$

5. Check your answers in the original equations.

Check:

$$-2(\tfrac{3}{2})(\tfrac{3}{2} - 4) \overset{?}{=} 2(\tfrac{3}{2})^2 + 3$$

$$-3(\tfrac{-5}{2}) \overset{?}{=} 2(\tfrac{9}{4}) + 3$$

$$\tfrac{15}{2} \overset{?}{=} \tfrac{9}{2} + 3$$

$$\tfrac{15}{2} \overset{\checkmark}{=} \tfrac{15}{2}.$$

Also,

$$-2(\tfrac{1}{2})(\tfrac{1}{2} - 4) \overset{?}{=} 2(\tfrac{1}{2})^2 + 3$$
$$(-1)(-\tfrac{7}{2}) \overset{?}{=} 2(\tfrac{1}{2}) + 3$$
$$\tfrac{7}{2} \overset{?}{=} \tfrac{1}{2} + 3$$
$$\tfrac{7}{2} \overset{\checkmark}{=} \tfrac{7}{2}.$$

D. How to use factoring to solve a quadratic equation in one unknown.

Method	Example
The object is to first transform the equation to the form $ax^2 + bx + c = 0$, and then to factor into a product of two linear factors, and then to set each factor equal to zero.	$-2x(x - 4) = 2x^2 + 3$
Steps 1 through 3 are the same as in the method above.	$-2x^2 + 8x = 2x^2 + 3$ $0 = 2x^2 + 3 + 2x^2 - 8x$ $0 = 4x^2 - 8x + 3$
4. Factor the nonzero side into a product of linear factors.	$0 = (2x - 3)(2x - 1)$
5. Set each of the factors equal to zero. (If the product of two numbers is zero, then at least one of the numbers must be zero.)	$2x - 3 = 0$ or $2x - 1 = 0$
6. Solve each of the linear equations (as in method **A** above) to obtain two solutions.	$2x = 3$ or $2x = 1$ $x = \tfrac{3}{2}$ or $x = \tfrac{1}{2}$
7. Check, as in method **C** above.	See **C** for the check.

Note that the method of factoring is usually the simpler way to solve a quadratic equation, provided that you are able to find the factors of $ax^2 + bx + c$. If you are not able to do so, then you can always rely on the quadratic formula.

E. How to use the substitution method to solve a simultaneous system of two linear equations in two unknowns.

Method	Example
The object is to reduce the solution of the system to the solution of one equation in one	(i) $x - 14y = -31$ (ii) $4x + 3y = -6$

unknown by substituting for one of the unknowns an expression in terms of the other.

1. Use either of the equations to express one of the unknowns in terms of the other.	Adding $14y$ to both sides of equation (i), we obtain (iii) $x = -31 + 14y$.
2. Substitute the expression thus obtained into the other equation.	Substituting $-31 + 14y$ for x in equation (ii), we obtain

$$4(-31 + 14y) + 3y = -6$$
$$-124 + 56y + 3y = -6$$
$$56y + 3y = -6 + 124$$
$$59y = 118$$
$$y = 2.$$

3. Solve the resulting equation as in method **A** above.*

4. Substitute the value obtained into the expression found in step 1. This gives the solution.

Substituting $y = 2$ in equation (iii)

$$x = -31 + 14(2)$$
$$= -31 + 28$$
$$= -3.$$

The solution is $x = -3$, $y = 2$.

5. Check your answer in the original equations.

Check:

(i) $-3 - 14(2) = -3 - 28$
$$\overset{\checkmark}{=} -31$$

(ii) $4(-3) + 3(2) = -12 + 6$
$$\overset{\checkmark}{=} -6.$$

F. How to use the method of elimination to solve a simultaneous system of two linear equations in two unknowns.

Method	*Example*
The object is to reduce the solution of the system to the solution of one equation in one	(i) $3x + 4y = 18$ (ii) $2x - 5y = -11$

---

* It is possible that in solving the equation in step 2 that we obtain an equation of the form $0y = k$, where k is a nonzero constant. In this case, we say that the system is inconsistent, and there are no solutions.

It is also possible that we obtain an equation of the form $0y = 0$. In this case, any value of y can be used to obtain a value of x. The equations are said to be dependent and there are infinitely many solutions.

In all other cases the solution is unique.

unknown by eliminating one of the variables.

1. Multiply both equations by the proper numbers so as to obtain two new equations in which the coefficients of x (or of y) are the negatives of each other.

Multiply (i) by 2 and (ii) by -3

$$2(3x + 4y = 18)$$

$$-3(2x - 5y = -11)$$

to obtain

(iii) $6x + 8y = 36$
(iv) $-6x + 15y = 33$.

2. Add the resulting equations to eliminate x (or y) and obtain an equation in one unknown.*
3. Solve as in method **A** above.
4. Substitute into either original equation to solve for the other unknown.

Adding,

$$23y = 69$$

$$y = 3.$$

Substitute in equation (i),

$$3x + 4(3) = 18$$

$$3x + 12 = 18$$

$$3x = 6$$

$$x = 2.$$

The solution is $x = 2$, $y = 3$.

5. Check your answer in the original equations.

Check:
(i) $3(2) + 4(3) = 6 + 12 \overset{\checkmark}{=} 18$
(ii) $2(2) - 5(3) = 4 - 15 \overset{\checkmark}{=} -11$.

G. How to solve a simultaneous system of three linear equations in three unknowns.

Method

The object is to transform the system of equations to the form

$$ax + by + cz = d$$

$$ey + fz = g$$

$$kz = m$$

and then to solve from the bottom up.

1. Rearrange the equations, if

Example

(i) $3x + 7y - 4z = -11$
(ii) $2x - 4y + z = -5$
(iii) $-5x + 2y - 8z = -12$

---

* If, when we add, the equation becomes $0x + 0y = a$, where $a \neq 0$, then the system is inconsistent and there are no solutions. If we obtain $0x + 0y = 0$, then the equations are dependent and there are infinitely many solutions. In all other cases, there is a unique solution.

necessary, so that x has a nonzero coefficient in the first equation.

2. If x has a nonzero coefficient in any of the other equations, use method **F** above to eliminate x from these equations.

Use (i) and (ii) to eliminate x from (ii)

$$2(3x + 7y - 4z = -11)$$
$$-3(2x - 4y + z = -5)$$

$$6x + 14y - 8z = -22$$
$$\underline{-6x + 12y - 3z = 15}$$
(iv) $\qquad\qquad 26y - 11z = -7.$

Use (i) and (iii) to eliminate x from (iii)

$$5(3x + 7y - 4z = -11)$$
$$3(-5x + 2y - 8z = -12)$$

$$15x + 35y - 20z = -55$$
$$\underline{-15x + 6y - 24z = -36}$$
(v) $\qquad\qquad 41y - 44z = -91.$

You obtain a new system of the form

$$ax + by + cz = d$$
$$ey + fz = g$$
$$hy + iz = j.$$

(i) $3x + 7y - 4z = -11$
(iv) $\qquad 26y - 11z = -7$
(v) $\qquad 41y - 44z = -91$

3. Use the last two equations to eliminate y from the last equation.

Use (iv) and (v) to eliminate y from (v)

$$41(26y - 11z = -7)$$
$$-26(41y - 44z = -91)$$

$$41(26)y - 451z = -287$$
$$\underline{-41(26)y + 1144z = 2366}$$
(vi) $\qquad\qquad 693z = 2079.$

The system has now been transformed into the desired form.

The system becomes
(i) $3x + 7y - 4z = -11$
(iv) $\qquad 26y - 11z = -7$
(vi) $\qquad\qquad 693z = 2079.$

4. Solve the last equation for the third unknown, as in method **A** above.

$$693z = 2079$$
$$z = \tfrac{2079}{693} = 3$$

5. Substitute this value into the middle equation and solve for the second unknown.

$$26y - 11(3) = -7$$
$$26y - 33 = -7$$
$$26y = -7 + 33 = 26$$
$$y = 1$$

6. Use the values found for both of the unknowns to find the first unknown by substituting into the first equation.

$$3x + 7(1) - 4(3) = -11$$
$$3x + 7 - 12 = -11$$
$$3x = -11 - 7 + 12 = -6$$
$$x = -2$$

The solution is

$$x = -2, \quad y = 1, \quad z = 3.$$

7. Check your answer in the original equations.

Check:

(i) $3(-2) + 7(1) - 4(3)$
$$= -6 + 7 - 12 \overset{\checkmark}{=} -11$$

(ii) $2(-2) - 4(1) + (3)$
$$= -4 - 4 + 3 \overset{\checkmark}{=} -5$$

(iii) $-5(-2) + 2(1) - 8(3)$
$$= 10 + 2 - 24 \overset{\checkmark}{=} -12.$$

Note that, again, it is possible for the equations to be inconsistent [if we obtain an equation $0 = a \ (\neq 0)$] or dependent (if an equation becomes $0 = 0$), in which cases there will be no solutions or infinitely many solutions respectively. In all other cases, there will be a unique solution.

An alternative approach to solving a system of three linear equations in three unknowns is to use the method of substitution (method **E** above) to reduce the system to two equations in two unknowns, and then proceed as in method **E** or **F**.

H. How to solve a system of two equations in two unknowns, where one equation is linear and the other is quadratic.

Method	*Example*
The object is to reduce the system to one quadratic equation in one unknown and then to solve as in method **C** or **D** above.	(i) $x^2 + 2x - y^2 = -1$
	(ii) $\qquad 2x - y = 7$
1. Use the linear equation to solve for one of the unknowns in terms of the other.	Use (ii) to solve for y in terms of x

$$2x - y = 7$$
$$2x = 7 + y$$
$$y = 2x - 7.$$

2. Substitute into the quadratic equation.

Substitute into (i)

$$x^2 + 2x - (2x - 7)^2 = -1.$$

3. Simplify and solve as in method **C** or **D**.

$$x^2 + 2x - (4x^2 - 28x + 49) = -1$$

Note that there will be zero, one, or two solutions.

$$x^2 + 2x - 4x^2 + 28x - 49 = -1$$
$$-3x^2 + 30x - 49 = -1$$
$$0 = 3x^2 - 30x + 48$$
$$0 = 3(x^2 - 10x + 16)$$
$$0 = x^2 - 10x + 16$$
$$0 = (x - 2)(x - 8)$$
$$x = 2 \quad \text{or} \quad x = 8$$

4. Use the equation in step 1 to find the other unknown.

$$y = 2x - 7$$
$$y = 2(2) - 7 = -3$$

or $y = 2(8) - 7 = 9$

There are two solutions:

$$x = 2 \text{ and } y = -3$$

or $x = 8$ and $y = 9$.

5. Check your answer in the original equations.

Check:

(i) $2^2 + 2(2) - (-3)^2$
$= 4 + 4 - 9 \overset{\checkmark}{=} -1$

(ii) $2(2) - (-3) = 4 + 3 \overset{\checkmark}{=} 7$.

(i) $8^2 + 2(8) - 9^2$
$= 64 + 16 - 81 \overset{\checkmark}{=} -1$

(ii) $2(8) - 9 = 16 - 9 \overset{\checkmark}{=} 7$.

PRACTICE PROBLEMS A

A.1. $\boxed{A}$ Solve for the unknowns:

(a) $x(x - 1) = 6$

(b) $4(x - \frac{1}{2}) + 3x = 2x - \frac{1}{3}$

(c) $x^2 + 3x - 7 = x(x - 5)$

(d) $2x^2 - 9x + 1 = 0$

(e) $3x - 4 < 7x + 1$

(f) $2x^2 - 4x + 3 = 0$

(g) $3x + 4y = 12$
    $2x + 3y = 7$

(h) $x^2 + y^2 = 25$
    $x - y = -1$

(i) $3x - 4y + \ z = 20$
    $x + 7y - 3z = -35$
       $3y - 2z = -19$

A.2. Solve for the unknowns:

(a) $7x - 4y = -9$
    $2x + 6y = 1$

(b) $x^2 - 8x + 16 = 0$

(c) $x(3x - 1) = 2x^2 - 7x + 1$

(d) $\ x + \ y + z = 4$
    $2x + \ y + z = 6$
    $3x + 2y - z = 1$

(e) $7x + 8 > \frac{1}{2} - 2x$

(f) $3x - 7(1 - x) = 10x + 2$

(g) $x^2 - y^2 = 9$
    $x + 2y = 13$

Appendix B

Mathematical Induction

Often in life, when we observe that some relationship has held true up to now, we conclude that it will continue to hold in the future. For example, even people who know nothing of the physical laws governing the motion of heavenly bodies "know" that the sun will rise tomorrow morning, because it has risen every morning in the past. Such reasoning is called inductive reasoning.

Although inductive reasoning plays a very important role in the physical and natural sciences, it can lead to mistakes. Consider the expression

$$x^4 - 10x^3 + 35x^2 - 50x + 24.$$

Observe that if we substitute the number 1, 2, 3, or 4 for x, the value of the expression becomes zero. We might conclude, by inductive reasoning, that the expression above will become zero no matter what integral value we substitute for x. But when we substitute 5, the value of the expression becomes 24, and so our conclusion is false.

You might argue that our induction is based on only four observations, hardly enough to draw any conclusions.

But, what is enough? Twenty? Forty?

Consider the expression

$$x^2 + x + 41.$$

When we substitute any integer from 1 to 20 for x, the value of this expression is a prime number. (See Chapter 4 for the definition of prime numbers.) In fact, the value of the expression is a prime whenever any integer from 1 to 40 is substituted for x. Can we conclude that the value of the expression will be prime no matter what integer is substituted for x? No. As a matter of fact, when $x = 41$, the expression takes the value 1763, which is not prime.

Does this mean that, no matter how many observations we make, we still cannot conclude that a result is true in general? The answer to this question is yes—observation alone, no matter how extensive, does not prove anything. If we observe that

$$1 + 2 + 2^2 + 2^3 + \cdots + 2^n = 2^{n+1} - 1$$

for each value of n, $1 \leq n \leq 100$ or $1 \leq n \leq 1000$, it is still conceivably possible that the result will fail to hold for some larger values of n.

Since we cannot make infinitely many observations, how can we hope to prove that the result above holds for all values of n? The answer lies in the method of mathematical induction.

The principle of mathematical induction is one of the defining properties of the positive integers. Its statement is as follows:

Let S_n be a statement about a general positive integer n. If

(i) S_1 is true (that is, the statement is true when $n = 1$)
and
(ii) whenever S_k is true, then S_{k+1} is also true (that is, for each positive integer k, the truth of S_k implies the truth of S_{k+1}), then S_n is true for all positive integers n.

That is, to prove that a statement holds for all positive integers, it is enough to show that it holds for 1 and that whenever it holds for a positive integer k, it also holds for $k + 1$.

A heuristic argument justifying the principle of mathematical induction is as follows: By (i), S_1 is true. Therefore, by (ii), S_2 must also be true (since S_1 is). But then, again by (ii), S_3 must be true (since S_2 is). Etc. Hence, S_n must be true for all positive integers.

Using a similar argument, we see that it is not always necessary to start our induction at the number 1. If we wish to prove that some statement S_n holds true for all integers n greater than or equal to five, we first show that S_5 is true, and then we show that S_k implies S_{k+1} for all positive integers $k \geq 5$.

Let us now illustrate the use of the principle of mathematical induction. We use it to prove that, for every positive integer n, $1 + 2 + 2^2 + \cdots + 2^n = 2^{n+1} - 1$. We present the proof on the right below, and explain the method on the left.

| Given S_n, we wish to show that S_n is true for all positive integers. | $S_n: 1 + 2 + \cdots + 2^n = 2^{n+1} - 1$ We wish to prove that S_n is true for all positive integers. |
| We must first show that S_1 is true. | (i) $S_1: 1 + 2 = 2^2 - 1$ This is certainly true ($3 = 3$). |

We next assume that S_k is true for some general positive integer k and, using this fact, we try to show somehow (using any tools of algebra, logic, etc.) that S_{k+1} must also be true. (Note that S_{k+1} is obtained by replacing n with $k+1$ in S_n.)

(ii) Assume S_k is true for some general positive integer k:

$$1 + 2 + \cdots + 2^k = 2^{k+1} - 1 \quad (1)$$

We try to show that S_{k+1} must then also be true. (Note,

$$S_{k+1}: 1 + 2 + \cdots + 2^k + 2^{k+1}$$
$$= 2^{k+2} - 1.)$$

Add 2^{k+1} to both sides of equation (1).

$$1 + 2 + \cdots + 2^k + 2^{k+1}$$
$$= 2^{k+1} - 1 + 2^{k+1}$$
$$= 2 \cdot 2^{k+1} - 1$$
$$= 2^{k+2} - 1$$

If we succeed, then we can conclude that S_n is true for all positive integers n.

But this is just S_{k+1}. Hence, by the principle of mathematical induction, S_n is true for all positive integers n.

We conclude this appendix with one more example of an inductive proof.

To present the result that we are about to prove, we introduce the symbol " ! " (read "factorial").

By $n!$, we mean the product of all the integers from 1 to n. That is,

$$n! = n(n - 1)(n - 2) \cdot \ldots \cdot 3 \cdot 2 \cdot 1.$$

For example,

$$4! = 4 \cdot 3 \cdot 2 \cdot 1 = 24.$$

Note that

$$(n + 1)! = (n + 1)n(n - 1) \cdot \ldots \cdot 3 \cdot 2 \cdot 1 = (n + 1) \cdot n!.$$

We are now ready to state the theorem we wish to prove.

Theorem For every integer $n \geq 4$, $2^n < n!$. That is,

$$S_n: 2^n < n!, \quad \text{for} \quad n \geq 4.$$

Since 4 is the smallest value for which S_n is claimed to be true, we start the induction at $n = 4$.

$$S_4: 2^4 < 4! \quad \text{is true since } 16 < 24.$$

Assume S_k is true. That is, $2^k < k!$.
(We must show that S_{k+1} is true. That is, $2^{k+1} < (k + 1)!$.)

Note that $2 < k + 1$, since $k \geq 4$.
Hence,

$$2 \cdot 2^k < (k + 1) \cdot 2^k < (k + 1) \cdot k! \quad \text{(since } S_k \text{ is true).}$$

Thus,

$$2^{k+1} < (k + 1)!$$

and so S_{k+1} is true. The proof is now complete.

PRACTICE PROBLEMS B

B.1. [S] Prove by mathematical induction:

$$1 + 2 + 3 + \cdots + n = \frac{n(n + 1)}{2}$$

B.2. Prove by mathematical induction:

$$1^2 + 2^2 + 3^2 + \cdots + n^2 = \frac{n(n + 1)(2n + 1)}{6}$$

B.3. [S] Consider a rectangle and n straight line segments, each of which begins on one edge of the rectangle and ends on another edge. Prove, by mathematical induction, that the resulting map can be colored using exactly two colors, so that bordering regions are of different colors. (See Chapter 6, page 198, for a discussion of map coloring.)

B.4. Use mathematical induction to prove that at least $n + 1$ integral weights are needed to be able to balance any integral weight not exceeding 2^n grams. (See Chapter 5, page 152.)

B.5. Prove by mathematical induction that there are 2^n cases in the truth table for a statement involving n distinct variables: $p_1, p_2, \ldots, p_n$.

B.6. Prove by mathematical induction that n guesses will suffice to determine a previously chosen positive integer that is between 1 and $2^n - 1$, provided that, with each incorrect guess, the guesser is told whether the chosen number is larger or smaller than the guess. (See Exercise 5.16.)

Appendix C

Probability

Although there are many games whose outcomes are completely determined by the skill of the players (see Chapter Seven), there are many other games in which chance is also an important factor. It was in considering certain gambling aspects of such games that Pascal and Fermat (in the seventeenth century) laid the foundation for the formal study of probability. (Cardano had done some work in the field earlier, but it is really Pascal and Fermat who developed the subject.)

According to Eves,[1] Chevalier de Mére, a gambler of the time, proposed the following problem to Pascal in 1654. (It should be noted that this problem had been posed in the literature as early as 1494 in Pacioli's Suma.)

Two equally skilled players are playing a game of chance where, at each play, one of the players wins a point. The game is to end when one player attains n points and, as a result, wins the total "pot." If the players must stop playing before either wins (e.g., when A has a points and B has b points), how should the money be distributed between the two players?

The general study of probability goes far beyond the scope of this book. However, we present a limited view of the subject which suffices to study most games of chance as well as certain other recreational problems. To introduce this view, we consider the following sample problems. Note that Problem C.6b is really a special case of the problem of Chevalier de Mére from which probability theory stems.

SAMPLE PROBLEMS

Problem C.1

Fenwick Featherhead has trouble making decisions, so he always carries around a pair of dice (one red and one white) to help him out. Today, at lunch, Fenwick had difficulty deciding whether to have a chocolate malted or a banana split for dessert. Taking out his trusty dice, he said, "I'll roll both dice. If either shows a six, I'll buy a malted; if the sum of the dice is either seven or eleven, I'll order the banana split; if both occur, I'll have two desserts; otherwise, I'll have no dessert at all." Assuming that the dice are honest, what is the probability that
    a) Fenwick has a banana split?
    b) Fenwick has a malted?
    c) Fenwick has both a malted and a banana split?
    d) Fenwick has no dessert at all?

Problem C.2

Before the era of the internet, fax machines, etc., all of the three-digit area codes used by the telephone companies in the United States satisfied the following rules:
    a) The first digit could not be 0 or 1.
    b) The second digit had to be 0 or 1.
    c) The third digit could not be 0.
    d) The third digit could be 1 only if the second digit is 0. How many possible area codes were there?

Problem C.3

At a track meet in which six teams participated, each team entered three runners in the one-mile run. Medals were awarded to the first three finishers: Gold for first, silver for second, and bronze for third.
    a) Assuming that there were no ties, in how many different ways could the medals have been distributed among the runners?
    b) If all eighteen runners are evenly matched, what is the probability that the three medal winners all belong to the same team?

[1] : H. Eves, <u>An Introduction to the History of Mathematics</u>, Rinehart, New York, 1960.

Problem C.4

You are "the shooter" in a game of Craps, which is being played according to the following rules: You roll two dice. If the total is seven or eleven you win immediately; if the total is two, three, or twelve, you lose immediately; if the total is any other number (referred to as "your point"), you continue rolling the dice until you obtain either a seven, in which case you lose, or your point, in which case you win.

a) What is the probability that you win?

b) Before rolling the dice, you place a bet. If you lose the game, then you lose the amount of the bet; if you win the game, then the amount of your winnings is determined as follows:

If you win by rolling a seven or eleven, you win the amount of your bet (even odds);

If your point is six or eight and you win, then again you win the amount of your bet (even odds);

But, if your point is four or ten and you win, you win twice the amount of your bet (odds of two to one); And, if your point is five or nine and you win, you win three halves times your bet (odds of three to two).

What is the expected value of this game for you, if you make a bet of $1 each time you play? (I.e., on the average, how much will you win or lose per game if you play the game many times?)

Problem C.5 [based on 9, Chapter 3, Problem 8]

Captain Damon "dead-eye" Dimwitty, the famed destroyer commander of World War II, was noted for his uncanny ability to sink enemy submarines. Based on a survey of his battle experiences, it was computed that each depth charge he fired had probability 1/2 of scoring a direct hit and thus sinking its target. It was also found that when a depth charge did not result in a sinking, it nevertheless caused less severe damage to an enemy vessel with probability 1/4 and that two damaging explosions are sufficient to sink a submarine. Finally, and this is the remarkable part, Captain Dimwitty only missed his adversary 1/4 of the time.

With these facts in mind, what is the probability that Captain Dimwitty would be able to sink an enemy sub by using no more than four depth charges?

Problem C.6

Two firemen, A and B, are playing the following game. Each antes up a certain amount of money to form a pot. A die is rolled. If the result is a 1 or a 6, A gets one point; otherwise B gets one point. The first player to attain four points wins the game and takes the money that is in the pot.

a) If they wish there to be $100 in the pot, what is the amount to the nearest penny that each should put in at the beginning, so that the game is a fair game (i.e., expectation = 0)?

b) After they have been playing for a while, the fire bell rings, and they must interrupt the game. If at this point, A is leading two points to one, how should the $100 be divided equitably between them?

➢ WHAT IS PROBABILITY?

Suppose we are "conducting an experiment" (such as playing a game of chance) in which a number of different outcomes are possible. If some of these outcomes are viewed as favorable and others are viewed as unfavorable, we may speak of the probability, or likelihood, that a favorable outcome will occur. For example, if we are about to roll two dice and wonder about the likelihood that at least one of the dice will show a six[2], then any outcome in which at least one die shows a six will be counted as favorable, and all the other outcomes will be considered as unfavorable. Alternatively, if we are interested in the probability that the sum of the numbers showing on the two dice will be seven, then any outcome for which

---

[2]: When we speak of the number showing on a die, we mean on the top face of a die (unless otherwise indicated); similarly, when we speak of the sum of two or more dice, we refer to the sum of the numbers on the top faces of the dice.

the sum of the two dice is seven will be viewed as favorable, and any other outcome will be regarded as unfavorable.

We sometimes refer to any particular subset of the set, S, of possible outcomes of an experiment as an _event_. Thus, in the example above, we may speak of the event that at least one die shows a six, or of the event that the sum of the dice is seven, and so on.

PRACTICE PROBLEM SET C.A

1. A a) If we toss a penny, a nickel, and a dime, what are the eight possible outcomes (considering heads and tails assuming a coin does not land on its edge)?
    b) List the outcomes in each of the following events:
        i)   The penny comes up heads.
        ii) The nickel and the dime come up the same.
        iii) There are more heads than tails.

2. a) If we toss a penny, a nickel, a dime and a quarter, what are the sixteen possible outcomes?
    b) List the outcomes in each of the following events:
        i)   The dime comes up tails.
        ii) Exactly 15 cents comes up heads.
        iii) At least 15 cents comes up heads.
        iv) There are more heads than tails.

3. A a) As an experiment, you administer a four question true-false examination to a subject. If you consider each answer to be either right or wrong, find two different outcome sets which could be used to represent the possible outcomes of the experiment.
    b) For each of these outcome sets, list the outcomes in each of the following events:
        i)   The subject gets 100% correct.
        ii) The subject gets 75% correct.
        iii) The subject gets 50% correct.

In this chapter, we will be concerned mainly with experiments for which there are only a finite number, n, of possible outcomes, all of which are equally likely.[3]

However, let us first digress to point out that this is not the most general situation. If three horses are entered in a race, there is no reason to assume that each of the horses is equally likely to win. One horse is usually the favorite, meaning that most people estimate its likelihood of winning to be greater than that of any other horse.

In general, given the finite outcome set $S = \{s_1, s_2, ..., s_n\}$, each outcome, s_i, may be assigned a nonnegative weight w_i (i.e., $w_i > 0$ for all i), called the probability of s_i, such that the sum of these assigned weights is 1. (I.e., $\sum_{i=1}^{n} w_i = 1$.) The probability of an event E is then the sum of the probabilities of the outcomes in the set E.

What weights should be assigned to each of the outcomes is not always an easy question. If Fleet of Foot, Runs Like the Devil, and Ton of Lead are racing each other, Benny the Bookie might decide to assign them weights of 3/4, 1/4, and 0 respectively. That is, he may feel that the probability that Fleet of Foot may win is 3/4, that the probability of Runs Like the Devil winning is 1/4, and Ton of Lead has no chance at all. Hardluck Harry may disagree, and he may assign the weights 5/12, 5/12, and 1/6 respectively.

In some situations, theoretical considerations and/or empirical evidence dictate what weights we should assign. Given an ordinary coin, there is no reason to suspect that, if it is flipped, a head is any more likely to appear than is a tail; we should think that the two outcomes are equally likely. This can be borne out by experimental observation. Hence, we usually assign the two outcomes equal weights. In general, if an experiment has n possible outcomes, all of which we consider, for one reason or another, to be equally likely, then all n outcomes should be assigned the same weight. Since the sum of the weights of all

---

(3): Roughly speaking, this means that, if the experiment were repeated a very large number of times, all possible outcomes would occur with approximately the same relative frequency.

outcomes must be 1, we must assign the weight $1/n$ to each outcome. Thus, for example, if we roll a die, we usually assign a probability of 1/6 to each of the six possible outcomes. We indicate this by saying that we roll an "honest" die. Similarly, we speak of a "fair" coin, if the probability of a head and the probability of a tail are both 1/2.

Since the probability of an event should be the sum of the probabilities of the outcomes in that event, we make the following definition:

Definition C.1: If f of the possible outcomes of an experiment are favorable for a particular event E and if the experiment has n possible outcomes which are equally likely, then the probability that the event E will occur (referred to as the probability of E) is given by

$$\Pr(E) = \frac{f}{n} = \frac{\text{the number of outcomes favorable for } E}{\text{the total number of possible outcomes}} \tag{C.1}$$

Note that this is consistent with our assigning each element of the outcome set a weight of $1/n$.

For example, when we roll a die, the set of possible outcomes is $\{1,2,3,4,5,6\}$. Therefore,

  i) the probability of rolling a number greater than four is $2/6 = 1/3$ (two outcomes, 5 and 6, are favorable for this event);
  ii) the probability of rolling a number less than seven is $6/6 = 1$ (all outcomes are favorable);
  iii) the probability of rolling a number larger than six is $0/6 = 0$ (no outcomes are favorable).

Note that, in general, $0 \le f \le n$, so that

$$0 \le \frac{f}{n} = \mathrm{pr}(E) \le 1$$

The only way that $\mathrm{pr}(E)$ could be 0 is if $f = 0$; i.e., no outcomes are favorable for E, and so E is impossible[4]. Similarly, if $\mathrm{pr}(E) = 1$, then $f = n$, and so all outcomes are favorable for E. Thus, E is certain. In any case, $\mathrm{pr}(E)$ cannot be less than zero (i.e., negative) or greater than 1.

Note also, that if E is any event for which f outcomes are favorable, and if E' is the complement of E (i.e., the event that E does not occur), then $n-f$ outcomes are favorable for E'. Hence,

$$\mathrm{pr}(E') = \frac{n-f}{n} = \frac{n}{n} - \frac{f}{n} = 1 - \mathrm{pr}(E) \tag{C.2}$$

Therefore, to compute the probability of an event, it is possible (and sometimes easier) to compute the probability of its complement and then to use formula (C.2).

Referring to the example of rolling one die, the probability of obtaining a number that does not exceed four is equal to 1 minus the probability of obtaining a number greater than four; i.e., it is $1 - (1/3) = 2/3$. (Note that four of the six possible outcomes are favorable.)

PRACTICE PROBLEM SET C.B

1. [A] If a fair penny, nickel, and dime are tossed, what is the probability of each of the following events?
     a) The penny comes up heads.
     b) The nickel and the dime come up the same.
     c) There are more heads than tails.

---

[4] : Sometimes the probability of an event is so small as to make it effectively zero for most practical purposes. Nevertheless, a favorable outcome is possible, although extremely unlikely. We often say that the event in question is "virtually impossible." A real life example of this occurs when the weather bureau says that the probability of rain is zero. As we have learned from experience, this does not guarantee that it will not rain. A special case of "effectively zero" occurs when there is an infinite number of equally likely possible outcomes and only one or two (or any finite number) are favorable for a particular event E. Then the probability of that event is less than any positive number ε, and hence may be said to be zero. In such a case, E is far less likely than the weather bureaus "no chance of rain"; but, even so, as a matter of principle, E is not absolutely impossible. We may be justified in betting heavily against such an event, but, practically speaking, the stake can be lost.

d) All three coins come up heads.

2. If you roll an honest die, what is the probability of each of the following events?
    a) An odd number is obtained.
    b) An odd number greater than three is obtained.
    c) The number obtained is less than two.

3. $\boxed{A}$ If E is an event and $pr(E) = 1/8$, find $pr(E')$.

➤ ODDS

Another way of expressing a statement about the probability of an event occurring is in terms of odds. Given an equally likely outcome set, if twice as many outcomes are favorable for an event to occur as are unfavorable, we frequently say that the odds are 2 to 1 in favor of the event. In general, given an equally likely outcome set, if f outcomes are favorable and $n-f$ outcomes are unfavorable, we say the odds are f to $n-f$ in favor of the event occurring, or that they are n-f to f against the event occurring.

We frequently express this fact by writing the odds in favor of E as the ratio $f : n - f$, or as the fraction $\dfrac{f}{n-f}$. Thus, for example, odds of $4 : 2$ $(4/2)$ are the same as odds of $2 : 1$ $(2/1)$.

In terms of probability, $pr(E) = \dfrac{f}{n}$ and $pr(E') = \dfrac{n-f}{n}$, so the odds in favor of E are

$$\frac{f}{n-f} = \frac{f/n}{(n-f)/n} = \frac{pr(E)}{pr(E')} = \frac{pr(E)}{1-pr(E)}$$

This suggests the following definition:

Definition C.2: If p represents the probability that an event E occurs, then the odds in favor of E are p to $1-p$ and the odds against E are $1-p$ to p.

If we roll an honest die, the odds against getting a number greater than four are $2 : 1$, since four outcomes are unfavorable and only two are favorable. Note that the probability of getting a number greater than four is 1/3 and $\dfrac{1-(1/3)}{1/3} = \dfrac{2/3}{1/3} = \dfrac{2}{1}$.

PRACTICE PROBLEM SET C. C

1. $\boxed{A}$ If we roll an honest die,
    a)   What are the odds in favor of getting a three?
    b)   What are the odds against getting a three?

2. On an honest roulette wheel, there are thirty-eight equally likely possible outcomes: 00, 0, 1, 2, 3, 4, ..., 32, 33, 34, 35, 36.
    a)   What are the odds against getting an odd outcome?
    b)   What are the odds against getting an outcome between 13 and 24 inclusive?
    c)   What are the odds in favor of getting a particular outcome, say 2?

Usually, mention of "odds" arises in connection with betting. If we are offered odds of r to s on a bet, this means that we will have to pay $s if we lose the bet, but we will receive $r if we win. This may not reflect the "true odds" of the bet as defined above. We will return in the section on expectation to discuss how the two uses of the word "odds" are related.

➤ EQUALLY LIKELY OUTCOMES

We emphasize that formula (C.1) for calculating probability is applicable only if all possible outcomes of the experiment are equally likely.[5]

Determining a set of equally likely outcomes may sometimes require careful thought. For example, suppose we throw two honest dice and ask, "what is the probability of getting a seven?" We might reason as follows: There are eleven possible outcomes (2,3,4,5,6,7,8,9,10,11, and 12), one of which, 7, is favorable; therefore, the probability of getting a seven is 1/11. Wrong! The reasoning is faulty. The eleven possible outcomes are not equally likely; in the long run, for example, 7 will appear more often than 2 (as we will see below).

The following attempt is better, but is still faulty: There are twenty-one possible outcomes (1 and 1, 1 and 2, 1 and 3, 1 and 4, 1 and 5, 1 and 6, 2 and 2, 2 and 3, 2 and 4, 2 and 5, 2 and 6, 3 and 3, 3 and 4, 3 and 5, 3 and 6, 4 and 4, 4 and 5, 4 and 6, 5 and 5, 5 and 6, and 6 and 6), three of which are favorable (1 and 6, 2 and 5, and 3 and 4); so the probability of getting a seven is 3/21 = 1/7. Wrong again!

The mistake here is again that the twenty-one possible outcomes are not equally likely. For example, 1 and 2 will appear twice as often as 1 and 1. To see this, suppose one die is red and the other is white. Then 1 and 1 will occur only if both dice have 1, whereas 1 and 2 will occur if the red die shows 1 and the white one 2, or if the red shows 2 and the white 1.

This then leads us to the proper solution of the problem. If we consider the dice as being red and white, then there are actually 36 possible outcomes, which are equally likely. If we let (r,w) represent the outcome in which the red die shows r and the white die w, then the possible outcomes are:

$(1,1), (1,2), (1,3), (1,4), (1,5), (1,6),$
$(2,1), (2,2), (2,3), (2,4), (2,5), (2,6),$
$(3,1), (3,2), (3,3), (3,4), (3,5), (3,6),$
$(4,1), (4,2), (4,3), (4,4), (4,5), (4,6),$
$(5,1), (5,2), (5,3), (5,4), (5,5), (5,6),$
$(6,1), (6,2), (6,3), (6,4), (6,5),$ and $(6,6).$

Of these, six are favorable ($(1,6), (2,5), (3,4), (4,3), (5,2),$ and $(6,1)$). Thus, the probability of getting a seven is 6/36 = 1/6.

(Note that, now that we have found the equally likely outcome set, we know what weights we should assign if we want to use the outcome set {2,3,4,...,12}, where the outcomes represent the sum of the dice. The outcome 2 should be assigned a weight of 1/36, since one of the thirty-six possible outcomes results in a sum of 2; the outcome 3 should be assigned the weight 2/36; etc.)

PRACTICE PROBLEM SET C.D

1. $\boxed{A}$ a) Toss three pennies; each lands heads or tails. What equally likely outcome set should we consider if we are interested in the probability that exactly two of the coins come up heads? What is the probability in question?

   b) What weights should be assigned if we wish to use the outcome set {0 heads, 1 head, 2 heads, 3 heads}?

---

[5]: However, probability theory can also handle the case of outcomes that are not equally likely, as in the example of the horse race. Although our discussion of the formulas and theorems presented in the remainder of the chapter will primarily deal with the equally likely outcome case, all formulas and theorems of the chapter hold for a more general discussion of probability (unless otherwise indicated as in the case of (C.1)). They may be applied in this broader context to some of the problems and exercises of the chapter, provided that we are told the underlying weights or probabilities to use. (See, for example, the solution of Problem C.5, later in the chapter.)

2. Two possible outcome sets for Problem 3 of Practice Problem Set 9.1 are: {(w,w,w,w), (w,w,w,r),...,
    (r,r,r,r) }, where w stands for "wrong" and r stands for "right", and {0,1,2,3,4}, where each outcome
    represents the number of correct answers.
    a) Assuming that the subject guesses randomly at each answer, which of the above sets contains
        equally likely outcomes?
    b) What weights should be assigned to the outcomes in the other outcome set?

3. [A] Randomly draw a card from an ordinary 52 cards deck of playing cards. What equally likely
    outcome set should we use if we are interested in the probability that the chosen card is a picture card?
    What is the probability in question?

The example above points out one of the difficulties in computing probabilities of the type we
consider – expressing the results of the experiment in terms of outcomes that are equally likely.

Once this has been done, the next difficulty is counting the number of possible outcomes and the
number of favorable ones.

You now should be able to solve Problem C.1. Try it again if you haven't already solved it.

SOLUTION TO PROBLEM C.1

In an experiment such as the one considered above, we may actually list all possible outcomes and
count how many there are and which are favorable.

For example, in Problem C.1, if E_a is the event that a seven or eleven occurs, then
$$E_a = \{(1,6),(2,5),(3,4),(4,3),(5,2),(5,6),(6,1),(6,5)\} \text{ , so pr}(E_a) = 8/36 = 2/9,$$ and Fenwick
has a banana split with probability 2/9. Similarly, if E_b is the event that at least one die shows a six, then
$$E_b = \{(1,6),(2,6),(3,6),(4,6),(5,6),(6,1),(6,2),(6,3),(6,4),(6,5),(6,6)\}.$$
Thus pr(E_b) = 11/36, and so Fenwick has a malted with probability 11/36.

The probability that Fenwick has both a malted and a banana split can be computed in a similar
manner:
$$\text{pr}(E_c) = \text{pr}(\{(1,6),(5,6),(6,1),(6,5)\}) = 4/36 = 1/9 \text{ .}$$
To compute the probability that Fenwick has no dessert, it is easier to first compute the probability
of the complementary event (that he has at least one dessert) and then to subtract from 1.
$$E_d' = \{(1,6),(2,5),(2,6),(3,4),(3,6),(4,3),(4,6),(5,2),(5,6),(6,1),(6,2),(6,3),(6,4),(6,5),(6,6)\} \text{ ,}$$
so pr(E_d') = 15/36 = 5/12, and pr(E_d) = 1 – 5/12 = 7/12.

➤ COUNTING: ADDITIVITY

In many experiments, there are just too many possibilities to list, and so we must develop other
techniques for counting. One helpful technique is to partition the outcomes into disjoint cases (i.e., cases
which have nothing in common).

Reconsider Problem C.1a. As we saw above, there are 36 possible outcomes when two dice are
thrown. We must count how many of these outcomes result in a seven or eleven. The answer is eight – six
which result in a 7 (as we saw above) and two which result in an 11 ((5,6) and (6,5)). In other words, the
event E_a, that a seven or an eleven occurs, may be broken up into two disjoint events – E_1, that a seven
occurs, and E_2, that an eleven occurs. The number f of outcomes favorable for E_a is the sum of the
number, f_1, of outcomes favorable for E_1, and the number, f_2, of outcomes favorable for E_2. In terms of
probabilities, pr(E_a) = f/n = $(f_1+f_2)/n$ = $(f_1/n) + (f_2/n)$ = pr(E_1) + pr(E_2). This holds whenever E can be broken up into disjoint events E_1 and E_2.
We state this result as a theorem.

Theorem C.1: If the event E is made up of two disjoint events E_1 and E_2, then
$\mathrm{pr}(E) = \mathrm{pr}(E_1) + \mathrm{pr}(E_2)$.[6]

$$(C.3)$$

Thus, in Problem C.1a,
$$\mathrm{pr}(E_a) = 6/36 + 2/36 = 8/36 = 2/9 \ .$$

We emphasize the importance in formula (C.3) of the events E_1 and E_2 being disjoint. In Problem C.lb, if E_1 is the event that the red die shows six, and E_2 is the event that the white die shows six, then we might be tempted to say that $\mathrm{pr}(E_b) = \mathrm{pr}(E_1) + \mathrm{pr}(E_2)$. Wrong!
$$E_1 = \{(6, 1),(6,2),(6,3),(6,4),(6,5),(6,6)\} \ ,$$

so
$$\mathrm{pr}(E_1) = 6/36 = 1/6;$$

and
$$E_2 = \{(1,6),(2,6),(3,6),(4,6),(5,6),(6,6)\},$$

so
$$\mathrm{pr}(E_2) = 6/36 = 1/6 \ ;$$

however, as we saw above,
$$\mathrm{pr}(E_b) = 11/36.$$
Obviously, $\mathrm{pr}(E_b) = 11/36 \neq 1/6 + 1/6 = \mathrm{pr}(E_1) + \mathrm{pr}(E_2)$. The reason for the error is that E_1 and E_2 are not disjoint; the outcome $(6,6)$ belongs to both and hence is counted twice in $\mathrm{pr}(E_1) + \mathrm{pr}(E_2)$.

In the case that E_1 and E_2 are not disjoint, we can compensate for our error by subtracting off outcomes that are counted twice. That is, we have the following theorem.

Theorem C.2: If E_1 and E_2 are events associated with the same experiment, and if E is the event consisting of those outcomes which belong to E_1 or E_2 (or both), then
$$\mathrm{pr}(E) = \mathrm{pr}(E_1) + \mathrm{pr}(E_2) - \mathrm{pr}(E_1 \cap E_2)$$
where $E_1 \cap E_2$ (called the intersection of E_1 and E_2) consists of those outcomes which are in both E_1 and E_2.

In Problem C.1b, $\mathrm{pr}(E_1) = 1/6$, $\mathrm{pr}(E_2) = 1/6$, and $\mathrm{pr}(E_1 \cap E_2) = 1/36$, so
$$\mathrm{pr}(E_b) = 1/6 + 1/6 - 1/36 = 11/36 \ .$$
If we are careful, we can always choose the manner in which we break up an event so that the smaller events are disjoint. For example, in Problem C.1b, let E_1 be the event that the red die shows six but the white die does not; let E_2 be the event that the white die shows six but the red die does not; and let E_3 be the event that both dice show six. Then E_1, E_2, and E_3 are pairwise disjoint (i.e., no two have any outcomes in common) and together form E. There are five outcomes in E_1 $((6,1),(6,2),(6,3),(6,4)$ and $(6,5)$), five outcomes in E_2 ($(1,6),(2,6),(3,6),(4,6)$ and $(5,6)$), and one in E_3 ($(6,6)$). Thus
$$\mathrm{pr}(E_b) = \mathrm{pr}(E_1) + \mathrm{pr}(E_2) + \mathrm{pr}(E_3) = 5/36 + 5/36 + 1/36 = 11/36.$$

PRACTICE PROBLEM SET C.E

1. A If a card is randomly selected from an ordinary fifty-two card playing deck, what is the probability of each of the following events?
   a) The selected card is red.
   b) The selected card is a picture card.
   c) The selected card is a red picture card.
   d) The selected card is red, but not a picture card.
   e) The selected card is either red or a picture card.
   Obtain the answer to e) in two different ways.

---

[6]: We remind you once more that, although Theorem C.1 and the remaining theorems of this chapter are only being presented for the equally likely outcome definition of probability, corresponding theorems hold for the more general definition of probability.

2. If four fair coins (one penny, two nickels and a dime) are tossed, what is the probability that
   a) the dime will land on heads?
   b) both nickels will land on heads?
   c) the value of the coins which land on heads is at least ten cents?

➤ THE MULTIPLICATION PRINCIPLE

Another helpful technique of counting is based on the Multiplication Principle, which was discussed in Chapter One. We restate it here to refresh your memory.

Theorem C.3 (The Multiplication Principle): Suppose that a task can be broken up into two steps, the first of which can be done in m ways and the second of which can be done in n ways (regardless of the result of the first step). The original task can be done in mn ways.

In the experiment of rolling two dice, we listed all possible outcomes and found that there were 36 in all. An alternate route to this same result would be the following: Suppose that, instead of rolling both dice simultaneously, we first rolled one and then rolled the other. There are six possibilities for the number showing on the first die, and then there are six possibilities for the number showing on the second. Hence, by the Multiplication Principle, there are 6 x 6 = 36 possible outcomes in all.

PRACTICE PROBLEM SET C.F

1. $\boxed{\text{A}}$ A box of anagram letters contains four S's, five T's, ten A's and four R's.
   a) In how many ways may we select and arrange letters from the box to spell the word STAR?
   b) In how many ways may we select and arrange letters from the box to spell the word START?

2. The library shelves currently contain six novels by Charles Dickens, three by Robert Louis Stevenson, and four by Mark Twain. If you wish to read one book by each of these authors, in how many different ways could you make the selection?

We are ready to consider Problem C.2. Try it again if you haven't already solved it.

SOLUTION TO PROBLEM C.2

There are eight possibilities (2,3,4,5,6,7,8, or 9) for the first digit, two possibilities (0 or 1) for the second digit, and nine possibilities (1,2,3,4,5,6,7,8, or 9) for the third digit. By the Multiplication Principle, we should expect 8·2·9 = 144 possible area codes. However, we must be careful. Rule (d) of the problem restricts our freedom of choice for the third digit; specifically, for some choices of the second digit, only eight choices (2,3,4,5,6,7,8, or 9) are possible for the third digit. We cannot use the Multiplication Principle directly; instead, we proceed in either of two ways.

We can divide possible area codes into disjoint sets – those with middle digit 0 and those with middle digit 1. By the Multiplication Principle, there are 8·1·9 = 72 possibilities in the first class and 8·1·8 = 64 in the second. By additivity, there are 72 + 64 = 136 area codes possible.

As another approach, of the 144 possibilities originally considered, only those which end in 11 must be eliminated. As there are eight such numbers $(211, 311, ..., 911)$, we have $(144 - 8 = 136)$ area codes.

➤ PERMUTATIONS

An important application of the Multiplication Principle occurs when we have a collection of objects and we wish to order them in some way. For example, we might want to assign post positions in a horse race, or we might wish to select Ms. America, 1st runner up, 2nd runner up, 3rd runner up and 4th runner up from a group of five finalists.

Given n objects that we wish to order, there are n ways in which to select the first. Once this selection is made, regardless of which object was chosen, there remain n - 1 possible choices for the second. After first and second are chosen, there remain n - 2 possible choices for the third. And so on. By

the time only two objects remain, there are two choices for next to last, and that leaves only one choice for last. Thus, by successive applications of the Multiplication Principle, there are $n(n-1)(n-2)...3 \cdot 2 \cdot 1$ possible ways in which the n objects can be ordered. We introduce the symbol ! (read "factorial") to help abbreviate our notation. Thus, for any positive integer n, $n! = n(n-1)(n-2)...3 \cdot 2 \cdot 1$.
For example,

$$5! = 5 \cdot 4 \cdot 3 \cdot 2 \cdot 1 = 120 ;$$
$$6! = 6 \cdot 5 \cdot 4 \cdot 3 \cdot 2 \cdot 1 = 6(5!) = 720.$$

For reasons which will become apparent later, we define 0! to be 1. In general, $n! = n \cdot (n-1)!$

PRACTICE PROBLEM SET C.G

1. $\boxed{\text{A}}$ Compute each of the following:

   a) 7!   b) 8!   c) $\dfrac{8!}{6!}$ d) $\dfrac{8!}{6!2!}$

2. Compute each of the following:

   a) 9!   b) $\dfrac{9!}{0!}$ c) $\dfrac{9!}{6!}$ d) $\dfrac{9!}{6!3!}$

Often, it is not necessary to order all objects in a set, but rather only some of them. For example, we may wish to select a president, vice president, treasurer, and secretary from among the membership of some organization. We are interested not only in selecting four individuals, but we must order them too – i.e., which one should be president, etc.

An arrangement or ordering of r objects is called a <u>permutation</u>. If we wish to arrange r out of n objects, then the number of ways in which this can be done is referred to as the <u>number of permutations of n objects taken r at a time</u>, and this number is denoted by $_nP_r$ or $P(n,r)$. By the Multiplication Principle, we have the following formula for computing $_nP_r$.

Theorem C.4: For any nonnegative integers $r, n, r < n$, $_nP_r = n(n-1)...(n-r+1) = \dfrac{n!}{(n-r)!}$.   (C.4)

That is,	For example,
The number of ways to select r objects in a specific order out of n is	The number of ways to select a president, vice president, and secretary from among twenty peo
$_nP_r$	$_{20}P_3$
The first object may be chosen in n ways; the second, in $n-1$ ways; the third, in $n-2$ ways, ... The rth object in $n-(r-1) = n-r+1$ ways	The president may be chosen in 20 ways; the vic president, in 19 ways; the secretary, in 18 ways.
	(Note the third person above can be chosen in $20-(3-1) = 20-2 = 18$ ways.

Therefore, by the Multiplication Principle

$_nP_r = n(n-1)(n-2)...(n-r+1)$	$_{20}P_3 = 20 \cdot 19 \cdot 18$

$_nP_r = n(n-1)(n-2)...(n-r+1) =$

$$\frac{n(n-1)...(n-r+1)(n-r)...2 \cdot 1}{(n-r)...2 \cdot 1}$$

$$= \frac{n!}{(n-r)!}$$

That is,

$$_nP_r = \frac{n!}{(n-r)!}$$

Note that

$$_{20}P_3 = 20 \cdot 19 \cdot 18$$

$$= \frac{20 \cdot 19 \cdot 18 \cdot 17 \cdot 16...2 \cdot 1}{17 \cdot 16...2 \cdot 1}$$

$$= \frac{20!}{17!}$$

For $r = n$, formula (C.4) becomes $_nP_n = \frac{n!}{0!} = n!$ which is consistent with our definition of 0!; i.e., we can select an ordered set of n objects from n given objects in $n!$ ways.

PRACTICE PROBLEM SET C.H

1. [A] Compute each of the following:

    a) $_{10}P_3$ b) $_{10}P_5$ c) $_{10}P_{10}$ d) $_{10}P_1$ e) $_{10}P_0$

2. Compute each of the following:

    a) $_8P_3$ b) $_8P_4$ c) $_8P_5$ d) $_8P_1$ e) $_8P_0$

We are now ready to consider Problem C.3. If you have not already solved it, try it again now.

SOLUTION OF PROBLEM C.3

    Part a) of the problem asks in how many ways may we select three runners, in a given order, from among the eighteen. This is just the number of permutations of eighteen objects taken three at a time. Thus, the medals may be awarded in $_{18}P_3 = \frac{18!}{15!} = 18 \cdot 17 \cdot 16 = 4896$ ways. That is, there are eighteen possibilities for the recipient of the gold medal; once first place is determined, there remain seventeen possibilities for the recipient of the silver medal, and so on.

    In part b) of the problem, we must determine how many of the 4896 possible outcomes are favorable for the event in question. There are still eighteen possibilities for the winner of the race; but, once the winner is determined, there are only two possibilities (the winner's teammates) for the second place finisher. Once first and second are determined, there remains only one possibility for third. Hence, by the Multiplication Principle, only $18 \cdot 2 \cdot 1 = 36$ of the possible outcomes are favorable.

    Thus, the probability that one team will sweep the event is $36/4896 = 1/136$.

PRACTICE PROBLEM SET C.I

1. [A] A box of anagram letters contains seven B's, ten O's, four K's, twelve E's, six P's, and eight R's. In how many ways may we select and arrange letters from the box to spell the word BOOKKEEPER?

2. A popular betting event at some race tracks is the trifecta, for which bettors must try to select the first three horses in a race in their exact order of finish. In a race in which eight horses are entered, how many different trifecta tickets are possible?

3. [A] Five slips of paper, with the numbers 1,2,3,4, and 5 respectively, are placed in a hat. A four-digit number is formed by randomly selecting one of the slips for the first digit, then selecting one of the

remaining slips for the second digit, and so on. What is the probability that the resulting number will exceed 2510?

➢ COMBINATIONS

So far, we have considered the number of ways in which we can select an ordered set of r objects from a given set of n objects. Sometimes, we merely wish to select r objects from the given n objects, without ordering them. For example, we want to select a committee of r people. In this case, it does not matter who is first, who is second, and so on; all that does matter is who the r people selected are.

The number of ways in which we can select r objects from a given set of n objects, without regard to order, is called the <u>number of combinations of n objects taken r at a time</u>. It is denoted by $_nC_r, C(n,r)$ or $\binom{n}{r}$. (Although the notation $_nC_r$ more closely parallels our notation for the number of permutations, the notation $\binom{n}{r}$ is more widely used in applications, and so we use it in the remainder of this chapter.)

To compute this number, observe that, in selecting an ordered set of r objects, we could first just select r objects and then order them. By the Multiplication Principle, the number of ways in which we may select an ordered set of r objects from n given objects is equal to the number of ways in which we may select r objects (without order) multiplied by the number of ways in which a selected set of r objects may be ordered. In other words,

$$_nP_r = \binom{n}{r}_rP_r$$

Since $_nP_r$ and $_rP_r$ are known, we get

$$\frac{n!}{(n-r)!} = \binom{n}{r} \cdot r!$$

or, equivalently, we have the following theorem.

Theorem C.5: For any nonnegative integers $r, n, r \leq n$,

$$\binom{n}{r} = \frac{n!}{(n-r)!\,r!}$$

For example, to select a president, vice president, and secretary from among 20 people, we could first select the three people (without regard to order) and then decide which of the three should be president, which should be vice president and which should be secretary.

The selection of three people can be done in $\binom{20}{3}$ ways.

And the ordering of the three selected people can be done in $_3P_3 = 3!$ ways.

Since, as we saw above, the president, vice president and secretary can be selected in $_{20}P_3$ ways, we have

$$_{20}P_3 = \binom{20}{3}_3P_3;$$

$$\frac{20!}{17!} = \binom{20}{3}3!.$$

Therefore,

$$\binom{20}{3} = \frac{20!}{17!\,3!}.$$

Note that $1 = \binom{n}{n} = \dfrac{n!}{n!\,0!}$ is again consistent with our definition of $0!$; there is only one way to choose n objects from a set of n objects – take them all.

Note also that

$$\binom{n}{n-r} = \frac{n!}{\left(n-(n-r)\right)!\,(n-r)!} = \frac{n!}{r!\,(n-r)!} = \frac{n!}{(n-r)!\,r!} = \binom{n}{r}$$

This is not surprising, as, whenever we select r objects from a given set of n objects, we simultaneously distinguish n-r objects – i.e., those objects not selected.

PRACTICE PROBLEM SET C.J

1.   Ⓐ Compute each of the following:

   a) $\binom{10}{3}$ b) $\binom{10}{7}$ c) $\binom{10}{10}$ d) $\binom{10}{1}$ e) $\binom{10}{0}$ f) $\binom{10}{5}$

2.   Compute each of the following:

   a) $\binom{8}{3}$ b) $\binom{8}{5}$ c) $\binom{8}{8}$ d) $\binom{8}{7}$ e) $\binom{8}{0}$

PROBLEM C.3b REVISITED

Let us reconsider Problem C.3b. Since the order of the first three finishers does not matter, but only whether or not they belong to the same team does, we may attack the problem using combinations. There are $\binom{18}{3} = \dfrac{18!}{15!\,3!}$ ways of selecting three medal winners (disregarding order) from the eighteen contestants. Of these, only six have all three winners on the same team (i.e., all three are on team A, or all three are on team B, etc.). Thus, the probability that all three winners are on the same team is

$$\frac{6}{\binom{18}{3}} = \frac{6}{\dfrac{18!}{3!\,15!}} = \frac{6 \cdot 3!\,15!}{18!} = \frac{6 \cdot 6}{18 \cdot 17 \cdot 16} = \frac{1}{136}$$

(Note that this is the same answer we obtained using permutations.)

As in Problem C.3b, it is sometimes (but not always) the case that a probability problem may be attacked using either permutations or combinations, depending on the point of view we adopt. If the problem may be viewed in such a way that order does not matter, we may use combinations; if we can also consider the problem so that order is relevant, then we may also use permutations. Whenever this choice of approaches exists, it is important to be consistent in counting total possible outcomes and favorable outcomes. We consider order in counting favorable outcomes if and only if we consider order in counting all outcomes.

PRACTICE PROBLEM SET C.K

1.   Ⓐ A box of anagram letters contains seven B's, ten O's, four K's, twelve E's, six P's, and eight R's. Letters are selected from the box and arranged to spell the word BOOKKEEPER. These letters are then returned to the box and thoroughly mixed. If we again select letters to spell the word BOOKKEEPER, what is the probability that we will choose exactly the same letters as we did before (although not necessarily in the same positions)?

2. a) Ten basketball players decided to play a game. In how many ways could they possibly form two teams of five players each?

b) Twelve basketball players decided to play a game. In how many ways could they possibly form two teams so that each team has at least five players and so that every player is on one of the two teams?

> **THE BINOMIAL THEOREM**

An important application of combinations is the Binomial Theorem, which deals with the expansion of $(x+y)^n$. When several polynomials are multiplied, each term of each polynomial is multiplied by each term of every other polynomial and then the results are added. Thus, in particular,

$$(x + y)^2 = (x + y)(x + y) = xx + xy + yx + yy = x^2 + 2xy + y^2$$

and

$$(x + y)^3 = (x + y)(x + y)(x + y) = x^3 + x^2y + xyx + xy^2 + yx^2 + yxy + y^2x + y^3$$
$$= x^3 + 3x^2y + 3xy^2 + y^3$$

Note that each term of the above products is of the form $cx^r y^{n-r}$ where c is a constant (called the coefficient of the term), and n is the power to which $x + y$ is being raised.

Note further that the coefficient of the $x^r y^{n-r}$ term is always $\binom{n}{r}$, the number of combinations of n things taken r at a time, as that is the number of different ways in which r x's may be chosen from n factors.

For example, the $x^2 y$ term in $(x + y)^3$ comes from choosing two x's and one y from the three factors $(x + y)$, $(x + y)$, and $(x + y)$. This may be done in $\binom{n}{r} = 3$ ways. I.e.,

$$(x + y)\ (x + y)\ (x + y) \qquad x^2 y$$
$$(x + y)\ (x + y)\ (x + y) \qquad xyx$$
$$(x + y)\ (x + y)\ (x + y) \qquad yx^2.$$

Similarly,

$$xy^2 + yxy + y^2x = 3xy^2 ;$$

whereas the x^3 and y^3 terms may each be obtained in only one way:

$$(x + y)(x + y)(x + y) \qquad x^3$$
$$(x + y)(x + y)(x + y) \qquad y^3$$

This same result holds true in general; that is, we have the following theorem:

Theorem C.6 (The Binomial Theorem): For any positive integer n,

$$\left(x + y\right)^n = \binom{n}{n}x^n + \binom{n}{n-1}x^{n-1}y + \binom{n}{n-2}x^{n-2}y^2 + ... + \binom{n}{1}xy^{n-1} + \binom{n}{0}y^n$$

(We sometimes abbreviate the right hand side of this equation by writing

$$\sum_{i=0}^{n}\binom{n}{i}x^i y^{n-i}$$

which is read "the sum, as i goes from 0 to n of, $\binom{n}{i}x^i y^{n-i}$.")

Since the combinatorial symbols are the coefficients in the Binomial Theorem, they are often referred to as the <u>binomial coefficients</u>.

One important consequence of Theorem C.6 which is sometimes useful in counting puzzles is obtained by setting $x = y = 1$ in the theorem. This gives

$$2^n = (1+1)^n = \sum_{i=0}^{n}\binom{n}{i}1^i 1^{n-i} = \sum_{i=0}^{n}\binom{n}{i}$$

In other words, we have the following corollary:

Corollary C.1: For any positive integer n

$$\binom{n}{0}+\binom{n}{1}+...+\binom{n}{n} = \sum_{i=0}^{n}\binom{n}{i} = 2^n$$

We can also obtain corollary C.1 in another way. For example: by the Multiplication Principle a student can answer a four question true false exam in $2\cdot 2\cdot 2\cdot 2 = 2^4 = 16$ ways.

We could also count in the following manner: count the number of ways in which none of the student's answers are 'true,' + the number of ways in which one of his answers is 'true' + the number of ways in which two of his answers are 'true' and so on. This gives

$$\binom{4}{0}+\binom{4}{1}+\binom{4}{2}+\binom{4}{3}+\binom{4}{4},$$

so that

$$\binom{4}{0}+\binom{4}{1}+\binom{4}{2}+\binom{4}{3}+\binom{4}{4} = 2^4.$$

A similar argument holds for general n.

PRACTICE PROBLEM SET C.L

1. $\boxed{A}$ Expand each of the following:

    a) $(x+y)^4$ b) $(x+y)^5$ c) $(x-y)^3$

2. Expand each of the following:

    a) $(x-y)^4$ b) $(1-x)^3$ c) $(2+y)^5$

3. $\boxed{A}$ Evaluate $\displaystyle\sum_{i=0}^{n}(-1)^i\binom{n}{i}$

➢ INDEPENDENCE AND CONDITIONAL PROBABILITY

At times, given two experiments (or steps in a procedure), the result of one affects the outcome of the other. Other times, this is not the case.

Definition C.3: When the outcome of each experiment (or step) of a two part experiment is not in any way influenced by the outcome of the other, the two experiments are said to be independent.

For example, given a hat containing ten slips of paper numbered one to ten, consider the following two step procedure.

Step one: Choose a number from the hat. Do not replace it.
Step two: Choose a second number from the hat.

Clearly the result of step one affects the outcome of step two. That is, if, for example, 6 is chosen in step one, then 6 cannot possibly be chosen in step two. Thus experiment two is dependent on experiment one.

If, on the other hand, the slip of paper chosen in step one is replaced in the hat before step two is performed, then the outcome of step two will in no way be influenced by the result of step one, and so the two experiments will be independent.

Just as we may speak of independent or dependent experiments, we may also speak of two events associated with the same experiment as being dependent or independent.

Definition C.4: Two events, are independent if the occurrence of one of them does not alter the probability that the other will occur.

For example, if we view the two step procedure described above (without replacing the slip chosen in step one) as a single experiment, and if E_1 is the event that 6 is chosen in the first step, then $\mathrm{pr}(E_1) = 1/10$, as there are ten equally likely possible outcomes, only one of which is favorable. Similarly, if E_2 is the event that 6 is chosen in the second step, and E_3 is the event that 7 is chosen in the second step, then, before the experiment is conducted, $\mathrm{pr}(E_2) = \mathrm{pr}(E_3) = 1/10$. (Again each of the ten numbers 1,2,3,...,10 has the same chance as each of the others of being selected in step two.) However, once E_1 has occurred, the set of possible outcomes for step two is no longer the set of ten equally likely outcomes that we used in step one. The outcome set of step two is now $\{1,2,3,4,5,7,8,9,10\}$, consisting of nine equiprobable cases; whence $\mathrm{pr}(E_3) = 1/9$, and $\mathrm{pr}(E_2) = 0$ (since E_2 cannot possibly occur). Thus the supposition that E_1 will occur alters the probability of E_2 occurring and also changes the probability of E_3 occurring. We say that E_1 and E_2 are dependent events, as are the events E_1 and E_3 .

On the other hand, if the experiment is conducted with replacement (i.e., the slip chosen in step one is replaced in the hat before step two is performed), the outcome set of step two is exactly the same as the outcome set of step one, and so $\mathrm{pr}(E_2) = \mathrm{pr}(E_3) = 1/10$, even if we suppose that E_1 occurs. Hence, in this case, E_1 and E_2 are independent events, as are E_1 and E_3.

In general, in performing an experiment, the probability that a particular event E occurs may be altered if we acquire knowledge (or if we suppose) that some other event F does occur.

Definition C.5: The new probability described above is called the <u>conditional probability of E given F</u> and is denoted by $\mathrm{pr}(E|F)$.

Thus, in the example above (without replacing the slip), $\mathrm{pr}(E_2|\,E_1) = 0$ and $\mathrm{pr}(E_3|\,E_1) = 1/9$.

Consider another example. In rolling two dice, the probability of getting a total of eleven is, as we saw in Problem C.1a, $2/36 = 1/18$.

Suppose, however, that we have already rolled the dice but do not yet know the full result, as the white die fell off the table. If the die that is still on the table shows a 1, we would no longer think that the probability that our total is eleven is 1/18, as, no matter what the white die shows, our total cannot possibly be eleven. (The new outcome set is $\{(1,1),(1,2), (1,3),(1,4),(1,5),(1,6)\}$.) That is, if E_1 is the event that the total is eleven, and E_2 is the event that the red die shows 1, then $\mathrm{pr}(E_1|\,E_2) = 0$.

Similarly, if the die remaining on the table shows a 5, then we would say that the probability that our total is eleven is 1/6, as exactly one of the six equally likely possible outcomes $\{(5,1), (5,2), (5,3), (5,4), (5,5)$ and $(5,6)\}$ yields a total of eleven. Thus, $\mathrm{pr}(E_1|\,E_3) = 1/6$, where E_3 is the event that the die remaining on the table shows a 5.

The technique used for computing $\mathrm{pr}(E_1|\,E_3)$ in the example above may be used to derive a general formula for computing $\mathrm{pr}(E|\,F)$. Knowing (or supposing) that F occurs alters the possible outcomes of the experiment that need be considered. Outcomes for which F does not occur are no longer possible, and, hence, may be discarded. The outcomes that remain are exactly those which are favorable for F. Of these, only outcomes which are also favorable for E result in E occurring. Hence, in the equiprobable case, we have the following (compare the formula with the computations in the examples above):

Theorem C.7: $\mathrm{pr}(E|\,F) = e/f$, where f is the number of possible (equally likely) outcomes favorable for F, and e is the number of these which are favorable for E – i.e., the number of outcomes which are favorable for both E and F.

If the original experiment has n possible outcomes, then $\mathrm{pr}(F)$ would be f/n . Furthermore, if $E \cap F$ represents the event that both E and F occur, then $\mathrm{pr}(E \cap F) = e/n$. Hence, we obtain $\mathrm{pr}(E \cap F)/\mathrm{pr}(F) = (e/n)/(f/n) = e/f = \mathrm{pr}(E|\,F)$.

More generally, we have the following theorem:

Theorem C.8: For any events E and F associated with the same experiment
$$\mathrm{pr}(E|\,F) = \mathrm{pr}(E \cap F)/\mathrm{pr}(F) \tag{C.5}$$
or, equivalently,
$$\mathrm{pr}(E \cap F) = \mathrm{pr}(F) \cdot \mathrm{pr}(E|F) \tag{C.6}$$
(Note that formula (C.5) makes no sense if $\mathrm{pr}(F) = 0$, as we cannot divide by 0.)

Reconsidering the example of the slips in the hat, without replacement,
$$\mathrm{pr}(E_3 \cap E_1) = \mathrm{pr}(E_1)\mathrm{pr}(E_3|\,E_1) = (1/10)\,(1/9) = 1/90$$
and, with replacement,
$$\mathrm{pr}(E_3 \cap E_1) = \mathrm{pr}(E_1)\mathrm{pr}(E_3|\,E_1) = (1/10)(1/10) = 1/100$$

PRACTICE PROBLEM SET C.M

1. $\boxed{A}$ Two honest dice are rolled. Find the conditional probability that the resulting sum is seven, given that neither die showed a 3.

2. Two honest dice are rolled, one red and one white. Find the conditional probability that the resulting sum is seven, given that the number showing on the red die is odd.

The definition of independence may be rephrased in terms of conditional probability: E and F are independent if the supposition that F occurs does not alter the probability that E will occur, i.e., if $\mathrm{pr}(E|F) = \mathrm{pr}(E)$. Substituting this into (C.6), we obtain the following:

Theorem C.9: If E and F are independent events associated with the same experiment, then
$$\mathrm{pr}(E \cap F) = \mathrm{pr}(E) \cdot \mathrm{pr}(F). \tag{C.7}$$

Thus, in the example above, with replacement,
$$\mathrm{pr}(E_3|\,E_1) = \mathrm{pr}(E_3) = 1/10$$
and
$$\mathrm{pr}(E_3 \cap E_1) = \mathrm{pr}(E_3) \cdot \mathrm{pr}(E_1) = (1/10)(1/10) = 1/100,$$
as we already discovered.

Equation (C.7) is frequently taken as the definition of independence. Note that the equation is symmetric with respect to E and F. I.e.,
$$\mathrm{pr}(E \cap F) = \mathrm{pr}(E) \cdot \mathrm{pr}(F) = \mathrm{pr}(F) \cdot \mathrm{pr}(E) = \mathrm{pr}(F \cap E).$$

PRACTICE PROBLEM SET C.N

1. $\boxed{A}$ Roll an honest die. Let E_1, E_2, E_3, and E_4, respectively be the events: the result is even, the result is greater than three, the result is odd, the result is greater than four.
   Which of the following pairs of events are independent?
   a) E_1 and E_2 b) E_1 and E_3 c) E_1 and E_4 d) E_2 and E_4

2. Select a card at random from an ordinary 52 card deck of playing cards. Let E_1, E_2, E_3, and E_4 respectively be the events: the card is a picture card, the card is a king, the card is black, the card is a diamond.
   Which of the following pairs of events are independent?
   a) E_1 and E_2 b) E_1 and E_3 c) E_1 and E_4 d) E_2 and E_4 e) E_3 and E_4.

We are now ready to consider Problem C.4a. If you have not already solved it, try it again now.

SOLUTION OF PROBLEM C.4a – THE GAME OF CRAPS

We begin by listing all the ways in which the shooter can win.

E_1: He rolls 7 or 11.
E_2: He rolls 4 and then rolls another 4 before rolling 7.
E_3: He rolls 5 and then rolls another 5 before rolling 7.
E_4: He rolls 6 and then rolls another 6 before rolling 7.
E_5: He rolls 8 and then rolls another 8 before rolling 7.
E_6: He rolls 9 and then rolls another 9 before rolling 7.
E_7: He rolls 10 and then rolls another 10 before rolling 7.

Since these are pairwise disjoint, the probability that the shooter will win is just the sum of the probabilities of the above events:

$$\text{pr(shooter wins)} = \text{pr}(E_1) + \text{pr}(E_2) + \ldots + \text{pr}(E_7) \ .$$

We know from Problem C.1 that $\text{pr}(E_1) = 8/36 = 2/9$.

In order for E_2 to happen, two independent events must both occur – the shooter must roll 4 on his first roll (call this event E_2^*) and, thereafter, the shooter must roll 4 before he rolls 7 (call this event E_2^{**}). (These events are clearly independent, as the first roll of the dice in no way affects later rolls.) Thus, $E_2 = E_2^* \cap E_2^{**}$, and so, by formula (C.6), $\text{pr}(E_2) = \text{pr}(E_2^*) \cdot \text{pr}(E_2^{**})$.

$\text{pr}(E_2^*)$ is easily seen to be $3/36 = 1/12$.

To compute $\text{pr}(E_2^{**})$, we may restrict our attention to the cases that either 4 or 7 occurs (as all other cases are irrelevant to E_2, since any other roll would be disregarded and the roll would be redone). That is, $\text{pr}(E_2^{**})$ is the conditional probability that 4 occurs, given that either 4 or 7 has occurred. Therefore, the relevant outcome set is $\{(1,3),(2,2),(3,1),(1,6),(2,5),(3,4),(4,3),(5,2),(6,1)\}$, and, since three of these give a total of four,

$$\text{pr}(E_2^{**}) = 3/9 = 1/3$$

and

$$\text{pr}(E_2) = \text{pr}(E_2^*) \cdot \text{pr}(E_2^{**}) = (1/12)(1/3) = 1/36 \ .$$

In a similar manner, $\text{pr}(E_3)$ is equal to the product of the probability that the shooter rolls a five and the conditional probability that a five occurs given that either a five or a seven has occurred. That is,

$$\text{pr}(E_3) = (4/36)(4/10) = 2/45 \ .$$

The remaining probabilities may be computed analogously:

$$\text{pr}(E_4) = (5/36)(5/11) = 25/396$$
$$\text{pr}(E_5) = (5/36)(5/11) = 25/396$$
$$\text{pr}(E_6) = (4/36)(4/10) = 2/45$$
$$\text{pr}(E_7) = (3/36)(3/9) = 1/36 \ .$$

Hence, the probability that the shooter wins is

$$2/9 + 1/36 + 2/45 + 25/396 + 25/396 + 2/45 + 1/36 = 244/495 \ \text{(slightly less than 1/2)}.$$

We can indicate the various possibilities and their associated probabilities in a tree diagram (Figure C.1):

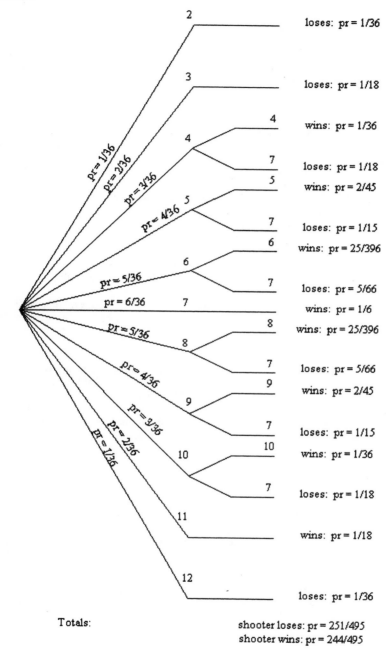

FIGURE C.1

Note that we read "and" along each of the tree branches and that Theorem C.9 applies, so that we multiply probabilities along the branches from start to finish to obtain the probability of the total branch.

➢ REPEATED EXPERIMENTS

In discussing Problem C.1, we saw that the act of rolling two dice could be viewed as if we first rolled one (red) die and then rolled another (white) die. In fact, we clearly could view the experiment as if we first rolled one die, noted the result, and then rolled the same die again. Thus, the experiment of rolling two dice is equivalent to a double performance of the experiment of rolling one die.

In general, any experiment may be repeated any number of times. In this section, we are interested in how the probability of events associated with repeated performances of an experiment can be obtained from the probabilities associated with the single experiment.

In the solution of Problem C.4a, we observed that the total obtained when we roll a pair of dice in no way affects the total obtained when we roll that same pair of dice a second time. I.e., the two performances of the experiment are independent. This is true in general. That is, in any sequence of repetitions of the exact same experiment, each performance of the experiment is independent of the others. Thus, in particular, if E_1 and E_2 are two events associated with an experiment, and if the experiment is performed twice, then the event that E_1 occurs during the first performance of the experiment and the event that E_2 occurs during the second performance of the experiment are independent. Hence, by formula (C.6), the probability that E_1 occurs during the first performance of the experiment and that E_2 occurs during the second performance is $\mathrm{pr}(E_1) \cdot \mathrm{pr}(E_2)$.[7]

This result can be generalized to any number of repetitions of the same or even different experiments, provided that the experiments are independent. Specifically, we have the following theorem.

Theorem C.10: If $E_1, E_2 \ldots, E_k$ are events respectively associated with k independent experiments $G_1, G_2 \ldots, G_k$, then the probability that E_1 will occur when G_1 is performed and that E_2 will occur when G_2 is performed, and so on, is $\mathrm{pr}(E_1)\mathrm{pr}(E_2)\ldots\mathrm{pr}(E_k)$, where $\mathrm{pr}(E_i)$ is the probability that E_i occurs when G_i is performed.

For example, suppose a manufacturer has three machines producing electrical switches. If the probability that the first machine produces a defective switch is 1/100, the probability that the second machine produces a defective switch is 1/150, and the probability that the third machine produces a defective item is 1/400, then, if one item is sampled from each machine, the probability that all three sampled items will be defective is (1/100)(1/150)(1/400) = 1/6000000. Similarly, the probability that none of the three sampled items are defective is (99/100)(149/150)(399/400) = 5885649/6000000.

PRACTICE PROBLEM SET C.0

1. $\boxed{\text{A}}$ if a red, a white, and a green die are rolled, what is the probability that the red die will show an odd number, the white die will show a three, and the green die will show a number greater than two?

2. Stu Dent is taking three courses this semester. He estimates that the probability of his getting a C or better in Math is .80, in English is .65, and in History is .90. Assuming that his grades in the three courses are independent of each other, what is the probability that Stu receives a C or better in all three courses?

We are now ready to consider Problem C.5. If you have not already solved it, try it again now.

---

[7]: This result also follows easily from the Multiplication Principle. If the experiment has n possible outcomes of which f_1 are favorable to E_1 and f_2 are favorable to E_2, then, by the Multiplication Principle, there are n^2 possible outcomes when the experiment is performed twice. Of these, $f_1 f_2$ are favorable to the event E (that E_1 occurs during the first performance of the experiment and E_2 occurs during the second). Hence

$$\mathrm{pr}(E) = f_1 f_2 / n^2 = (f_1/n)(f_2/n) = \mathrm{pr}(E_1) \cdot \mathrm{pr}(E_2)$$

SOLUTION TO PROBLEM C.5

Here, we are not dealing with an equally likely outcome model of probability (e.g., each depth charge is more likely to sink a submarine than it is to miss it). However, we are given all the relevant underlying probabilities, and so we can still proceed, using the theorems of the chapter. Since the captain fires the depth charges in some sequence (i.e., first one blows, then a second, and so on), he is essentially conducting four performances of the experiment of setting off a single depth charge. If the first charge sinks the submarine, then we need not consider what would have happened when the other three charges exploded. Similarly, if the first charge damages the ship and the second charge either scores a direct hit or damages it again (and hence sinks it), then the third and fourth charges do not matter. And so on. We draw a tree diagram (Figure C.2) to indicate all the various possible outcomes.

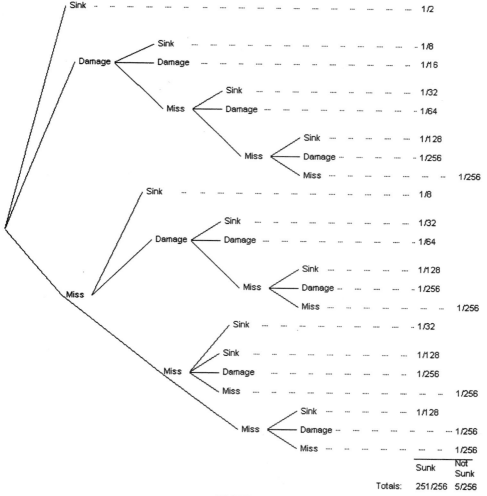

FIGURE C.2

In the diagram, the probability with which each case occurs is indicated at the right. It is computed by using Theorem C.10 and the assumption that the probability of success of a subsequent depth charge is not affected by the knowledge that a previous charge did not sink the ship. For example, the probability of the case that the first charge causes damage (probability = 1/4), the second charge misses (probability = 1/4) and the third charge is a direct hit (probability = 1/2) is (1/4)(1/4)(1/2) = 1/32. If we now add up the probabilities of all the branches on which the submarine eventually sinks, we obtain the desired probability. Thus, the probability that Captain Dimwitty will sink the submarine using no more than four depth charges is

$1/2 + 1/8 + 1/16 + 1/32 + 1/64 + 1/128 + 1/256 + 1/8 + 1/32 + 1/64 + 1/128 + 1/256 + 1/32 + 1/128 + 1/256 + 1/128$

$$= 1/2 + 2/8 + 1/16 + 3/32 + 2/64 + 4/128 + 3/256$$
$$= (128 + 64 + 16 + 24 + 8 + 8 + 3)/256 = 251/256.$$

(Actually, there is a simpler method of solving Problem C.5, to which we will return shortly.)

So far we have considered the question, "in two performances of an experiment, what is the probability that E_1 occurs the first time and E_2 occurs the second?" We emphasize that this is not the same as the question, "in two performances of an experiment, what is the probability that E_1 and E_2 both occur?" The latter question does not consider when E_1 and E_2 occur, only whether or not they both occur. This question is more difficult, especially if E_1 and E_1 are not disjoint. We shall not attempt to answer this question here, but the discussion above and in the rest of this section should enable the reader to handle any problem in which the answer to such a question is called for. Basically, the technique involved is to break the event under consideration into disjoint events whose probabilities can be found, and then to use additivity. The difficulty, if there is one, arises in making sure that the events into which the original event is broken up are disjoint.

This same technique also applies to the solution of the following question: In n repetitions of an experiment, what is the probability that the event E occurs exactly k times?

The answer may be found in using a tree diagram. We illustrate the case $n = 4$ (see Figure C.3), using S (success) to denote the possibility that E occurs during a particular performance of the experiment and F (failure) to denote the possibility that E does not occur. S occurs with probability $p = \mathrm{pr}(E)$ and F occurs with probability $q = 1 - p$.

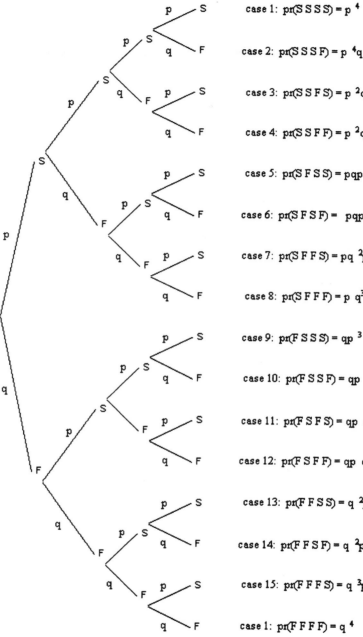

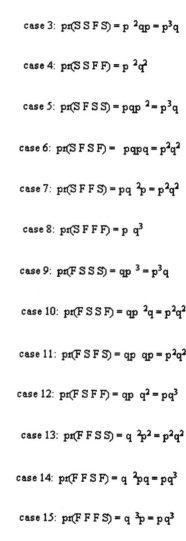

case 1: $pr(S\,S\,S\,S) = p^4$

case 2: $pr(S\,S\,S\,F) = p^4 q$

case 3: $pr(S\,S\,F\,S) = p^2 qp = p^3 q$

case 4: $pr(S\,S\,F\,F) = p^2 q^2$

case 5: $pr(S\,F\,S\,S) = pqp^2 = p^3 q$

case 6: $pr(S\,F\,S\,F) = pqpq = p^2 q^2$

case 7: $pr(S\,F\,F\,S) = pq^2 p = p^2 q^2$

case 8: $pr(S\,F\,F\,F) = p\,q^3$

case 9: $pr(F\,S\,S\,S) = qp^3 = p^3 q$

case 10: $pr(F\,S\,S\,F) = qp^2 q = p^2 q^2$

case 11: $pr(F\,S\,F\,S) = qp\,qp = p^2 q^2$

case 12: $pr(F\,S\,F\,F) = qp\,q^2 = pq^3$

case 13: $pr(F\,F\,S\,S) = q^2 p^2 = p^2 q^2$

case 14: $pr(F\,F\,S\,F) = q^2 pq = pq^3$

case 15: $pr(F\,F\,F\,S) = q^3 p = pq^3$

case 1: $pr(F\,F\,F\,F) = q^4$

FIGURE C.3

Note, for example, that there are $\binom{4}{3} = 4$ cases with three S's and one F. (There are $\binom{4}{3}$ ways to select three slots out of four in which to put the S's.) Each of these occurs with probability

$p^3 q = p^3 (1 - p)$. Hence, the probability of exactly three S's (i.e., E occurs exactly three times) is $\binom{4}{3} p^3 (1 - p)$. Similarly,

$$\text{pr}(E \text{ does not occur}) = \binom{4}{0} q^4 = q^4 = (1 - p)^4$$

$$\text{pr}(E \text{ occurs exactly once}) = \binom{4}{1} pq^3 = 4pq^3 = 4p(1 - p)^3$$

$$\text{pr}(E \text{ occurs exactly twice}) = \binom{4}{2} p^2 q^2 = 6p^2 q^2 = 6p^2 (1 - p)^2$$

$$\text{pr}(E \text{ occurs exactly four times}) = \binom{4}{4} p^4 = p^4.$$

In general, if an experiment is repeated n times, there are $\binom{n}{k}$ ways for E to occur exactly k times. For each of these possible cases, E occurs during a particular performance with probability p and does not occur with probability $1 - p$. Hence, we get k p's and $n - k$ $(1 - p)$'s for each choice. We thus have the following theorem:

Theorem C.11: In n repetitions of an experiment, the probability that a particular event E occurs exactly k times is

$$\binom{n}{k} p^k (1 - p)^{n-k}$$

where p is the probability that E occurs in one performance of the experiment.

For example, if we toss ten fair coins, the probability that we obtain exactly eight heads is

$$\binom{10}{8} \left(\frac{1}{2} \right)^8 \left(\frac{1}{2} \right)^2 = \frac{10!}{8!\,2!\,2^{10}} = \frac{10 \cdot 9}{2 \cdot 1 \cdot 2^{10}} = \frac{45}{2^{10}} = \frac{45}{1024}.$$

If we are interested in the probability that E occurs at least k times, then we must add the probability that E occurs exactly k times to the probability that E occurs exactly $k + 1$ times, and so on. Thus $\text{pr}(E$ occurs at least k times$)$

$$= \binom{n}{k} p^k (1 - p)^{n-k} + \binom{n}{k+1} p^{k+1} (1 - p)^{n-k-1} + \dots + \binom{n}{n} p^n. \tag{C.8}$$

Thus, if we toss ten fair coins, the probability of at least eight heads is

$$\binom{10}{8} (1/2)^8 (1/2)^2 + \binom{10}{9} (1/2)^9 (1/2) + \binom{10}{10} (1/2)^{10} = \left[\binom{10}{8} + \binom{10}{9} + \binom{10}{10} \right] (1/2)^{10}$$

$$= (45 + 10 + 1)(1/2)^{10} = 56/1024 = 7/128.$$

PRACTICE PROBLEM SET C.P

1. $\boxed{\text{A}}$ On a roulette wheel (see Problem 2 of Practice Problem Set C.C), the 0 and 00 are green. Half of the remaining 36 numbers are red, and the other half are black.

   If the wheel is spun four times in succession, what is the probability of each of the following events?
   a) All four spins come up black.
   b) At least two spins come up black.
   c) More blacks than reds.

2. If five honest dice are tossed, what is the probability of each of the following events?
   a) Five 6's.
   b) The same number showing on all five dice.
   c) Three odd numbers and two even numbers.
   d) At most three 6's.

PROBLEM C.5 REVISITED

Let us now return to Problem C.5. Instead of computing the probability that Captain Dimwitty sinks the submarine, using no more than four depth charges, we compute the probability that he does not sink it.

The sub does not sink only if all four depth charges miss, or if three miss and the fourth only damages the sub. The probability that all four miss is $\binom{4}{4}(1/4)^4 = 1/256$. The probability that three miss and the fourth just causes damage is $\binom{4}{3}(1/4)^3(1/4) = 4/256$. Hence, the probability that the ship is not sunk is $1/256 + 4/256 = 5/256$.

Thus, the probability that Captain Dimwitty does sink the ship is $1 - 5/256 = 251/256$.

PRACTICE PROBLEM SET C.Q

1. $\boxed{A}$ If ten fair coins are tossed, what is the probability of obtaining at least one head?

➢ EXPECTATION

When playing a game of chance, the probability of winning is not the only important consideration; the potential payoff and the expected payoff are other important factors which should be considered in deciding whether or not to play.

Potential payoff refers to the amounts of money that could possibly be won or lost. Few of us would be willing to risk $1000 against the possibility of winning a nickel, even if our chances of winning are very high; on the other hand, many of us would be willing to risk a nickel for the opportunity to win $1000, even if our chances of winning are very low. We will return to this point later.

Expected payoff, on the other hand, refers to the amount we theoretically expect to win or lose if we play the game a large number of times. For example, consider the following game: We roll a die. If we get a two, the bank pays us $4; otherwise, we pay the bank $1. (I.e., the bank is giving us four to one odds that we do not roll a two.) If we play this game six hundred times, then, since we expect each face of the die to show up approximately equally often, we anticipate rolling a two one hundred times. We will therefore expect to win $4 one hundred times and lose $1 five hundred times, for a net loss of $100. That is, our expected payoff if we play the game six hundred times is –$100, which makes our average expected payoff per game –$100/600 (approximately –$.17). (Equivalently, we expect to win $4 one sixth of the time and lose $1 five sixths of the time so our expected payoff per game is $(1/6) \cdot \$4 - (5/6)\$1 = -\$.17$.)

This does not mean that we will lose $.17 every time we play the game – the only possible amounts that could be won or lost at a given play are $4 or $1. Rather, it means that, if we play the game a large number of times, we expect, in the long run, to lose money; in fact, we expect to lose an amount approximately equal to $.17 multiplied by the number of games we play. Obviously, if we are lucky, we could conceivably win (or, if we are unlucky, we could lose larger amounts), but the laws of probability tell us that it would be unwise for us to play the game in question.

Similarly, if we are offered odds of s to r that some event E does not occur (i.e., we win $\$s$ if E occurs and lose $\$r$ otherwise), and if the probability of E occurring is p, then our expected payoff is $sp - r(1 - p)$. If this is negative, then we expect to lose money in the long run, and we say the odds are not favorable to us. If it is positive, we expect to win money in the long run, and say the odds are favorable to us. Finally, if $sp - r(1 - p) = 0$, then, in the long run, we expect to break even, and we say the odds are fair. In this case, $sp = r(1 - p)$, and so

$$\frac{r}{s} = \frac{p}{1 - p}$$

Note that fair odds correspond to the theoretical definition of odds given at the beginning of the chapter (Definition C.2).

PRACTICE PROBLEM SET C.R

1. [A]   If someone hands you a pair of dice and offers to bet that you do not roll a double (two like numbers), what odds should you be offered in order for the bet to be a fair bet?

2.   A roulette wheel contains each of the numbers 1 to 36 as well as 0 and 00. If you bet on a number and it turns up, you win thirty five times the amount of your bet, i.e., odds of 35 to 1. Are these odds fair? If not, what would fair odds be?

In general, a game of chance may have more than two possible payoffs. Suppose that the possible outcomes of a game are grouped into k disjoint events, $E_1, E_2 ..., E_k$, which occur with probabilities $p_1 p_2 ...,$ p_k respectively $(p_1 + p_2 + ... + p_k = 1)$- Suppose further that these events respectively require payoffs of $m_1, m_2 ..., m_k$. (I.e., if E_1 occurs, then you win m_i if m_i is positive, lose m_i if m_i is negative, and neither win nor lose if $m_i = 0$.)

Definition C.6: In the situation described above, the <u>expectation</u> or <u>expected value</u> of the game is
$$e = m_1 p_1 + m_2 p_2 + ... + m_k p_k.$$
(The rationale behind this definition is essentially the same as in the case of two payoffs. Try to supply the details yourself.)

For example, suppose we have the following arrangement: We roll one die. If we roll a one, we win \$1; if we roll a three, we win \$3; for a five, we win \$5; otherwise, we lose \$2. Then the expected value of the game is $1(1/6) + 3(1/6) + 5(1/6) - 2(1/2) = 1/2 = \$.50$, since the probability of rolling a one is 1/6, the probability of rolling a three is 1/6, the probability of rolling a five is 1/6, and the probability of rolling something else is 1/2.

We are now ready to solve Problem C.4b. Try it now if you haven't already solved it.

SOLUTION OF PROBLEM C.4b

We distinguish five events.
    E_1: The shooter loses.
    E_2: The shooter wins by rolling 7 or 11.
    E_3: The shooter wins by making his point, which is 6 or 8.
    E_4: The shooter wins by making his point, which is 5 or 9.
    E_5: The shooter wins by making his point, which is 4 or 10.

From the solution of Problem C.4a, the probability of each of these events is easily found. (See Figure C.4.)

Event	Probability	Payoff
E_1		−1
E_2		1
E_3		1
E_4		3/2
E_5		2

FIGURE C.4

Hence, the expected value of the game is
$$2(2/36) + (3/2)(4/45) + 1(50/396) + 1(2/9) - 1(251/495)$$
$$= 85/990 \approx .086.$$

That is, in the long run, the shooter expects to win an average of about 8.6 cents for each dollar he bets.

PRACTICE PROBLEM SET C.S

1. [A] Select a card at random from an ordinary 52 card deck of playing cards. If you select a spade, you win $3; if you select a red picture card, you win $5; otherwise, you lose $1. What is the expected value of the game?

2. In a game of roulette, you place a $1 bet on red and a $1 bet on 11 (which happens to be a black cell). If a red number comes up, you win $1. If 11 comes up, you win $35 dollars; otherwise you lose $2. What is the expected value of the game?

We reiterate that we do not win or lose the expected value of a game each time we play. However, if we play the game often, then we expect our net gain (or loss) to be approximately the expected value times the number of games we play.

If the expectation of a game is positive, we expect to win in the long run, and we say that the game is favorable; if it is negative, we expect to lose, and say that the game is unfavorable; if the expectation is zero, we expect to break even. In the latter case, we say that the game is a fair game. That is, we have

Definition C.7: A <u>fair game</u> is a game with expected value zero.

Any gambling game may be converted to a fair game by requiring one of the players to pay for the privilege of playing. For example, if a player's expectation is +$2 and if he pays $2 for the privilege of playing, then the game becomes fair.

PRACTICE PROBLEM SET C.T

1. [A] You toss three fair coins and win a number of dollars equal to the number of heads you obtain. What should you pay for the privilege of playing this game in order to make it a fair game?

We emphasize that it is possible to lose at a favorable game and to win at an unfavorable game (otherwise, no one would gamble). However, the more often a game is played, the less likely it is for this to happen.

We are now ready to solve Problem C.6. Note that part b) of the problem is just a special case of the problem which led Pascal and Fermat to develop the theory of probability.

Now it is your turn to try to solve the problem using the theory that has been developed in the chapter.

SOLUTION OF PROBLEM C.6

In this problem, there are several different probabilities in which we are interested. There is the probability that A will win a point at any given roll of the die. This probability remains constant throughout the game (successive rolls are independent) and is easily seen to be 1/3 (two of the six outcomes are favorable).

Similarly the probability that B will win a point at any given roll of the die is 2/3.

In addition, since a fair distribution of the money in the pot will depend on the relative probabilities that A and B will be the ultimate winner, we are interested in these probabilities. These probabilities change as the game progresses and the score changes.

In part a) of the problem, we are interested in the probability before the game begins that A will be the eventual winner. Let P_A denote this probability. Define P_B similarly.

Let S_A and S_B denote the sums of money anted by A and B respectively. Then A's expectation is $P_A S_B - P_B S_A$ and B's expectation is $P_B S_A - P_A S_B$.

Since the game is to be a fair game $P_A S_B - P_B S_A = 0$, or, equivalently,

$$P_A S_B = P_B S_A$$

Since $S_A + S_B = 100$ and $P_A + P_B = 1$,

$$P_A (100 - S_A) = (1 - P_A) S_A$$
$$100\, P_A - P_A S_A = S_A - P_A S_A$$
$$100\, P_A = S_A$$

Therefore, we will know S_A if we can find P_A.

How can A be the winner? He can win by a score of 4–0, 4–1, 4–2, or 4–3.

In the first case he must win the first four points. This occurs with probability $\left(\dfrac{1}{3}\right)^4$.

If he wins by a score of 4 to 1, he must win the fifth point and three out of the first four points (losing the remaining one). This occurs with probability $\dbinom{4}{3}\left(\dfrac{1}{3}\right)^4\left(\dfrac{2}{3}\right)$.

Similarly he wins by a score of 4–2 with probability $\dbinom{5}{3}\left(\dfrac{1}{3}\right)^4\left(\dfrac{2}{3}\right)^2$

and he wins by a score of 4–3 with probability $\dbinom{6}{3}\left(\dfrac{1}{3}\right)^4\left(\dfrac{2}{3}\right)^3$. Thus

$$P_A = \left(\dfrac{1}{3}\right)^4 + \dbinom{4}{3}\left(\dfrac{1}{3}\right)^4\left(\dfrac{2}{3}\right) + \dbinom{5}{3}\left(\dfrac{1}{3}\right)^4\left(\dfrac{2}{3}\right)^2 + \dbinom{6}{3}\left(\dfrac{1}{3}\right)^4\left(\dfrac{2}{3}\right)^3$$

$$= 379/2187 \approx .1733$$

Therefore

$$S_A = 100\, P_A = \$17.33$$

That is, A should ante $17.33 and B should put up $82.67.

For part b) of the problem, since A has two points and B has one, A needs only two points to win and B needs three. Thus the probability that A would have won had the game continued is the probability that A would have gotten two more points before B gained three more. Again this could happen in several ways:

A wins the next two points – probability $= \left(\dfrac{1}{3}\right)^2$

A wins the third point and one of the next two – probability $= \dbinom{2}{1}\left(\dfrac{1}{3}\right)^2\left(\dfrac{2}{3}\right)$

A wins the fourth point and one of the next three – probability $= \dbinom{3}{1}\left(\dfrac{1}{3}\right)^2\left(\dfrac{2}{3}\right)^2$

Thus, had the game continued, the probability that A would have won is

$$\left(\dfrac{1}{3}\right)^2 + 2\left(\dfrac{1}{3}\right)^2\left(\dfrac{2}{3}\right) + 3\left(\dfrac{1}{3}\right)^2\left(\dfrac{2}{3}\right)^2 = 11/27 \approx .4074.$$

Therefore, the probability that B would have won is $16/27 \approx .5926$. Hence the pot should be divided in the proportion 4074 to 5926. That is, A should get \$40.74 and B should get \$59.26. (Note that A wins \$23.41.)

➤ ENTERTAINMENT VALUE AND POTENTIAL PAYOFF

In general, most games of chance operated by gambling casinos or racetracks are slightly unfavorable to the player –i.e., the player's expectation is slightly negative. This is necessary in order for the gambling establishment to stay in business. Under these circumstances, why do many intelligent people choose to gamble anyway? The answer seems to involve at least two factors.

Firstly, participation in a game of chance is a form of entertainment. Just as one might be willing to spend money to attend a show, concert, movie, sporting event, etc., or to eat at a fancy restaurant, one might also be willing to spend a certain amount of money simply for the enjoyment of gambling. Since, in the long run, a player should expect to lose an average of "e" for each time he plays a game having (negative) expectation, –e, we say that e is the entertainment value of the game. That is, the player is essentially willing to invest e per game, just for the fun of playing.

The second, and probably more important, factor which influences people to gamble is the possibility, no matter how unlikely, of winning a big payoff. The person who invests one dollar per week in lottery tickets usually figures that losing a dollar a week is not going to have a great effect on his or her standard of living; but, imagine what would happen to his (her) standard of living if he (she) should, by some miracle, win one million dollars.

Some of us apply the same principle, on a smaller scale, to sweepstakes sponsored by magazines and mail order houses. In this case, the only investment required of the player is a postage stamp (at the time of this writing \$.13). However, chances of winning the grand prize are usually so small that even a \$.13 investment is not warranted unless the potential payoff is sufficiently large. To our mind, prizes of \$5000 or even \$10000 are not sufficiently tempting[8] to be worth the investment of \$.13, chances of winning being as small as they are. (This is not to say that we would not be thrilled to win even \$1000.) However, we find prizes of \$100,000 or more sufficiently enticing to justify the investment of \$.13, even though we know in our heads and hearts that we will not win.

➤ THE CHAPTER IN RETROSPECT

Our object in this chapter has been to introduce you to the notions of probability, expectation, odds, etc., not only so that you can gamble more intelligently but also since these words are becoming more and more part of our daily vocabulary.

Although we have basically considered probability as defined in the case of equally likely outcomes, many of the results hold in general. Remember though, that, in the general case, care is needed in deciding on how the outcomes should be weighted.

It is extremely important to be able to count cases properly, and so we have considered several types of counting techniques. Make sure you understand thoroughly when each type can be used. Try now to apply these ideas in the exercises which follow.

---

[8]: We never could understand the thinking of a bank robber in movies which have him steal \$10,000 or even \$50,000 with the expectation of escaping to South America and being able to live the rest of his life like a king.

EXERCISES

Introductory Problems

C.1 [H] [A] a) Remove nine cards (AS, 2S, 3S, ..., 9S) from a deck of playing cards.

Place them in a number of piles so that the sum of the face values of the cards in each pile is the same. Disregarding the order of the piles, in how many ways can this be done?

b) What if ten cards (AS, ..., 10S) are used?

C.2. [H] [A] How many triangles (of all sizes) are there in Figure C.5?

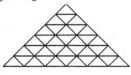

FIGURE C.5

C.3. [A] Winnie Gamble loves to play the slot machines. She has nineteen silver dollars and one slug (which will be rejected by the machine) in her purse.
a) If Winnie randomly selects one coin from her purse, what is the probability that it will be the slug?
b)   What is the probability it will be a good coin?
c) If Winnie selects four coins from her purse, what is the probability that at least one of them will be the slug?

C.4.[A]   In every four hundred year period, the thirteenth of a month falls on a Friday 688 times and the first day of a century falls once on Monday, once on Wednesday, once on Friday, and once on Saturday (see Exercise 4.36).

a) If a year and a month are selected at random, what is the probability that the thirteenth of that month falls on a Friday in the selected year?

b) What is the probability that the first day of a century falls on a Sunday?

c) What is the probability that the first day of a century falls on a Monday?

C.5.   [A] Jack and Jill each selects two distinct odd numbers between three and seventeen inclusive. They then simultaneously throw three dice each. Whoever first rolls his or her dice and obtains a sum equal to one of his or her own selected numbers is the winner, unless his or her

opponent is also successful, in which case th game starts over again.

Jack and Jill have selected distinct pairs (odd numbers so that they have equal chances (winning.

If Jack has selected seven as one of hi numbers, what numbers has Jill selected, an what is Jack's other number?

Counting, Using The Multiplication Principle, Permutations, Combinations And Additivity

C.6. [A] Don and Dinah went to a diner for dinne The menu is shown in Figure C.6.

Appetizer	Entree	Dessert
Fruit juice	Filet mignon	Chocolate mo
Soup du jour	Breaded veal cutlets	Ice cream
Melon	Fried flounder	Sherbet
	Beef stew	Cheese Cake
		Fresh Fruit

FIGURE C.6

a)   How many different complete dinne (including one appetizer, one entree, and o dessert) are available?

b) Dinah could not make up her mind abou what to order, so she told the waiter to surpris her. Assuming that the waiter randomly selecte one dish in each category, what is the probabili that Dinah had a liquid appetizer, a meat entre and either ice cream or sherbet for dessert?

C.7   [H] [A] The repair lady for the telephor company has been given a list of five phor numbers which are in need of repair (see Figu C.7).

Phone Number	Signature
527 – 5263	
468 – 4486	
643 – 3546	
266 – 7269	
267 – 8666	

FIGURE C.7

After each phone is repaired, she mu obtain the signature of each of the people whose homes the phones are located. Assumir that no two names are the same, what is th probability that, when the signatures a obtained, the last names of the people will be in alphabetical order?

C.8. [A] There was one other number that the repair lady was supposed to fix, but she forgot to bring the number with her. However, she does recall that the last four digits of the number were all distinct.

What is the probability that the last four digits of the number begin with a 2 and contain a 1?

C.9. [A] In how many different ways can a cube be placed in a box of the correct size?

C.10. [S] [A] Given six colors, in how many ways can the faces of a cube be colored so that each face is a different color? (Two colorings are considered to be the same if the resulting cubes can be so placed so that the same colors face the same directions.)

C.11. [A] Mrs. Cooke is having a big dinner party and has decided to start the dinner with stone soup. Her recipe calls for two large stones, one cup each of any five vegetables, diced, and one quarter pound of meat, for each two gallons of water. (Note, you must never mix two varieties of meat in the same pot of stone soup!)

a) If Mrs. Cooke has already placed two stones and the meat in two gallons of boiling water, and if she has twelve varieties of vegetables to choose from, in how many ways could she choose which five vegetables to add to the soup?

b) After the soup was finished, Mrs. Cooke cooked up another potfull. (She could not have doubled the recipe in the first place, because she did not own a pot that would be large enough.) She had six stones remaining to choose from, as well as seven varieties of vegetable and three different kinds of meat.

In how many different ways could she make the second pot of soup?

C.12. [A] Sandra, Jessie, and Veronica bowl together every Thursday night. This Thursday, however, each, unbeknownst to the others, decided that she would rather attend their college basketball game. Not wishing to offend her friends, each of the girls told the others that she was sick in bed and could not bowl. Fortunately, they were all sitting in different locations in the basketball arena, so they did not notice each other. However, at the exact same moment, all three decided to go to the powder room. If there are three powder rooms in the building, what is

the probability that none of the girls will find out that anyone else lied?

C.13. [H] [A] At the Nationalist Party's fundraising dinner, there were nine tables arranged in a square. There are three red, three white and three blue tablecloths.

In how many ways can they be placed on the tables so that no two tablecloths of the same color are in the same row or the same column?

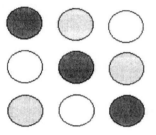

C.14. [H] [A] John, Joan and Jack Smith, John, Joan, and Jack Brown, and John, Joan and Jack Jones all had tickets to the same performance at the opera. Their seats were in rows D, E, and F, and were numbers 101, 102, and 103 in each row.

a) In how many ways can the tickets be distributed so that no members of the same family have tickets with the same number or letter?

b) In how many ways can the tickets be distributed so that not only do no two members of the same family have tickets with the same letter or the same number, but also the same is true of people with the same first name.

c) Assuming that the conditions referred to in part b are realized, what is the probability that Jack Smith is sitting immediately to the right of Joan Brown?

**C.15. [H] [A] The four Bridges, Manny, Brooke, William, and Queenie, play bridge together four evenings a week. If, at the beginning of each week, they make a seating plan so that each person sits in a different direction (North, East, South, or West) each of the four evenings that week, how many weeks could pass without their having to repeat a complete weekly seating plan?

C.16. [A] Mr. and Mrs. B. A. Scout are going on a camping trip with their baby. They figure that they will need ten jars of baby food for the trip. Their pantry is stocked with eight jars of baby fruits, twelve jars of baby vegetables, and six jars of baby meats. If they select ten jars at random, what is the probability that

a) they select five jars of vegetables, three of fruit, and two of meat?

b) all ten jars are the same (i.e., all fruit, all vegetables, or all meat)?

c) at least one type of food is missing (i.e., there is no fruit, or there is no vegetable, or there is no meat)?

**C.17. [H] [A] Given the six colors and the cube of Exercise C.10, how many colorings are possible if more than one face may receive the same color? E.g., the entire cube may be colored red, etc. Remember that two colorings are considered to be the same if the resulting cubes can be so placed that the same colors face the same directions.

C.18. [A] a) How many different three letter words (meaningful or not) can be written using the English alphabet?

b) (H) How many are there if the middle letter must be a vowel and the last must be a consonant, and if no more than one "y" is to be used in any word? ("y" is to be considered as a consonant unless there is no other vowel in the word.)

C.19. [S] [A] a) How many four digit numbers may be formed using only the odd digits (1,3,5,7,9)?

b) What is the sum of all these numbers?

c) If these numbers are arranged in numerical order from smallest to largest, what is the 314th number on the list?

*C.20. [H] [A] a) How many numbers of no more than four digits may be formed using only the odd digits?

**b) What is the sum of all these numbers? If these numbers are arranged in numerical order, what is the 314th number on the list?

*C.21. [H] [A] a) How many four digits numbers are there whose digits are distinct?

**b) What is the sum of all these numbers?

c) If these numbers re arranged in numerical order, what is the 314th number on the list?

C.22. [A] In how many ways can three red and four black checkers be arranged in a straight line? (Consider checkers of the same color to be indistinguishable from each other.)

C.23. [H] [A] In how many ways can four red and four black checkers be arranged in a straight line so that the nth red checker (counting from the left) is not to the left of the nth black checker, for n = 1, 2, 3, and 4?

C.24. [S] [A] The elections committee was holding an election, by secret ballot, for chairperson of the committee. The final vote was 4 for Della Gate and 6 for Victor Riaz. The votes were counted one at a time.

a) What is the probability that Della was ahead when the fifth vote was counted?

b) What is the probability that Della was ever ahead in the counting of the ballots?

C.25. [H] [A] Ima Gardner has enough space to plant five fruit trees in a row on her property. She has been given two apple, two pear, two peach, two plum, one apricot and one cherry tree from which she may choose any five.

How many different arrangements are possible? (Trees of the same type are to be considered as indistinguishable.)

C.26. [H] [60, volume 1, page 85] a) In how many ways can you place two kings, one white and one black, in a legal position (i.e., in nonadjacent cells) on an 8 x 8 chessboard?

b) In how many ways can you place two rooks (white and black) so that they do not attack each other? (I.e., they are not in the same row or file.)

c) In how many ways can you place two bishops (white and black) so that they do not attack each other? (I.e., they are not on the same diagonal.)

d) In how many ways can you place two queens (white and black) so that they do not attack each other? (I.e., they are not in the same row, file, or diagonal.)

e) In how many ways can you place two knights (white and black) so that they do not attack each other? (See Chapter Six for a description of the knight's move.)

C.27 [H] [A] Using the dots in a 6 x 6 array (Figure C.8) as vertices,

a) How many squares having horizontal and vertical sides can be drawn?

b) How many rectangles which are not square can be constructed having horizontal and vertical sides?

*c) How many squares can be drawn with sides which are not horizontal or vertical?

FIGURE C.8

.28. H A How many rectangles (including
quares) having horizontal and vertical sides can
e drawn on an n x m array of dots?

.29. H A A man has three dice. On one of
hem, the one has been changed to a second three;
n another, the six has been rubbed out and the
ace is blank.

a) Counting the blank as zero, how many
ifferent numbers can be made by turning the
ice any way you like and placing the top faces
ogether. For example, the dice in Figure C.9 are
howing the number 305.

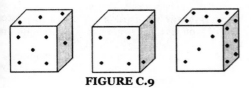

FIGURE C.9

b) If the dice are rolled, what is the
robability that the top three faces show a three, a
ive, and a six (regardless of order)?

.30. A a) In how many different ways can ten
eople line up in a row?

b) In how many different ways can ten people
ne up in a row if two particular people, say,
urtis and Clementine, must be next to each
ther?

c) H How many different circular
rrangements of ten people are there?

d) How many circular arrangements are
here if two particular people, say, Curtis and
lementine, must be next to each other?

.31. A How many different necklaces can be
nade from seven beads, each of a different color?
Two necklaces are considered the same if they
an be turned or flipped over so that they appear
dentical.)

.32. H A How many different necklaces can be
nade so that each contains four red, three blue
nd two yellow beads?

.33. H A Given a supply of four red, four green,
nd four white beads, how many different six
ead necklaces could be made?

C.34. A If a bridge hand (thirteen cards) is dealt
rom an ordinary fifty-two card deck of playing
ards, find an expression for the probability that

a) all thirteen cards are of the same suit
(clubs, diamonds, hearts, or spades)?

b) S at least eight cards are in the same suit?

c) at least seven cards are in the same suit?

**d) S at least six cards are in the same suit?

e) exactly two different suits are represented?

f) two suits contain ten or more cards between
them, but no suit contains seven or more cards?

g) S there are four cards in one suit and
three cards in each of the others?

h) there are four cards in each of three suits
and one card in the fourth suit

i) there are no more than four cards in any
suit?

j) there are five cards in one suit and no more
than three cards in any other suit?

k) the hand contains four or more picture
cards (jacks, queens, and kings)?

l) the hand contains all four aces?

m) the hand contains no card lower than a
ten? (Consider aces to be higher than a ten.)

n) the hand contains no card higher than a
ten?

*C.35 A A hand containing five cards is dealt
from an ordinary fifty-two card deck of playing
cards. What is the probability that

a) the hand is a flush? (I.e., all five cards are
in the same suit.)

b) no two cards are in the same suit?

c) S the cards form a straight? (I.e., five
cards which can be arranged, regardless of suit, to
form a consecutive sequence such as A,2,3,4,5;
2,3,4,5,6; ...; 9,10,J,Q,K; or 10,J,Q,K,A.)

d) the cards form a straight flush? (I.e., they
are in the same suit and can be arranged to form a
consecutive sequence.)

e) four cards in the hand have the same face
value. (Such a hand is referred to as four of a
kind.)

f) S the hand is a full house? (I.e., three of
the cards have the same face value, and the other
two cards have the same face value as each other.
The face value of the two cards is obviously
different from that of the three cards.)

g) S three of the cards have the same face
value, but the other two have different, distinct
face values. (Such a hand is referred to as three of
a kind.)

h) two of the cards have the same face value,
and the other cards have different distinct values?
(Referred to as two of a kind, or one pair.)

i) two of the cards have one face value, two
other cards have the same face value as each other
(different from that of the first two cards), and the

remaining card has a third face value? (This is referred to as two pairs.)

j) $\boxed{H}$ no two cards have the same face value and the hand is not a straight nor is it a flush?

*C.36. $\boxed{H}$ $\boxed{A}$ In poker, a straight flush (see Exercise C.35) beats four of a kind, which beats a full house, which beats a flush, which beats a straight, which beats three of a kind, which beats two pair, which beats one pair, which beats no pair. In a game of draw poker, you are dealt the following five cards: 10S, JS, QS, KS, KH. You are allowed to throw away up to four cards and to replace them by new cards drawn from the deck.

a) If you keep only the two kings, what are the probabilities that you will obtain a straight flush, four of a kind, etc?

b) If you keep the four spades, what is the probability of each type of hand?

**C.37. $\boxed{S}$ $\boxed{A}$ In a game of five card stud poker, two of your opponents are currently showing the following hands: (You do not know what the face down cards are.)
$\boxed{}$,$\boxed{}$, AS,3H,2D, and $\boxed{}$,$\boxed{}$,6H,8H,9H,
Two other players have already folded (i.e., given up), having shown KS,6D,5S, and 2H between them.

You have 5D,5H,5C,7S,10C.

What is the probability you hold the winning hand? (Refer to Exercises C.35 and C.36 for the various types of hand and their relative values.) Note that three aces beats three kings which beats three queens, etc.

C.38. $\boxed{S}$ $\boxed{A}$ Three people participate in a grab bag. Each brings a gift. These are then placed in a bag, and each person selects one of the gifts, sight unseen

a) How many different outcomes are possible, if no one is to receive the gift that he or she brought?

b) How many outcomes are possible if four people participate and if no one receives the gift that he or she brought?

c) Five people?

*d) Six people?

e) If six people participate, and if no one receives the gift that he or she brought, what is the probability that Ross receives the gift brought by Sanford and that Sanford receives the gift brought by Ross?

*C.39. $\boxed{H}$ $\boxed{A}$ The ace of spades, two of spades, ..., ten of spades are all removed from a deck of cards, shuffled, and placed face down on the table.

You turn over the first card, saying "one," turn t next card, saying "two," and so on. What is t probability that the ace will come up when you s one, or that the two will come up when you s two, etc? I.e., that you name at least one ca correctly as you turn it?

Conditional Probability And Independen

C.40. $\boxed{A}$ During a game of draw poker, Co Sharpe "accidentally" dropped one of his fi cards on the table. It was a deuce (two). What the probability that Cord holds at least one mc deuce in his hand?

C.41. $\boxed{A}$ In a game of Craps, your point is s What is the probability that
a) you win in no more than three additior rolls?
b) you lose in no more than three additior rolls?

C.42. $\boxed{A}$ When Una and Zvi Dreier got marrie they decided to have exactly three childre Assuming that each child is as likely to be a boy it is to be a girl and that there were no multip births except as indicated,
a) what is the probability that the Dreie have three boys?
b) what is the probability that the Dreie have three boys, given that the first born is a boy
c) what is the probability that the Dreie have three boys, given that at least one of tl children is a boy?
d) what is the probability that the Dreie have exactly three boys, given that Una gave bir to identical twins at her third delivery?

C.43. $\boxed{S}$ $\boxed{A}$ Referring to Exercise C.14, a assuming that the conditions referred to in part of the problem are realized, what is tl probability that Jack Smith is sitting immediate to the right of Joan Brown?

C.44. $\boxed{S}$ $\boxed{A}$ In a game of Blackjack (Twenty On using a single ordinary deck of fifty two cards, yc have a nine and a six, totaling fifteen. The deal has a nine showing.
a) What is the probability that you can dra another card so that your total will not exce twenty one? (Picture cards count as ten; aces one or eleven, as you wish. If you exceed 21, yc lose and the game ends.)
b) By looking around the table ar remembering cards which have already bee played, you can account for nine picture card

two tens, all four nines (counting yours and the dealer's), two eights, one seven, two sixes (including yours), three fives, all four fours, three threes, all four twos, and all four aces. In this case, what is the probability that, if you draw another card, your total will not exceed twenty one?

c) The rules of the game state that the dealer must continue to draw cards until his total is seventeen or more, at which point he may not draw any more cards. If his total exceeds twenty one, he loses.

Using the information given in part b), what is the probability that, if you stand pat at fifteen (i.e., take no more cards) then the dealer will beat you (i.e., he will have a total between seventeen and twenty one, inclusive)?

*d) Should you draw or not? (In the event that you and the dealer have the same total which is less than twenty one, then no one wins.)

C.45. H A Repeat Exercise C.44 in the case that the dealer has a six showing rather than a nine. (You can now account for three nines and three sixes rather than four nines and two sixes. Everything else is as in Exercise C.44.)

C.46. A Eight people wish to choose up sides to play a game of football. They decide to choose by the method often referred to as "the four same fingers." That is, they all chant, "The four same fingers are together," and, on the word "together," each person extends either one or two fingers on the right hand. If four people have extended one finger and the other four have extended two, then the two teams are determined; otherwise, the process is repeated again. Assuming that each person is just as likely to extend one finger as two,

a) what is the probability that the method will be successful on the first try?

b) what is the probability that Dan and Eliot will be on the same team?

C.47. S A A second method for choosing up two four person teams out of eight players is as follows: One of the eight is separated from the rest. The remaining seven then choose as in Exercise C.46, with the "four same fingers" constituting one team and the designated player then belonging to the other team with the "three same fingers."

a) What is the probability that this method will be successful on the first try?

b) If Dan is the designated player, what is the probability that Dan and Eliot are on the same team?

c) If neither Dan nor Eliot is the designated player, what is the probability that they will both be on the same team?

*C.48. A A third method of choosing teams from eight players is to have two designated players and to have the remaining players choose as in Exercise C.46. If there are "four same fingers," then the two designated players are on the same team with the "two same fingers." If there are two groups of "three same fingers," then the designated players flip a (fair) coin to see who is on which team. If the choosing process comes up five to one or six to zero, then it is repeated.

a) What is the probability that the method will be successful on the first try
    i) without requiring a coin toss?
    ii) requiring a coin toss?

b) If Dan and Eliot are the two designated players, what is the probability they end up on the same team?

c) If Dan is one of the designated players but Eliot is not, what is the probability that they will end up on the same team?

d) If neither Dan nor Eliot is a designated player, what is the probability they will end up on the same team?

C.49. H A If there are twenty five people in a room, what is the probability that at least two of them have the same birthday? (Assume that none have birthdays on February 29, and that all other birthdates are equally likely.)

Repeated Trials

C.50. A A student is taking a ten question multiple choice exam, each question offering four choices. If the student answers the questions completely at random, what is the probability that
    a) he gets everything right?
    b) he gets everything wrong?
    c) he gets eight or more right?

C.51. A Senta Dead, Cy Bull, and Mark Hitda were sent behind enemy lines to blow up the bridge on the Kly River. They had planted their explosives and were about to blow up the bridge when they discovered that the detonator was not working properly. Their only hope of a successful mission was to shoot the dynamite, which had been placed in three key locations. Unfortunately, they each had only one bullet remaining.

Fortunately, all were excellent markspeople. At the distance from which they had to shoot, Cy would hit his target 95% of the time, Senta would not miss more than 10%, and Mark had an 80% probability of success. They decided that they would fire simultaneously, each at a different one of the key dynamite placements. Assuming that the success of each of the three is independent of the success of each of the others,

a) what is the probability that all three hit their targets?

b) what is the probability that at least two hit their targets? (This would probably be enough to bring down the whole bridge anyway.)

c) what is the probability that they at least damage the bridge? (I.e., they have at least one hit.)

C.52. |A| Ten slips of paper, numbered 0 to 9 respectively, are placed in a hat. A slip is selected, its number noted, and then it is replaced. The procedure is repeated until a total of five numbers have been noted.

a) What is the probability that all five numbers are the same?

b) What is the probability that all five numbers are different?

c) What is the probability that at least two numbers are the same?

C.53. |A| Ty M. Waster spends all his free time playing with a game of the type pictured in Figure C.10. A metal ball is inserted through the opening at the top of the game, and, as the ball falls, it hits the metal nails and either falls to the right or left, eventually landing at the bottom.

a) How many different possible routes are there that the ball could follow on its way down?

b) For each of the six bottom slots, how many routes end in that slot?

c) What is the probability that a ball placed at the top of the game will land in slot number two?

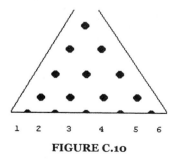

1   2   3   4   5   6

FIGURE C.10

*C.54. |H| |A| Last time I went to the theater noticed that there were ten seats in the row which I was seated and that, although the sea were filled one at a time, at no time were the vacant seats between seats that were alrea occupied.

In how many different ways could this occur

C.55. |A| The Women's Auxiliary of t Amalgamated Plumbers of Watertown is holdi its annual gala affair, and each of the ni members has been asked to bake either an ange food cake, a devil's food cake, or a marble cake. each woman is just as likely to bake any one of t three as she is to bake any of the others, what the probability that

a) there are five angel's food cakes, thr devil's food cakes, and one marble cake?

b) there are five or more angel's food cakes?

c) the number of angel's food cakes excee the number of devil's food cakes?

Odds And Expectation

C.56. |A| At Upson Downs, four horses are enter in today's feature race. The odds on Far Behi are 5 to 1; on Dead Heat, they are 2 to 1; Running Last, 11 to 1. If these are fair odds (i. the track is not taking a percentage), what are t odds on the fourth horse, Favorite Son?

C.57. |A| In the game of Chuckluck (also known Bird Cage) there are three dice in a cage. The ca is rotated, acting as a shaker.

a) What is the probability that the sum of t top faces of the three dice will be thirteen?

b) What is the probability that at least two the dice will show a two?

c) Before each roll, money can be bet on a of the numbers 1, 2, 3, 4, 5, or 6. If you bet on number and if that number shows up on one the dice, then you win the amount of your bet; if shows up on two dice, you win twice the amou of your bet; if it shows up on all three dice, y win three times the amount of your be Otherwise, you lose.

What is the expected value of this gam (Assume one dollar bets.)

C.58. |A| Raphael Loff was in need of $200, b his paycheck was only in the amount of $175. H therefore decided on the following plan. He hel a raffle, selling two hundred raffle tickets at $ each, and offering his paycheck as the prize. Tekka Shantz bought one ticket, what was h expectation?

.59. A Rudolph, Thomas and Neil each has hree denominations of U.S. currency in his ocket. Rudolph has one $5 bill, one $20 bill, and ne $100 bill; Thomas has one bill each of $1, 50, and $500; Neil has one bill each of $2, $10, nd $1000.

a) If Rudolph and Thomas each randomly elects one bill from his pocket, who is more likely ɔ select the bill with the higher denomination?

b) What if Thomas and Neil each selects one ill?

c) What about Neil and Rudolph?

d) Each of the three selects one bill, at andom, from his pocket, and they then compare ills. The one who has the bill with the highest enomination wins all three bills. What is the robability that Rudolph will win? Thomas? Jeil?

e) What is the expected value for each of the hree players of the game described in d)?

.60. A A man plays the following game. He lips a coin. If it lands on heads, then he wins $1. f the coin lands on tails, then the man flips again. f the coin comes up heads on the second flip, hen he wins $2; otherwise he flips again. Ie continues flipping in this manner until he inally obtains a head. If he obtains his first head n the nth toss, then he wins 2^{n-1} dollars.

a) What would be a fair price for the man to ›ay for the privilege of playing the game, if he is llowed at most six tosses?

b) What if he is allowed to continue tossing ntil a head comes up, regardless of how long that akes?

C.61. A Tickets for the Pennsylvania "Big Fifty" ottery contain a three digit red number, a two ligit white number, and a one digit blue number. drawing is held weekly in which the winning ed, white, and blue numbers are randomly elected. The following prizes are awarded[9]:

$50,000 if your red, white, and blue umbers all match the corresponding winning umbers;

$1,000 if your red and white numbers ›oth match the winning numbers;

$100 if your red and blue numbers both natch;

$25 if your white and blue numbers both natch;

$25 if your red number matches;

$5 if your white number matches.

You may only win one prize per ticket.

In addition, if you do not win one of the above prizes and if your blue number matches the winning blue number, then your ticket is entered into a special drawing, which is held after twenty million tickets have been sold in all. This drawing awards one prize in each of the following amounts: $1,000,000, $25,000, $20,000, $15,000, and $10,000.

If you buy one ticket for $.50, what is your expected value?

Miscellaneous

C.62. S A In how many ways may the numbers l, 2, 3, ..., 8 be placed in the eight cells in Figure C.11 so that no number is directly to the left of or directly above a smaller number?

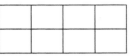

FIGURE C.11

C.63. H A In how many ways may the numbers 1, 2, ..., 9 be placed in the nine cells in Figure C.12 so that no number is to the left of or above a smaller number?

FIGURE C.12

C.64. H A The butcher left his shop for only a few moments. However, while he was gone, Beggar, the neighborhood mutt, wandered in and found nine sausages hanging as pictured in Figure C.13.

In how many different ways could Beggar possibly eat all nine sausages, one at a time, if he never eats a sausage higher on a string unless he has finished all sausages below it.

---

⁹): There is also an additional chance of four in one nillion of winning a Cadillac or Lincoln; but, to simplify he problem, we ignore this possibility.

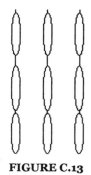

FIGURE C.13

C.65. [A] How many routes are there from A to B in the diagram in Figure C.14 if you can only travel toward the South or the East?

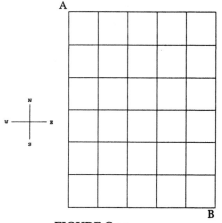

FIGURE C.14

C.66. [S] [A] Our local mailman is only willing to walk in a southerly, easterly or southeasterly direction. In the diagram in Figure C.15 (which represents our town), how many routes are there for the mailman to walk
    a) from A to B?
    b) from A to C?
    c) from A to D?
    d) from A to E?
    e) from A to F?

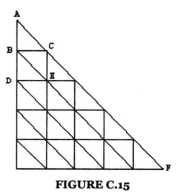

FIGURE C.15

C.67. [H] [A] Andy Hardly lives on the outskirt (the corner of A Street and First Avenue) of th town of Planesville. He attends Planesville Hi which is located at the intersection of two of th streets pictured in Figure C.16. Because And likes variety, he plans to follow a different rou each day for as long as he can. However, to avo backtracking, he always walks in a southern easterly, or south easterly direction. He h computed that he will be able to select differe route in this manner for exactly 377 days. Where is the school located?

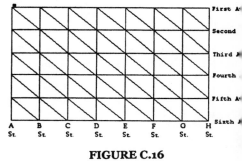

FIGURE C.16

C.68. ⊞Ⓐ Using the array in Figure C.17 count the number of ways in which you could start at a C, move to an adjacent O, etc., to spell the word COUNT.

```
              C
             COC
            COUOC
           COUNUOC
          COUNTNUOC
           COUNUOC
            COUOC
             COC
              C
```

FIGURE C.17

C.69. ⊞Ⓐ When the first man on Earth was told that a companion for him would be created, he made the sign. shown in Figure C.18 – both to introduce himself and to test Eve's intelligence.

a) In how many ways can you spell "MADAM M ADAM" by starting at an exterior M, moving horizontally or vertically to an adjacent A, etc., and ending at an exterior M?

b) What if you are allowed to start at any M and end at any M?

```
              M
             MAM
            MADAM
           MADADAM
          MADAMADAM
         MADAMIMADAM
          MADAMADAM
           MADADAM
            MADAM
             MAM
              M
```

FIGURE C.18

C.70. ⊞Ⓐ In how many ways can you start at an S in the array in Figure C.19 and spell SPHINX by moving from letter to adjacent letter. (Two letters are adjacent if they are connected by a line segment.)

```
        S–P–H–I–N–X
       P–H–I–N–X–S–P
      H–I–N–X–S–P–H–I
     I–N–X–S–P–H–I–N–X
    N–X–S–P–H–I–N–X–S–P
   X–S–P–H–I–N–X–S–P–H–I
    P–H–I–N–X–S–P–H–I–N
     I–N–X–S–P–H–I–N–X
      X–S–P–H–I–N–X–S
       P–H–I–N–X–S–P
        I–N–X–S–P–H
```

FIGURE C.19

Bibliography

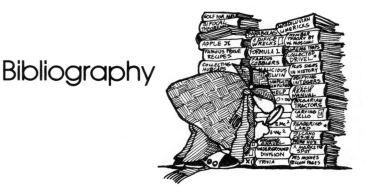

1. *American Mathematical Monthly*, Problems and Solutions section, Vol. 1– (1894–), the Mathematical Association of America.
2. Appel, K., and Haken, W., "The Solution of the Four Color Map Problem," *Scientific American*, October 1978.
3. Ball, W. W. R., *Mathematical Recreations and Essays*, 11th ed. Macmillan, New York, 1960.
4. Beck, A., Bleicher, M., and Crowe, D., *Excursions into Mathematics*, Worth, New York, 1969.
5. Beiler, A. H., *Recreations in the Theory of Numbers—The Queen of Mathematics Entertains*, Dover, New York, 1964.
6. Berlekamp, E., Conway, J. H., and Guy, R., *Winning Ways*, to be published by Academic Press, London.
7. Biggs, N., Lloyd, E. K., and Wilson, R., *Graph Theory 1736–1936*, Clarendon Press, Oxford, 1976.
8. Carroll, L., *Mathematical Recreations of Lewis Carroll*, Dover, New York, 1958.
9. Conway, J. H., *On Numbers and Games*, Academic Press, New York, 1977.
10. Degrazia, J., *Math Is Fun*, Emerson Books, New York, 1954.
11. Dudeney, H. E., *536 Puzzles and Curious Problems*, Scribner's, New York, 1954.
12. Dudeney, H. E., *Canterbury Puzzles*, Dover, New York, 1958.
13. Dudeney, H. E., *Amusements in Mathematics*, Dover, New York, 1970.
14. Eves, H., *An Introduction to the History of Mathematics*, 3rd ed., Holt, Rinehart, and Winston, New York, 1969.

15. Friedland, A. J., *100 New Recreations, Puzzles in Math and Logic,* Dover, New York, 1970.
16. Gardner, M., "Mathematical Games" column, *Scientific American,* Vol. 196– (1957–).
17. Gardner, M., *Mathematics, Magic, and Mystery,* Dover, New York, 1956.
18. Gardner, M., *The Scientific American Book of Mathematical Puzzles and Diversions,* Simon and Schuster, New York, 1959.
19. Gardner, M., *The Second Scientific American Book of Mathematical Puzzles and Diversions,* Simon and Schuster, 1961.
20. Gardner, M., *New Mathematical Diversions from Scientific American,* Simon and Schuster, New York, 1966.
21. Gardner, M., *The Numerology of Dr. Matrix,* Scribner's, New York, 1967.
22. Gardner, M., *The Unexpected Hanging and Other Mathematical Diversions,* Simon and Schuster, New York, 1969.
23. Gardner, M., *Martin Gardner's Sixth Book of Mathematical Games from Scientific American,* W. H. Freeman and Company, San Francisco, 1971.
24. Gardner, M., *Mathematical Carnival,* Knopf, New York, 1975.
25. Gardner, M., *Mathematical Magic Show,* Knopf, New York, 1977.
26. Golomb, S., *Polyominoes,* Scribner's, New York, 1965.
27. Gottlieb, A., "Puzzle Corner" column, *Technology Review,* Vol. 69– (1966–).
28. Graham, L. A., *Ingenious Mathematical Problems and Methods,* Dover, New York, 1959.
29. Greenes, C. E., "Function Generating Problems: The Row Chip Switch," *Arithmetic Teacher,* Vol. 20. (1973).
30. Grosswald, E., *Topics from the Theory of Numbers,* Macmillan, New York, 1966.
31. Hunter, J. A., *Mathematical Brain Teasers,* Dover, New York, 1976.
32. Hunter, J. A., and Madachy, J. S., *Mathematical Diversions,* Dover, New York, 1975.
33. *Journal of Recreational Mathematics,* Vol. 1– (1968–).
34. Kemeny, J. G., Snell, J. L., and Thompson, G. L., *Introduction to Finite Mathematics,* 3rd ed., Prentice-Hall, Englewood Cliffs, N.J., 1974.
35. Kraitchik, M., *Mathematical Recreations,* Dover, New York, 1953.
36. Leblanc, H., and Wisdom, W., *Deductive Logic,* 2nd ed., Allyn and Bacon, Boston, 1976.
37. Lindgren, H., *Recreational Problems, Geometric Dissections and How to Solve Them,* Dover, New York, 1972.
38. *Litton's Problematical Recreations,* compiled and edited by Angela Dunn, Van Nostrand Reinhold, New York, 1971. Some of the problems are being republished in *Mathematical Bafflers* by Angela Dunn, Dover, New York, 1980.

39. Loyd, S., *Mathematical Puzzles of Sam Loyd*, Dover, New York 1959.
40. Loyd, S., *More Mathematical Puzzles of Sam Loyd*, Dover, Nev York, 1960.
41. *Mathematics Magazine*, Mathematical Association of America, Vo 21– (1947–).
42. *Mathematics Teacher*, Vol. 1– (1908–).
43. Meyers, J., *Puzzle Quiz and Stunt Fun*, Dover, New York, 1956
44. *More Problematical Recreations*, compiled and edited by Angel Dunn, Litton, Beverly Hills, California.
45. Mott-Smith, G., *Mathematical Puzzles for Beginners an Enthusiasts*, Dover, New York, 1954.
46. *National Mathematics Magazine*, Vol. 9–20 (1934–1945).
47. Niven, I., and Zuckerman, H., *An Introduction to the Theory o Numbers*, 3rd ed., John Wiley and Sons, New York, 1972.
48. Ore, O., *Theory of Graphs*, American Mathematical Society Providence, Rhode Island, 1962.
49. Patton, T., *Card Tricks Anyone Can Do*, Castle Books, New York 1968.
50. Perlman, A., *Damnable Puzzles from Intellectual Digest*, Communi cations/Research/Machines, 1973.
51. Phillips, H. ("Caliban"), *My Best Puzzles in Logic and Reasoning* Dover, New York, 1961.
52. *Problematical Recreations*[11], compiled and edited by Angela Dunn Litton, Beverly Hills, California, 1970.
53. Read, R. C., *Tangrams, 330 Puzzles*, Dover, New York, 1965.
54. Schaaf, W., *A Bibliography of Recreational Mathematics*, Vol. 1 National Council of Teachers of Mathematics, Reston, Virginia 1970.
55. Schaaf, W., *A Bibliography of Recreational Mathematics*, Vol. 2 National Council of Teachers of Mathematics, Washington, D.C. 1970.
56. Schaaf, W., *A Bibliography of Recreational Mathematics*, Vol. 3, National Council of Teachers of Mathematics, Reston, Virginia 1973.
57. Schuh, F., *The Master Book of Mathematical Recreations*, Dover, New York, 1968.
58. Silverman, D., *Your Move*, McGraw-Hill, New York, 1971.
59. Smith, D. E., *History of Mathematics*, Vol. 2, Dover, New York, 1953.
60. *Sphinx: Revue Mensuelle des Questions Récréatives*, Vol. 1–9 (1931–1939).
61. Summers, G., *New Puzzles in Logical Deduction*, Dover, New York, 1968.
62. Summers, G., *Test Your Logic*, Dover, New York, 1972.

63. Trigg, C. W., *Mathematical Quickies*, McGraw-Hill, New York, 1967.
64. Wilson, R., *Introduction to Graph Theory*, 2nd ed., Longman Group Limited, Harlow, Essex, England, 1979.
65. Wylie, C. R., Jr., *101 Puzzles in Thought and Logic*, Dover, New York, 1957.

Hints and Solutions

CHAPTER 1

Exercises

1.1. Hint: Sue's mother must be one of the five women whose names are given.

1.4. Solution: The first step is to make a chart, listing the players' initials in the left column and the positions along the top. We then use the given information to place √'s and X's in the chart. Using the clues, we can immediately place X's in some boxes of the chart. For example, clue (a) enables us to place an X in the box where Allen's row meets the catcher's column. To indicate that clue (a) was used to place this X, we use a subscript (X_a). We continue eliminating boxes in this manner:

    1. Allen is not the catcher (clue a).
Ed is not the second baseman (clue b).
Harry is not the third baseman (clue d).
Neither Paul nor Chuck is the pitcher (clue e).
Ed is not an outfielder (clue f).
Chuck, Harry, and Allen are not outfielders (clue h).
Sam is not an infielder or in the battery (clue h).
Neither Paul nor Allen is the shortstop (clue i).

Neither Paul, Harry, nor Bill is the second baseman or the catcher (clue j).

Neither Ed, Paul, nor Jerry is the right fielder or the center fielder (clue m).

Bill is not the shortstop or the third baseman (clue n).

At this point, our chart appears as follows:

	p	c	1b	2b	ss	3b	rf	cf	lf
Allen		X_a			X_i		X_h	X_h	X_h
Bill		X_j		X_j	X_n	X_n			
Chuck	X_e						X_h	X_h	X_h
Ed				X_b			X_f	X_f	X_f
Harry		X_j		X_j		X_d	X_h	X_h	X_h
Jerry							X_m	X_m	
Mike									
Paul	X_e	X_j		X_j	X_i		X_m	X_m	
Sam	X_h	X_h	X_h	X_h	X_h	X_h			

We cannot tell yet who plays what position, so we return to the clues to make a list of who is married and so on:

The second baseman is engaged (b).
The pitcher is married (g).
Sam is married (k).
The catcher and the third baseman are not achelors (l).
Ed, Paul, Jerry, the right fielder, and the center elder are bachelors and everyone else is married (m).
From these observations, we can conclude that:

2. Ed, Paul, and Jerry are not the pitcher, the catcher, or the third baseman. (Place X_2 in the boxes where rows Ed, Paul, and Jerry meet columns p, c, and 3b.)

3. Sam is not the right fielder or the center fielder. (Place X_3's.)

4. The only possible position remaining for Sam is left field. (Place $\checkmark_4$.)

5. No one else can be the left fielder (X_5's).

6. The only possibility remaining for Paul is first base ($\checkmark_6$).

7. No one else can be the first baseman (X_7's).

8. The only possibility for Ed is now shortstop ($\checkmark_8$).

9. No one else can be the shortstop (X_9's).

10. The only possibility remaining for Harry is pitcher ($\checkmark_{10}$).

11. No one else can be the pitcher (X_{11}'s).

12. The only possible position remaining for Jerry is second base ($\checkmark_{12}$).

13. No one else can be the second baseman (X_{13}'s).

14. The only position remaining for Allen is third base ($\checkmark_{14}$).

15. No one else is the third baseman (X_{15}'s).

16. The only position remaining for Chuck is catcher ($\checkmark_{16}$).

17. No one else is the catcher (X_{17}).

18. This leaves Mike and Bill as the right fielder and the center fielder. From clues (c) and (o), Mike must be the right fielder, leaving Bill to be the center fielder.

This completes the chart:

	p	c	1b	2b	ss	3b	rf	cf	lf
Allen	X_{11}	X_a	X_7	X_{13}	X_i	$\checkmark_{14}$	X_h	X_h	X_h
Bill	X_{11}	X_j	X_7	X_j	X_n	X_n	X_{18}	$\checkmark_{18}$	X_5
Chuck	X_e	$\checkmark_{16}$	X_7	X_{13}	X_9	X_{15}	X_h	X_h	X_h
Ed	X_2	X_2	X_7	X_b	$\checkmark_8$	X_2	X_f	X_f	X_f
Harry	$\checkmark_{10}$	X_j	X_7	X_i	X_9	X_d	X_h	X_h	X_h
Jerry	X_2	X_2	X_7	$\checkmark_{12}$	X_9	X_2	X_m	X_m	X_5
Mike	X_{11}	X_{17}	X_7	X_{13}	X_9	X_{15}	$\checkmark_{18}$	X_{18}	X_5
Paul	X_e	X_j	$\checkmark_6$	X_j	X_i	X_2	X_m	X_m	X_5
Sam	X_h	X_h	X_h	X_h	X_h	X_h	X_3	X_3	$\checkmark_4$

Note that we have entered all the information about the problem in this one chart, with subscripts indicating the sequence in which the various $\checkmark$'s and X's were entered. Without these subscripts, there would be no way to reconstruct our reasoning from the final chart alone. Even with the subscripts, reconstructing the argument from the chart is a tedious task; thus the verbal description of the steps in our solution is important for others to be able to follow our argument. Bear this in mind if you ever write up a solution for someone else to look at. You should include a verbal description of your procedure as well as charts showing several intermediate points along the way.

1.5.   Hint: Make a chart showing what happens each day. Whom can St. Jacques play on the fourth day?

1.6.   Hint: Make a chart with columns as the nine days of the vacation and the rows as the three noise-makers.

1.7.   Hint: Make a chart in which the column headings are the various events and the row headings are the names of the teams. Include a column for total points earned by each team. How many points were scored altogether? How many could each team have obtained?

1.8.   Hint: Note that a total of ten points is scored in each game. What were the scores of the three games in which Alice participated? (Caution—Alice outscored her opponents by 22 points, but this does not mean that she scored only 22 points.)

1.9. Comment: Note that each of clues 1–4 tells us only that the particular resorts mentioned have particular features. They do not imply that the other resorts do not have these features. Thus, for example, we cannot conclude from clue 1 that the Shangri La does not have a swimming pool. Etc.

1.11. Hint: First consider the characteristic of knowing the odds. Since everyone is sitting next to someone who does know the odds, and yet four of the five are sitting next to someone who does not know the odds, the only possible seating arrangement is to have three people who know the odds sitting next to each other, as this figure indicates:

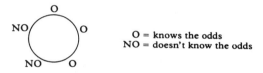

O = knows the odds
NO = doesn't know the odds

Do the same for each characteristic. Then consider Edie first and Cleo next.

1.13. Hint: Make two charts—one with three columns representing the caller, the person he intended to call, and the person who was actually reached; and the other plotting the names of the men against their occupations.

1.14. Solution: He said to himself: "If my face were clean, then Jack would realize that his own face is dirty, for why else would Jill be laughing. Since Jack did not draw this conclusion, my face must be dirty too."

1.15. Hint: See the solution of Exercise 1.14.

1.16. Hint: Don't forget that Carla knows that John did not receive what he ordered. Also, see the solution of 1.17 below.

1.17. Solution: Each cup contains five coins. There are only six possibilities for the contents of a cup: (Let N represent one nickel and D represent one dime.)

NNNNN	NNNND	NNNDD
(25¢)	(30¢)	(35¢)
NNDDD	NDDDD	DDDDD
(40¢)	(45¢)	(50¢)

Since there are six cups, each of the above possibilities must actually occur—one in each cup.

We begin by making a chart showing what each of the first two men could have felt (based on their statements), the possible incorrect labels that could have led them to these conclusions, and what their cups could actually have contained (see the figure).

	felt	label on cup	actual contents
first oil man	NNND	30¢	NNNDD =35¢
	NNDD	35¢	NNDDD = 40¢
	NDDD	40¢	NDDDD = 45¢
second oil man	NNNN	30¢	NNNNN = 25¢
	DDDD	50¢	NDDDD = 45¢

For example, if the first oil man felt NNND, the only way he could know that the fifth coin was a dime is if the label said 30¢.

Now consider the third man. On the basis of feeling only two coins, no matter what they were, there are still four possibilities for what his cup could contain. He could eliminate one of these by knowing the label on his cup, but the only way that he can eliminate more than one is if he can figure out what the first two have in their cups.

If the label on the third man's cup said 50¢, he could know all about the second man's cup (it would be labeled 30¢ and would contain NNNNN), but he would not be able to tell anything about the first's.

Similarly, if the third man's label said 45¢, 40¢, 35¢, or 25¢, he would not have been able to make his claim.

However, if his label actually said 30¢, then the second cup must have been labeled 50¢ and must have contained NDDDD. But then, the first cup could not have contained NDDDD and could not have been labeled 30¢. It must therefore have been labeled 35¢ and must have contained NNDDD = 40¢, as this figure shows:

oil man	label	actual contents
first	35¢	NNDDD = 40¢
second	50¢	NDDDD = 45¢
third	30¢	?

If the third man felt *DN*, he could now conclude that he has *NNNDD*. This is the only case that works.

We are now ready to consider the fourth leader of the oil industry. As he felt only one coin, there seem to be five possibilities for the contents of his cup. However, he is capable of the same reasoning we used above, and so he can deduce the labels and contents of the cups of the first three fellow students. Using the fact that the label on his cup is incorrect, he may be able to eliminate four of the five possibilities.

If he felt a dime, then no matter what his own label read, he would only have been able to eliminate three of the possibilities. (Check this.) But to know what he has, he must be able to eliminate four. Hence, he must have felt a nickel. The only way he would have been able to come to a conclusion is if his cup were labeled 25¢, in which case it would contain *NNNND*. We obtain the chart below:

oil man	label	actual contents
first	35¢	*NNDDD* = 40¢
second	50¢	*NDDDD* = 45¢
third	30¢	*NNNDD* = 35¢
fourth	25¢	*NNNND* = 30¢

Thus, the two remaining cups must be labeled 40¢ and 45¢; one must contain five nickels and the other five dimes. They thus contain 75¢ between them.

1.20. Hint: Set up a chart with five columns representing the competitor's name, school, entry number, prior standing, and position in the high jump. Enter as much information as possible, using letters, x, y, z, etc., to represent unknown quantities. For example, since Manners' position in the high jump was one number higher than his prior standing, make the following entry in one row of the chart:

competitor	school	entry number	prior standing	high jump
Manners			x	$x + 1$

Be sure you use different letters to represent different unknown quantities.

1.21. Hint: Which of the five houses is the red house? Consider cases.

1.22. Hint: First find out which show is at which theater. Make a chart showing all possible pairings as the row headings, and the shows and days of the week as the column headings. (There are twenty-five rows and ten columns.) Fill in what you can, then go back to the beginning and go through the clues again.

1.23. Hint: Hook A and B together, C and D together, and proceed as in Problem 1.6 (page 22). Then hook B, L, and C together and proceed as in Problem 1.6.

CHAPTER 2

Exercises

2.3. Hint: What is the largest number who could possibly be Truthfuls?

2.4. Solution: How many Truthfuls could there be? None? No, because then the fourth troll would be telling the truth and hence would be a Truthful. One? Possibly, if the fourth troll is the only Truthful. Two? Possibly, if the third and fifth trolls are Truthfuls. Three? Possibly, if the first, second, and fifth trolls are Truthfuls. Four? No, since the first three would be lying. Five? No, since the first three would be lying.

Consider, now, what the fifth troll might have said. Remember that his answer left no doubt in Silas' mind as to the clan affiliation of each.

If he said "three," then Silas would still have been in doubt because the fourth troll could have been the only Truthful or the first, second, and fifth trolls could be telling the truth.

Similarly, if the fifth troll said "two," two possible cases could remain (only the fourth is a Truthful, or the third and fifth are both Truthfuls).

If the fifth says anything else, then it is no longer possible that there are either two or three Truthfuls (see the cases above). Thus, the fourth must be the only Truthful.

2.6. Hint: What kind of troll could say about himself that he is a Liar?

2.8. Hint: Consider the possible alternatives for Hocus (he is a Truthful; he is a Liar; he is an Alternator who first tells the truth; he is an Alternator who first tells a lie). In each case, see what Pocus can be.

2.11. Hint: Ask a hypothetical question within a question. For example, "If I were to ask ..., what would you say?"

2.15. Solution: We will refer to the statements as Jeeves 1, Jeeves 2, etc. Jeeves made 3 statements (1, 2, 4) implying his innocence. At least one of these must be true. Therefore, Jeeves didn't do it.

Similarly, Jessica 1, 3, 4 imply she was innocent.

We now know that certain statements are true and certain are false. These are shown in the chart below:

Jeeves	Fifi	Julia	Jessica
1 2 3 4	1 2 3 4	1 2 3 4	1 2 3 4
T F		F	T

Since everyone refers to Jessica's being black-mailed, let's consider the possibilities: She was being blackmailed, or she wasn't being blackmailed.

Case 1   Suppose she wasn't being blackmailed.

Then her first two statements are true and therefore the last two are false. Therefore Fifi is innocent (Jessica 4 is F) and Julia must be the murderess. But in this case, Julia 1, 2, 3 are F. This contradicts the fact that each person makes two true and two false statements.

Case 2   Jessica was being blackmailed. This enables us to further fill in the chart:

Jeeves	Fifi	Julia	Jessica
1 2 3 4	1 2 3 4	1 2 3 4	1 2 3 4
T F	T	T F	T F

Now make a secondary assumption considering whether or not Jessica was in Chicago. If she was, then Jessica 1 and 3 are true, forcing Jessica 4 to be false, so that Fifi is innocent and Julia is the murderess. But then, Julia 1, 3, and 4 are false, again giving a contradiction.

Thus Jessica was not in Chicago. Our chart is now

Jeeves	Fifi	Julia	Jessica
1 2 3 4	1 2 3 4	1 2 3 4	1 2 3 4
T F	T	T F	T F F T

Note Jessica 4 must be true, and Fifi is the murderess. (We still have to check that this answer is consistent with the given information.)

We can fill in the chart as follows:

Jeeves	Fifi	Julia	Jessica
1 2 3 4	1 2 3 4	1 2 3 4	1 2 3 4
T F F	F F T T	T T F F	T F F T

Jeeves 3 must therefore be true and then everything fits. Hence, Fifi is the murderess.

2.20. Solution: Label the statements as Earl 1, Earl 2, etc. From Earl 2, Earl is not the architect since, if he were, this statement could not possibly be true.

Similarly, Luis is not the architect by Luis 1, and Randy is not the mason by Randy 2.

Since we now know that neither Earl nor Luis is the architect, we have two possibilities: Case 1, Moe is the architect; Case 2, Randy is the architect.

Case 1   In this case, Moe 2 must be true so that Luis is the mason. This leaves Earl and Randy to be the plumber and the carpenter (not necessarily in that order).

However, in this case, Earl 1 and Randy 1 would both have to be true; but they contradict each other. Thus, this case leads to a contradiction.

Therefore Case 2 holds: Randy is the architect. (Randy's statements will now be ignored as neither need be true. He does not mention the architect.) Moe cannot be the mason since if Moe 2 were true, Luis would have to be the mason—contradiction.

Therefore Moe is the carpenter or the plumber and Moe 1 must be true. Therefore Earl 1 is false and Earl cannot be the carpenter or the plumber, so Earl is the mason.

Similarly, Luis 2 is false, so Luis is not the carpenter. This leaves Luis to be the plumber and Moe to be the carpenter.

23. Hint: Can there be more red cards in the top
alf than there are black cards in the bottom half?

24. Solution: We begin by representing the state-
ients symbolically. Let g, h, and f represent the
»llowing statements:

> g: Martians are green.
> h: Martians have three heads.
> f: Martians can fly.

'hen the astronaut's statement, which we are told is
ue, can be represented as

$$[g \to (h \lor (\sim f))] \to [(g \leftrightarrow f) \land (\sim h)].$$

y Observation 9 (page 48), either $[g \to (h \lor (\sim f))]$
 false or $[(g \leftrightarrow f) \land (\sim h)]$ is true.
 Consider cases:

`ase 1` $[g \to (h \lor (\sim f))]$ is false. Then, by Ob-
ervation 10, g must be true and $(h \lor (\sim f))$ must
»e false. That is, g is true and h is false and f is
rue (by Observation 5).

`ase 2` $[(g \leftrightarrow f) \land (\sim h)]$ is true. Then, by Observa-
ions 3 and 11, h is false and g and f have the same
ruth values. However, since Martians have at least
»ne of the three characteristics, it is not possible for
, f, and h all to be false. Thus, g and f must both
»e true, with h false.

  Since both cases lead to the same conclusion, we
ind that Martians do not have three heads, they are
;reen, and they can fly.

2.28. Hint: Consider whether Archie owns the
Chrysler or the Chevy. Caution—if Archie does own
:he Chevy, the second clue tells us nothing about the
:olor of the car.

2.30. Hint: Express this problem symbolically.

2.31. Hint: Express this problem symbolically.

2.32. Hint: Express this problem symbolically.

CHAPTER 3

Exercises

3.2. Solution: The easiest way to solve the problem
is to work backward. However, we will take the
algebraic approach:

Let M = the amount of dollars they started with.
Dinner cost $\frac{1}{3}M$.
This left them with $M - \frac{1}{3}M = \frac{2}{3}M$.
The two theater tickets cost \$19.60.
This left them with $\frac{2}{3}M - 19.60$.
The taxi cost $\frac{1}{4}(\frac{2}{3}M - 19.60)$.
This left them with $\frac{3}{4}(\frac{2}{3}M - 19.60) = \frac{1}{2}M - 14.70$.
The nightclub cost \$23.10.
This left them with $\frac{1}{2}M - 14.70 - 23.10 = \frac{1}{2}M - 37.80$.
The cab fare home cost $\frac{1}{2}(\frac{1}{2}M - 37.80) = \frac{1}{4}M - 18.90$.
This left them $\frac{1}{4}M - 18.90$.
After the \$1 tip they are left with $\frac{1}{4}M - 18.90 - 1$,
which must $= 4.10$.
Therefore we get the equation

$$\frac{1}{4}M - 19.90 = 4.10$$
$$\frac{1}{4}M = 24.00$$
$$M = 96.$$

Since they started with \$96 and ended with \$4.10,
they spent \$91.90.

3.4. Hint: Let C = the amount of candy.
  How much did Samuel take and what was left?
  How much did Mildred take and what was left?

3.5. Solution: Each share is 4 bottles, so Tom
gives Mack 3 bottles and Don gives Mack 1. Hence
they should split Mack's money in the ratio of 3 : 1.

$$3x + x = 8.40$$
$$x = 2.10$$

Tom gets $3x$ = \$6.30 and Don gets x = \$2.10.

3.7. Hint: s = the number of sacks of wheat seed,
s is also the price per sack, so that altogether Farmer
Grey spent s^2 on wheat seed; etc.

3.10. Solution: Assuming that x is well defined,
then, squaring both sides,

$$x^2 = 1 + \sqrt{1 + \sqrt{1 + \sqrt{\cdots}}}$$

But the right side is just $1 + x$

$$x^2 = 1 + x$$
$$x^2 - x - 1 = 0$$
$$x = \frac{1 \pm \sqrt{5}}{2} \text{ by the quadratic formula.}$$

But $\dfrac{1 - \sqrt{5}}{2}$ is negative and x is positive, so the solution is $x = \dfrac{1 + \sqrt{5}}{2}$.

3.11.   Hint: See the solution of Exercise 3.10.

3.14.   Hint: Let $n =$ the number of $.03 screws that Calvin bought. Observe that n is also equal to the number of $.04 screws, etc. Let $y =$ total amount that each person spent.

How much did Wendy spend on $.03 screws? How many $.03 screws did she buy? Do the same for the $.04 and $.05 screws.

3.15.   Solution: Define A, B, C, D, E, and F in the obvious manner:

(1) $A + B = 647$
(2) $B + C = 675$
(3) $C + D = 599$
(4) $D + E = 583$
(5) $F = 370$
(6) $A + B + C + D + E + F = 1927$

From (5), $A + B + C + D + E = 1557$.

From (1) and (3),

$$(A + B) + (C + D) = 647 + 599 = 1246$$
$$E = 1557 - 1246 = 311.$$

Now substitute back in (4) to find D, and substitute back in (3), to find C, etc., getting $A = 299$, $B = 348$, $C = 327$, $D = 272$, $E = 311$.

3.25.   Solution: Let $S =$ Sue's present age
$\phantom{\textbf{3.25.}\quad\text{Solution: Let }}C =$ Chin's present age
Make a chart as shown, with row headings for Sue and Chin, and column headings for present, past, and future. Since Sue is 3 times as old as Chin was when Sue was as old as Chin is now, we can put

	Present	Past	Future
Sue	S		
Chin	C		

C in the Sue-past box and $\dfrac{S}{3}$ in the Chin-past box.

Also, since Sue will be 56 when Chin is as old as Sue is now, we can put 56 in the Sue-future box and S in the Chin-future box. The result is:

	Present	Past	Future
Sue	S	C	56
Chin	C	$\frac{S}{3}$	S

Since the difference between Sue's and Chin's ages remains constant,

$$S - C = C - \frac{S}{3} = 56 - S.$$

We obtain the following system of equations:

(1) $S - C = C - \dfrac{S}{3}$

(2) $S - C = 56 - S$.

(1′) $3S - 3C = 3C - S$

giving

(1″) $2S - 3C = 0$

(2) $2S - C = 56$

Subtracting, $-3C + C = -56$

$\phantom{\text{Subtracting, }}2C = 56$

$\phantom{\text{Subtracting, }}C = 28, \quad S = 42.$

3.26.   Hint: See the solution of Exercise 3.25.

3.28.   Hint: See the solution of Exercise 3.25.

3.32.   Solution: Imagine a flag at the front of each train. At the time the trains meet, how far are the flags apart? Answer, 0 ft. At the time the trains just pass each other, how far are the flags apart? Answer $270 + 192 = 462$ ft, as the figure shows.

Therefore the total distance traveled is 462 ft. At what rate are the flags separating? In each hour, one flag moves 45 mi and the other moves 60 mi. So they separate at the rate of 105 mph. To maintain consistency of units, we convert to feet per second:

$$105 \text{ mi/hr} = \frac{88}{60} \cdot 105 \text{ ft/sec.}$$

Using the formula $d = rt$ or $t = \dfrac{d}{r}$ we get

$$t = \frac{462 \text{ ft}}{\dfrac{88 \cdot 105}{60} \dfrac{\text{ft}}{\text{sec}}} = 3 \text{ sec.}$$

A slightly different approach is as follows (note that the time interval is the same for both trains):

For the first train, $d_1 = \dfrac{45 \cdot 88}{60} t$

For the second train, $d_2 = \dfrac{60 \cdot 88}{60} t$

here $d_1 + d_2$ = the distance between the flags
           = 462 ft.

we add the two equations to obtain

$$462 = d_1 + d_2 = \dfrac{45 \cdot 88}{60} t + \dfrac{60 \cdot 88}{60} t$$

$$462 = \dfrac{105 \cdot 88}{60} t$$

$$t = 3 \text{ sec.}$$

33. Hint: If d = the distance between Topeka nd Santa Fe, what are the rates of the two trains?

35. Hint: What is the relationship between the istance from Miami to the point at which the trains eet and the distance from Washington to that same oint? What is the ratio of the rates of the two ains?

37. Hint: When will the trains collide?

.38. Solution: Picture the action as shown in the gure.

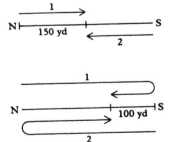

(a) Solution by two methods:

Let r_1 = the rate of the first cyclist
   r_2 = the rate of the second cyclist
   d = the length of track
   t_1 = the time from the start until they meet the first time

t_2 = the time from start until they meet the second time.

Method I

$$\frac{r_1}{r_2} = \frac{r_1 t_1}{r_2 t_1} = \frac{150}{d - 150}$$

$$\frac{r_1}{r_2} = \frac{r_1 t_2}{r_2 t_2} = \frac{d + 100}{2d - 100}$$

$$\frac{150}{d - 150} = \frac{d + 100}{2d - 100}$$

$$300d - 15{,}000 = d^2 - 50d - 15{,}000$$

$$d^2 - 350d = 0$$

$$d(d - 350) = 0$$

$$d = 0,\ 350.$$

(Note: 0 is extraneous.)

Method II

$$(r_1 + r_2)t_1 = d$$

$$(r_1 + r_2)t_2 = 3d$$

$$\frac{(r_1 + r_2)t_1}{(r_1 + r_2)t_2} = \frac{d}{3d} = \frac{1}{3}$$

$$3t_1 = t_2$$

$$d + 100 = r_1 t_2$$

$$= 3r_1 t_1$$

$$= 3 \cdot 150$$

$$= 450$$

$$d = 350.$$

(b) $\dfrac{r_1}{r_2} = \dfrac{150}{d - 150} = \dfrac{150}{200} = \dfrac{3}{4}$

$$r_1 = \frac{3}{4} r_2$$

Therefore $r_2 > r_1$ and the second cyclist must win.
(c) Observe that in time t_1 the first cyclist goes 150 yd and the second goes 200 yd; so in each time interval t_1, the second cyclist gains 50 yd. To overtake his opponent, the faster cyclist must gain a total of 350 yd. This will take 7 time intervals t_1. At the end of the seventh time interval t_1, the first cyclist will have gone $7 \cdot 150 = 1050$ yd $= 3$ laps of the track; thus, the winner will overtake the loser at the southern tip of the track.

3.39. Hint: Let n = the number of chairs showing on each side of the lift at one time.

Express the time of the skier's trip in two different ways: once using the up chairs; once using the down.

3.41. Hint: The movement of the boat relative to the driftwood is independent of the rate of the current. Thus, you can find the rate of the current even though you do not have enough information to determine the rate of the boat.

3.42. Solution: Let s = the number of steps showing at one time, t = the time it took Irv to take 3 steps (and Liz 2), and r = the number of steps that disappeared during time t.

Since Liz reached the top in 24 steps, it took her $12t$ seconds. During this time, $12r$ steps disappeared, so that Liz actually covered $s - 12r$ steps. That is, $24 = s - 12r$.

Similarly, consider the number of steps Irv climbed: $30 = s - 10r$.

Solving simultaneously,

$$6 = 2r$$
$$3 = r$$
$$s = 60.$$

3.43. Hint: See the solution of Exercise 3.42.

3.44. Solution: Let t be the total time in hours th Flog rides and let T be the total time that he walk Then he covers a total distance of $56t + 4T$ in a tot time of $t + T$. Since each of his companions mu cover the same distance in the same time, each rid for time t and walks for T. During the first pas senger's ride, the distance covered by the fir passenger is $56t$ km, while the walkers cover distance of $4t$ km (see Figure 1).

During the motorcycle's return trip, the cycli and the walkers are approaching each other at a rat of $56 + 4 = 60$ km/hr. As they are $56t - 4t = 52t$ k apart, the return trip of the motorcycle takes $\dfrac{52t}{60}$ h During this time, all walkers (including the firs passenger) cover a distance of $4\left(\dfrac{52t}{60}\right)$ km (Figure 2 The distance between the two groups of walker remains $52t$ km.

The second passenger now rides for time t, exactl overtaking the first passenger. The motorcycl returns $\left(\text{time } \dfrac{52t}{60}\right)$. The third passenger rides fo time t; the motorcycle returns $\left(\text{time } \dfrac{52t}{60}\right)$ for th

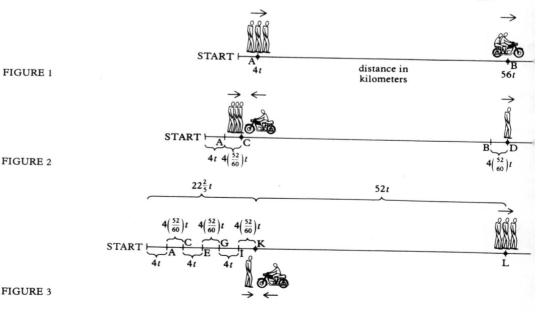

FIGURE 1

FIGURE 2

FIGURE 3

al walker. At this point the situation is as indicated
Figure 3.

The last passenger rides for t hours and catches up
$(78\frac{2}{5})t$ km; the total time for this trip is

$$4t + 3 \cdot \frac{52t}{60} = \frac{33}{5}t.$$

Since the trip must be completed in $1\frac{2}{3}$ hr $= \frac{5}{3}$ hr,

$$\frac{33}{5}t \le \frac{5}{3}$$

$$t \le \frac{25}{99}.$$

n the other hand, since the total distance must be
least 19.3 km,

$$78\frac{2}{5}t \ge 19.3$$

$$t \ge \frac{19.3}{(78\frac{2}{5})} = \frac{193}{784}.$$

hat is, t must satisfy

$$\tfrac{193}{784} \le t \le \tfrac{25}{99}.$$

here are values of t that satisfy both inequalities,
or example,

$$t = \tfrac{1}{4}.$$

herefore, the trip is possible. For instance, if
$= \frac{1}{4}$ hr $= 15$ min, each person walks for

$$3t + 3\left(\frac{52t}{60}\right) = 45 + 39 = 84 \text{ min}$$

nd rides for 15 min, giving a total trip of 99 min.
n this time they cover

$$84(\tfrac{4}{60}) + 15(\tfrac{56}{60}) = 19\tfrac{3}{5} = 19.6 \text{ km}$$

more than is needed).

3.45.   Hint: Let $n =$ the number of stations at which
the EE stops from the time it leaves Continental
Avenue until it reaches 34th St. (including 34th St.).
Let $m =$ the number of stations after Continental
Avenue up to and including Queen's Plaza. Let
$R_F =$ the rate of the F train (in stations per minute)
and $R_{EE} =$ the rate of the EE train (in stations
per minute).

3.52.   Hint: Let $A =$ the initial amount of grass in
the field (volume),

$C =$ the number of days required
for a cow to eat A,
$H =$ the number of days required
for a horse to eat A,
$S =$ the number of days required
for a sheep to eat A,
$G =$ the number of days required
for uneaten grass to grow A.

3.54.   Hint: If $S =$ the number of shirts that Leroy
bought, then how much did each shirt cost and what
was the total amount of money that he spent on
shirts?

3.55.   Solution: Although this can be done algebraically, it is quickly answered if you observe that the
amount of vinegar that was added to the oil must
exactly equal the net amount of oil that had been
removed. Since the oil that was no longer in the oil
bowl was now in the other bowl, the amount of
vinegar in the oil and the amount of oil in the
vinegar were equal.

3.60.   Hint: What is the total amount that the five
spent? How much did Bruce and Cookie spend
together? How much did Chuck spend?

3.61.   Hint: Let $L =$ the length of the left arm and
$R =$ the length of the right arm. Let $T =$ the weight
of 6 tomatoes, $B =$ the weight of the bag of beans.

Note that you will be able to determine $\dfrac{R}{L}$ but not

R and L separately.

3.71.   Hint: $(x - y)(x + y) = x^2 - y^2$.

3.72.   Hint: Let $n =$ the selected number.
   After the deck is shuffled, there are n cards face up
and $52 - n$ cards face down.
   In the pile of n cards taken from the top of the
shuffled deck, let $f =$ the number of cards that are
face up. How many cards in the packet are face down?
How many cards in the rest of the deck are face up?

3.73.   Solution: In a pile of cards constructed as
in this problem, if the first card has value V, how
many cards are in the pile? To reach 12 we must add
$12 - V$ more cards. So altogether there are $12 - V$
(added cards) $+ 1$ (the original card) $= 12 - V +
1 = 13 - V$. We can also look at this in another
way: The pile has cards corresponding to numbers
V up to 12; the numbers from 1 to $V - 1$ are
omitted. Therefore, there are $12 - (V - 1) =$

13 − V cards in the pile. These two approaches are illustrated below:

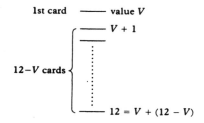

total is
1 + (12 − V) = 13−V cards

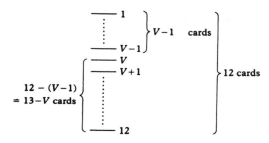

The trick can now be explained as follows:

Suppose there are n piles and the top cards have values $V_1, V_2, \ldots, V_n$.

Then the first pile has $13 − V_1$ cards,
the second pile has $13 − V_2$ cards,
$\vdots$
the nth pile has $13 − V_n$ cards.

Therefore there are $(13 − V_1) + \cdots + (13 − V_n) = 13n − (V_1 + V_2 + \cdots + V_n)$ cards on the table. There are R cards left over and altogether this will equal 52. That is,

$$52 = 13n − (V_1 + V_2 + \cdots + V_n) + R$$

or equivalently,

$$V_1 + V_2 + \cdots + V_n = 13n + R − 52.$$

Since $52 = (13)(4)$,

$$V_1 + V_2 + \cdots + V_n = 13(n − 4) + R.$$

In the particular problem, $n = 6$ and the number of cards left over is 5

$$V_1 + V_2 + \cdots + V_6 = 13(6 − 4) + 5$$
$$= 26 + 5 = 31.$$

3.74. Hint: Could the selected card have been placed anywhere in the deck, or does the magician make sure that it is in a certain position? What position?

See also the solution to Exercise 3.73.

3.76. Solution: Let $x =$ the number of cards taken by the first person, $y =$ the number of cards taken by the second person. Note that $y > x$ and $x + y < 26$. The magician is left with $52 − x − y$ cards.

In the face up pile of 26 cards dealt by the magician on the table, the card noted by the first subject is the xth from the bottom and the card noted by the second subject is the yth from the bottom. See the illustration.

Cards face up

When the 26 cards are turned face up on the table, the magician has $52 − x − y − 26 = 26 − x − y$ cards left in his hand. When the cards are placed face down at the bottom of his packet, the first subject's card is at the $26 − x − y + x = 26 − y$th position from the top. Similarly, the second subject's card is at the $26 − x$th position (see the illustration at the top of page 351). Altogether, the magician is holding $52 − x − y$ cards, hence the $26 − x$th card from the top is the $52 − x − y − (26 − x) + 1$st card = the $27 − y$th card from the bottom.

Throwing away the bottom card (which is the bottom card of the bottom half of the deck) leaves the selected cards at the $26 − y$th position from the top and the $26 − y$th position from the bottom. Hence, when the bottom half is turned over, the two selected cards will both appear at the $26 − y$th position.

Note that since $x < y$, $−x > −y$, and $26 − x > 26 − y$.

Cards face down

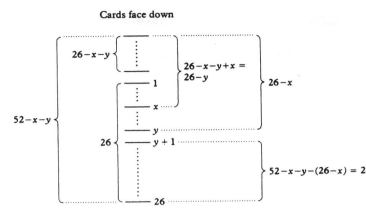

Therefore $26 - y <$ half of the deck in the magician's hand.

.77. Hint: Let $x =$ the number of cards that are removed, $y =$ the chosen number. Use a counting argument similar to the one given above for the solution of Exercise 3.76 to determine the position of the selected card.

3.78. Hint: Determine the position of the selected card by an argument similar to the one used above in the solution of Exercise 3.76.

3.79. Hint: Did the magician really select the four of diamonds? Where was the four of diamonds when the trick started?

Work backward and find out, then justify your answer.

CHAPTER 4

Practice Problems

4.M 5. Solution: Since

$$ABCDEF = A \cdot 10^5 + B \cdot 10^4 + C \cdot 10^3 + D \cdot 10^2$$
$$+ E \cdot 10 + F \cdot 1,$$

$$ABCDEF \equiv F \pmod{10}.$$

Therefore, by Theorem 4.9,

$$ABCDEF \equiv F \pmod 2.$$

In general, the remainder when a number is divided by 2 is the same as the remainder when the last digit is divided by 2—it is 1 if the last digit is odd and 0 otherwise.

Similarly,

$$ABCDEF \equiv F \pmod 5.$$

Thus, the remainder when ABCDEF is divided by 5 is the same as the remainder when F is divided by 5.

For example,

$$2387 \equiv 7 \equiv 2 \pmod 5.$$

To consider remainders upon division by 3, we use the fact that any number is congruent modulo 9 to the sum of its digits, and, by Theorem 4.9, this is also true modulo 3. In particular,

$$ABCDEF \equiv A + B + C + D + E + F \pmod 9,$$

so

$$ABCDEF \equiv A + B + C + D + E + F \pmod 3.$$

This enables us to "cast out threes." Thus, the remainder when a number is divided by 3 is the same as the remainder when the sum of the digits is divided by 3.

For example,

$$283147 \equiv 2 + 8 + 3 + 1 + 4 + 7 = 25 \equiv 2 + 5$$
$$= 7 \equiv 1 \pmod 3,$$

so the remainder when 283147 is divided by 3 is 1.

To find the remainder when ABCDEF is divided by 4, observe that $10^2 = 100 \equiv 0 \pmod 4$ and hence that $10^k \equiv 0 \pmod 4$ for $k \geq 2$. Therefore

$$ABCDEF \equiv 10E + F = EF \pmod 4.$$

That is, any number is congruent modulo 4 to its last two digits. Hence, to find the remainder when a number is divided by 4, it is only necessary to find the remainder when the number formed by the last two digits is divided by 4.

For example, $1873954618 \equiv 18 \equiv 2 \pmod 4$; so the remainder is 2 when 1873954618 is divided by 4.

The rule for 8 is obtained similarly:

$$10^3 = 1000 \equiv 0 \pmod 8,$$

so

$$ABCDEF \equiv DEF \pmod 8.$$

That is, the remainder when a number is divided by 8 is the same as the remainder when the number formed by the last three digits is divided by 8.

The rule for the number 11 is obtained in a manner similar to the rule for 9:

$$10 \equiv -1 \pmod{11}$$
$$10^2 \equiv 1 \pmod{11}$$
$$10^3 \equiv -1 \pmod{11}$$
$$\text{etc.}$$

Therefore

$$ABCDEF = A \cdot 10^5 + B \cdot 10^4 + C \cdot 10^3 + D \cdot 10^2$$
$$+ E \cdot 10 + F$$
$$\equiv -A + B - C + D - E + F \pmod{11}$$
$$= (F + D + B) - (E + C + A).$$

That is, to find the remainder when a number is divided by 11, first find the sum of the 1st, 3rd, 5th, etc. digits from the right and then subtract the sum of the 2nd, 4th, 6th, etc. digits from the right. The number obtained in this manner will have the same remainder upon division by 11 as the original number.

For example,

$$1879543765 \equiv (5 + 7 + 4 + 9 + 8)$$
$$- (6 + 3 + 5 + 7 + 1) \pmod{11}$$
$$= 33 - 22 = 11 \equiv 0 \pmod{11}.$$

So the remainder is 0 when 1879543765 is divided by 11; that is, 11 divides 1879543765.

Similarly,

$$29083 \equiv (3 + 0 + 2) - (8 + 9) = 5 - 17 = -12$$
$$\equiv 10 \pmod{11}.$$

That is, the remainder is 10 when 29083 is divide by 11.

The rule for the number 7 is somewhat mor complicated. It is based on the fact that $1001 = 7 \cdot 11 \cdot 13 \equiv 0 \pmod 7$. Thus

$$1000 \equiv -1 \pmod 7$$
$$1000000 = (1000)^2 \equiv 1 \pmod 7$$

and so on.

In particular,

$$ABCDEF = A \cdot 10^5 + B \cdot 10^4 + C \cdot 10^3 + D \cdot 10^2$$
$$+ E \cdot 10 + F$$
$$= (A \cdot 10^2 + B \cdot 10 + C) \cdot 10^3 + D \cdot 10^2$$
$$+ E \cdot 10 + F$$
$$= (ABC) \cdot 1000 + DEF$$
$$\equiv (ABC)(-1) + DEF \pmod 7$$
$$= DEF - ABC.$$

The general rule for finding the remainder when number is divided by 7 is to subdivide the number in blocks of 3 digits (counting from the right). Add the rightmost block, subtract the next block, add the next block, and so on. Divide the resulting number by 7 to obtain the desired remainder.

For example, to find the remainder when 8137496023 is divided by 7,

$$8137496023 \equiv 023 - 496 + 137 - 8 \pmod 7$$
$$= (23 + 137) - (496 + 8)$$
$$= 160 - 504$$
$$= -344 \equiv -1 \equiv 6 \pmod 7.$$

Therefore, the remainder is 6 when 8137496023 is divided by 7.

Exercises

4.3. Hint: Is 49,999 a prime? What would you have to do to show this?

4.6. Hint: If a number ends in a 0, what can be said about the prime factorization of this number?
What if it ends in two 0's?
What if it ends in three 0's? etc.
In the prime factored form of 50!, to what power is 2 raised? To what power is 5 raised?

4.7. Solution: Let B = the number of boys, G = the number of girls, and Y = the number of years the parents have been married. Each is a prime number and they are all distinct.

$$(B + G)B = 24 + Y$$

Suppose that B is even, that is, $B = 2$.

Then the left side is even and so Y is even, and must be 2 (the only even prime is 2). This is impossible since B and Y must be distinct. Thus $B \neq 2$ and must be odd.

If G were also odd, then again the left side would be even and $Y = 2$, leaving

$$B(B + G) = 26 = 2 \cdot 13.$$

In this case, B would have to be 2, which cannot be.

Therefore, G is even ($G = 2$). The equation becomes

$$B(B + 2) = Y + 24$$
$$B^2 + 2B - 24 = Y$$
$$(B - 4)(B + 6) = Y.$$

But Y is a prime and each factor is an integer since B is. Therefore, one factor must = 1. Since $B - 4 < B + 6$

$$B - 4 = 1$$
$$B = 5$$
$$Y = 11.$$

4.8. Hint: What is the relationship between the number of times that the kth switch is reversed and the number of divisors of k? How many divisors must k have for the kth light to end up on? Which lights end up on? (You might want to test your theory on 10 lights using cards to represent the lights.)

4.10. Solution: Let $a = p_1^{r_1} \cdots p_k^{r_k}$ be the prime factored form of a.

$$a^3 = p_1^{3r_1} \cdots p_k^{3r_k}.$$

Since $a^3 = b^2$ is a perfect square, each exponent must be even, that is, $3r_1$ is even, $3r_2$ is even, etc.

$$r_1 \text{ is even, } r_2 \text{ is even, etc.}$$
$$a \text{ is a perfect square, that is, } a = e^2.$$

Similarly c is a perfect square and $c = f^2$.

$$25 = c - a = f^2 - e^2$$
$$= (f - e)(f + e)$$

To solve this either $\begin{vmatrix} f - e = & 1 \\ f + e = & 25 \end{vmatrix}$ or $\begin{vmatrix} f - e = & 5 \\ f + e = & 5 \end{vmatrix}$

giving
$$2f = 26 \quad \text{or} \quad 2f = 10$$
$$f = 13 \qquad\qquad f = 5$$
$$e = 12 \qquad\qquad e = 0$$
$$\qquad\qquad\qquad \text{extraneous}$$

$$a = 12^2 = 144 \quad \text{and} \quad c = 13^2 = 169.$$

Finally, $b^2 = a^3 = 12^6$; therefore, $b = 12^3 = 1728$. Similarly, $d = 13^3 = 2197$.

4.11. Hint: Let x, y, and z be the three numbers. Then

$$x + y + z = 25$$
$$xyz = 540 = 2^2 \cdot 3^3 \cdot 5.$$

Therefore, x, y, and z are divisors of 540 which are less than 25. Note that exactly one of them, say x, is divisible by 5.

That is, $x = 5$ or 10 or 15 or 20.

In each of these cases, what can be said about y and z?

4.12. Hint: Considering that the census taker knows the number on the door, why did he have to ask whether the youngest is a twin?

4.13. Hint: If two numbers A and B have the same remainder when divided by d, what happens when their difference is divided by d?

4.17. Hint: Write two equations in three unknowns and eliminate one of the unknowns to obtain one linear Diophantine equation in two unknowns.

4.19. Solution: Let N = the price (in cents) of one notebook, L = the price (in cents) of one pencil, and P = the price (in cents) of one pen ($P > 10$). Then the equation representing the amount Garrett spent is:

$$695 = 6N + 4L + 5P \tag{1}$$

and the equation representing the amount Beth spent is:

$$285 = 2N + 5L + 3P. \tag{2}$$

Eliminate N by the addition subtraction method,

that is, three times equation 2 minus equation 1 gives

$$160 = 11L + 4P. \qquad (3)$$

This is a Diophantine equation in two unknowns and can be solved by the method discussed in the chapter. Convert equation 3 to a congruence modulo 4.

$$0 \equiv 3L \;(\text{mod } 4)$$

Since $\gcd(3, 4) = 1$, we may divide by 3 getting

$$L \equiv 0 \;(\text{mod } 4)$$

or, equivalently,

$$L = 4K$$

(where K is a positive integer since L must be positive). Substituting $4K$ for L in equation 3 we get

$$160 = 11(4K) + 4P$$
$$P = 40 - 11K.$$

Since P must be greater than 10, K is restricted to 1 or 2. Therefore, we have two possible cases:

When $K = 1$, $L = 4$, $P = 29$;
when $K = 2$, $L = 8$, $P = 18$.

For each case, we substitute the corresponding values for L and P in equation 1 and find N:

Case 1 $6N = 695 - 16 - 145 = 534$

$$N = 89$$

Case 2 $6N = 695 - 32 - 90 = 573$

$$N = 95\tfrac{1}{2}.$$

Since N must be a whole number, the case 2 answer does not satisfy the conditions of the problem. So $L = 4$, $P = 29$, $N = 89$, and Wilma paid

$$1(89) + 1(4) + 1(29) = 122 \text{ cents}$$
$$= \$1.22.$$

4.20. Hint: Set up the equation and solve it by considering cases—by considering how many grandfather clocks could have been sold.

4.23. Hint: $3^3 \equiv 2 \;(\text{mod } 5)$.

4.24. Hint: Let $x =$ the first number the subject selected, and let $y =$ the number the subject announced. (In the actual example, $y = 4$.)

If the subject counts "x" as the magician taps the number y on the clock (the first tap), how many taps will the magician have made at the time that the subject counts N (for a general number N)? At this time, to what number on the clock will the magician be pointing? (Note that the answer to this question must be expressed in terms of x, y, and N.) What should N be so that this indicated number will be "x" on the clock?

4.25. Hint: Use algebra and digital roots.

4.27. Hint: Use algebra and digital roots.

4.28. Hint: Use digital roots.

4.29. Hint: What is the largest value S could possibly have?
What is the digital root of S?
What is the largest value T could have?
What is the digital root of T?
What is the sum of the digits of T?

4.30. Hint: Let $G =$ the number of pieces of gum in the machine. From the information about Kevin, $G \equiv 1 \;(\text{mod } 4)$. Therefore, $G = 4K + 1$ for some positive integer K. Kevin took $K + 1$ balls, leaving $3K$ balls. From the information about Esteban, $3K \equiv 1 \;(\text{mod } 4)$. Therefore, $K \equiv 3 \;(\text{mod } 4)$, and so $K = 4J + 3$. Esteban took $3J + 3$ gum balls, leaving $9J + 6$ gum balls, etc. Remember we are looking for the smallest possible answer.

4.31. Solution: Let $p =$ the price per plum; then $3p =$ the price per apple.
Let $A =$ the total number of apples purchased, and $P =$ the total number of plums. Then

$$Pp = 1188, \qquad A(3p) = 1188.$$

Therefore $P = 3A$, and so $A + P = 4A$. That is, $A + P \equiv 0 \;(\text{mod } 4)$.
The sum of all the baskets $= 289 \equiv 1 \;(\text{mod } 4)$.
Therefore the number of fruits in the remaining basket is congruent to 1 modulo 4. Of all the baskets, only two (the one containing 17 fruits and the one containing 25 fruits) have a number of fruits congruent to 1 modulo 4.
Consider the two cases:

Case 1   The basket with 25 fruits remains:

$$A + P = 289 - 25 = 264$$

$$4A = 264$$
$$A = 66$$

The only sum of baskets that gives 66 is $30 + 17 + 19$.

Case 2 The basket with 17 fruits remains:

$$A + P = 289 - 17 = 272$$
$$4A = 272$$
$$A = 68$$

No sum of baskets can equal 68, so this case cannot occur.

Therefore the basket remaining contains 25 apples, 66 apples cost \$11.88 (so each apple costs \$.18), and the remaining basket is worth \$4.50.

4.32. Hint: See the solution to Problem 5 in Practice Problems 4.M (page 351).

4.33. Hint: See again the solution to Problem 5 in Practice Problems 4.M (page 351). Observe that saying that $n \mid m$ is the same as saying that $m \equiv 0 \pmod{n}$.

Note also that a number is divisible by 6 if and only if it is divisible by 2 and by 3; by 12, if and only if it is divisible by 4 and by 3; by 14, if and only if it is divisible by 2 and by 7; by 15, if and only if it is divisible by 3 and by 5.

The rule for 13 is similar to the rule for 7, since $1001 \equiv 0 \pmod{13}$. The rule for sixteen is analogous to the rules for two, four, and eight.

4.34. Hint: $99 = 9 \cdot 11$.

4.35. Hint: Find the prime factorization of 792.

4.36. (a) Solution: The number of days in 400 years is:

$400(365) + 97$ days, where 97 is the number of leap years in 400 years.

$400(365) + 97 \equiv 1 \cdot 1 + 6 \equiv 0 \pmod{7}$.

Thus, 400 years contains an exact number of weeks, so the day following the 400 year period is the same day of the week as the first day of the period.

(c) Solution: Label the days of the week as Sunday—0, Monday—1, Tuesday—2, Wednesday—3, Thursday—4, Friday—5, Saturday—6. In order to do this problem, we need a reference day with known date and known day of the week. We select our reference day to be Sunday, January 1, 1978. Since it is a Sunday, we represent it by 0. (You can select any day you like—the argument will have to be modified accordingly).

Let x be the number representing the day of the week on which January 1, 1901 fell. Then we can write a congruence modulo 7 representing the number of days between January 1, 1901 and January 1, 1978.

$$\underbrace{0}_{\substack{\text{Jan 1,} \\ 1978}} \equiv \underbrace{x}_{\substack{\text{Jan 1,} \\ 1901}} + \underbrace{77(365)}_{\substack{\text{number} \\ \text{of years}}} + \underbrace{19 \cdot 1}_{\substack{\text{no. of} \\ \text{leap yr}}} \pmod{7}$$

$$0 \equiv x + 0 \cdot 1 + 5 \cdot 1 \equiv x + 5 \pmod{7}$$

Therefore

$$x \equiv 2 \pmod{7}.$$

So, January 1, 1901 fell on a Tuesday.

(d) Hint: Use the statements in parts (b) and (c) of the exercise.

(e) Hint: Again, use parts (b) and (c).

(f) Hint: How many days are there in a century from January 1, 1901 to January 1, 2001? From January 1, 2001 to January 1, 2101? Etc. Use parts (a) and (c) of the exercise.

(g) Hint: Use parts (b) and (c). If a year contains 53 Sundays, what is the first day of the year?

CHAPTER 5

Exercises

5.2. Hint: Consider congruence modulo 10. (Note that a number is congruent modulo 10 to its last digit.)

5.3. Solution: Let n^2 = the number. n can be written in the form $n = 10k + r$ (where r is the last digit, $0 \leq r \leq 9$)

$$n^2 = (10k + r)^2 = 100k^2 + 20kr + r^2.$$

$100k^2$ does not affect the last two digits.

$20kr$ does not affect the last digit; it makes an even contribution to the next to last digit. Therefore, the parity (evenness or oddness) of the next to last digit of n^2 is the same as the parity of the next to last digit of r^2.

Consider the possibilities for r^2:

$$0^2 = 0 = 00, \quad 4^2 = 16, \quad 7^2 = 49,$$
$$1^2 = 1 = 01, \quad 5^2 = 25, \quad 8^2 = 64,$$
$$2^2 = 4 = 04, \quad 6^2 = 36, \quad 9^2 = 81.$$
$$3^2 = 9 = 09,$$

The only cases in which the next to last digit of r^2 is odd, are $r = 4$ and $r = 6$. In both of these cases, the last digit of r^2 is 6. Conversely, if the last digit of r^2 is 6, then $r = 4$ or 6 and so the next to the last digit of r^2 is odd.

5.4. Hint: See Exercises 5.2 and 5.3. It is also helpful to consider what the digital root of a perfect square could be.

5.6. Solution: If $T =$ the tens digit and $U =$ the units digit of n, then $n = 10T + U$. The product of the digits is TU. Therefore TU divides $(10T + U)$. Therefore T divides $(10T + U)$ and therefore T divides U (Theorem 4.1, page 105).

Similarly U divides $(10T + U)$ and therefore U divides $10T$. We make a chart of the possible values of U and T in which T divides U and U divides $10T$, as follows:

U	1	2	2	3	4	4	5	5	6	6	7	8	8	9
T	1	1	2	3	2	4	1	5	3	6	7	4	8	9

We thus obtain numbers 11, 12, 15, 22, 24, 33, 36, 44, 48, 55, 66, 77, 88, and 99. Of these, only 11, 12, 15, 24, and 36 satisfy the conditions of the original problem.

5.8. Hint: $10T + U - (10U + T) = 9(T - U)$; $k(T + U) = 9(T - U)$ for each possible value of k, what can T and U be?

5.9. (b) Solution: Let the number $= 10T + U$; then its reversal $= 10U + T$.

$$(10U + T) - (10T + U) = 9U - 9T$$
$$= 9(U - T) = k^2$$

Therefore, since 9 is a perfect square, $U - T$ must be a perfect square. Since $U - T < 9$, we get $U - T = 0, 1$ or 4.

We get the following possible values: 11, 12, 15, 22, 23, 26, 33, 34, 37, 44, 45, 48, 55, 56, 59, 66, 67, 77, 78, 88, 89, 99.

5.11. Hint: If a square $= 99k = 9 \cdot 11k$, k must be divisible by 11 and $k/11$ must be a perfect square. Also, if 11 divides $(T + U)(T - U)$, then either 11 divides $T - U$ or it divides $T + U$.

5.13. Solution: Write the number as $100H + 10T + U$; its reversal is $100U + 10T + H$.

If $H > U$, the difference is $100(H - U) + (U - H)$.

Since $H > U$, we write this as $100(H - U - 1) + 90 + (U - H + 10)$ so that all the terms will be positive.

Note that the three digits of this number are

$$H - U - 1, \ 9, \ U - H + 10;$$

reversing the digits, we get $U - H + 10$, 9, $H - U - 1$; and adding

$$\begin{aligned} &100(H - U - 1) + 90 + U - H + 10 \\ + \ &100(U - H + 10) + 90 + H - U + 1 \end{aligned}$$
$$\text{we get} \quad 100 \cdot 9 \quad + 180 + 9 = 1089.$$

5.15. Solution: One way of proceeding is by asking questions that divide the remaining possibilities in half. Or we can use the binary system and with each question can determine another digit in the base two representation of the number. That is, we may begin by asking, "Is the number less than 512?" $((1000000000)_{two})$.

If the answer is yes, then the base two representation of the number has fewer than ten digits, and we next ask, "Is the number less than 256?" Etc.

If the answer to the question about 512 is no, then the number has ten digits in its base two representation, and the leftmost digit is 1. We determine the second digit in the base two representation by asking whether the number is smaller than 768 $((1100000000)_{two})$. Each successive question determines the next digit in the base two representation of the number. Since there are ten possible digits, ten questions may be required to pinpoint the number exactly. As an example, suppose the number is 379. We then end up asking the following questions:

Q: Is it less than 512?
A: Yes (there are at most 9 digits).
Q: Is it less than 256?
A: No [the number is $(1--------)_{two}$].

Q: Is it less than 384 (256 + 128)?
A: Yes [the number is $(10\text{-------})_{two}$].
Q: Is it less than 320 (256 + 64)?
A: No [the number is $(101\text{------})_{two}$].
Q: Is it less than 352 (320 + 32)?
A: No [the number is $(1011\text{-----})_{two}$].
Q: Is it less than 368 (352 + 16)?
A: No [the number is $10111\text{----})_{two}$].
Q: Is it less than 376 (368 + 8)?
A: No [the number is $101111\text{---})_{two}$].
Q: Is it less than 380 (376 + 4)?
A: Yes [the number is $(1011110\text{--})_{two}$].
Q: Is it less than 378 (376 + 2)?
A: No [the number is $(10111101\text{-})_{two}$].
Q: Is it less than 379 (378 + 1)?
A: No [the number is $(101111011)_{two} = 379$].

5.16. Hint: See the solution of Exercise 5.15.

5.18. Hint: Note that card A contains all numbers up to 31 that have a 1 in the units digit (the last digit) in their base two representations; card B contains those numbers that have a 1 in the two's digit (the next to last digit); etc.

5.19. (a) Solution: On the first deal, all numbers having 0 as the units digit in their base two representations are placed, in their original order, on top of zero, and all numbers having units digit 1 are placed on one. When the deck is reassembled, it is in the order shown at the left below.

The cards are now sorted according to their last digits. The next deal sorts the cards according to the next to last digit, as at the right below.

$\text{---}0$	$\text{--}00$	$\text{--}10$
$\vdots$	$\vdots$	$\vdots$
$\text{---}0$	$\text{--}00$	$\text{--}10$
$\text{---}1$	$\text{--}01$	$\text{--}11$
$\vdots$	$\vdots$	$\vdots$
$\text{---}1$	$\text{--}01$	$\text{--}11$

When the deck is reassembled, we obtain the situation shown at the left at the top of the next column.

The next deal sorts according to the third digit from the right. After reassembly, the result is as shown at the right.

The final deal completes the sorting, with the original order restored.

$\text{--}00$	-000
$\vdots$	-001
$\text{--}01$	-010
$\vdots$	-011
$\text{--}10$	-100
$\vdots$	-101
$\text{--}11$	-110
$\text{--}11$	-111

0000	
0001	
$\vdots$	
1111	

5.20. Hint: At the time that the magician was blindfolded, how many matchsticks remained on the table?

If A represents the number of sticks originally given to the person who ended up with the apple, B represents the number of sticks originally given to the person who ended up with the banana, and P represents the original number of sticks given to the person who ended up with the peach, what is the physical significance in this problem of the number whose base three representation is $(PBA)_{three}$?

5.22. (b) Solution: We follow the card's position in the deck as the magician performs the trick. When the magician first reassembles the deck, after having dealt out the three piles, how many cards are above the selected card? Since each pile contains 9 cards, if the selected pile was placed on top, then there are $0 \cdot 9 + r$ cards above it where $0 \le r \le 8$; if the selected pile is placed in the middle, then there are $1 \cdot 9 + r$ cards above it where $0 \le r \le 8$; if the selected pile is placed on the bottom, there are $2 \cdot 9 + r$ cards above it where again $0 \le r \le 8$. That is, there are $9x + r$ cards above the selected card where $x = 0$, 1, or 2, depending on whether the selected pile was placed on the top, middle, or bottom.

When the deck is redealt face up, how many cards on the table are below the selected card? Since the $9x + r$ cards that were previously above the selected card are distributed into three piles, there will be $\frac{1}{3}(9x + r) = 3x + s$ cards below the selected card, where $0 \leq s \leq 2$ since $0 \leq r \leq 8$ (there cannot be a fraction of a card).

When the deck is reassembled, there will be $9y + 3x + s$ cards above the selected card, where $y = 0, 1,$ or 2, depending on whether the selected pile was placed on the top, middle, or bottom.

Similarly, when the cards are dealt face up for the third time into three piles, there will be $3y + x + (0)$ cards below the selected card in its pile [note $+ (0)$ because $0 \leq s \leq 2$, and we can't have a fraction of a card].

When the deck is reassembled this time, there will be $9z + 3y + x$ cards above the given card where $z = 0, 1,$ or 2 depending on whether the selected pile is placed on the top, middle, or bottom.

Therefore, if we wish to select the card to be at the Nth position, we must have

$$9z + 3y + x = N - 1.$$

That is, (z, y, x) is the base three representation of the number $N - 1$.

To apply this finding to part (a) of the problem if $N = 23$,

$$N - 1 = 22 = (211)_{three}.$$

Therefore we place the selected pile first in the middle (1), next in the middle (1), and lastly on the bottom (2).

5.24. Solution: In the base three representation of a number, the digits 0, 1, and 2 are used. In a sense, though, we could use 0, 1, -1, since

$$2 \cdot 3^k = (3 - 1)3^k = 1 \cdot 3^{k+1} - 1 \cdot 3^k.$$

Thus, any number can also be expressed as a sum of powers of 3 with coefficients 0, 1, and -1. For example,

$$\begin{aligned}
23 &= 2 \cdot 3^2 + 1 \cdot 3^1 + 2 \cdot 3^0 \\
&= 2 \cdot 3^2 + 1 \cdot 3^1 + (3 - 1)3^0 \\
&= 2 \cdot 3^2 + 1 \cdot 3^1 + 1 \cdot 3^1 - 1 \cdot 3^0 \\
&= 2 \cdot 3^2 + 2 \cdot 3^1 - 1 \cdot 3^0 \\
&= 2 \cdot 3^2 + (3 - 1)3^1 - 1 \cdot 3^0 \\
&= 2 \cdot 3^2 + 1 \cdot 3^2 - 1 \cdot 3^1 - 1 \cdot 3^0 \\
&= 3 \cdot 3^2 - 1 \cdot 3^1 - 1 \cdot 3^0 \\
&= 1 \cdot 3^3 - 1 \cdot 3^1 - 1 \cdot 3^0 \\
&= 1 \cdot 3^3 + 0 \cdot 3^2 - 1 \cdot 3^1 - 1 \cdot 3^0.
\end{aligned}$$

This representation of a number N is the basis of this trick. Card A contains those numbers whose last coefficient (coefficient of 3^0) is $+$ or -1; (-1 is indicated by an $\star$); card B contains those numbers whose next to last coefficient (coefficient of 3^1) is $+$ or -1 (-1 is indicated by a $\star$), etc.

Therefore, to recapture a number, add the proper powers of 3 corresponding to cards containing the selected number without an $\star$, and subtract that power of 3 if the number on the card has an $\star$.

For example, if $N = 8$, which appears on Card A with an $\star$ and on card C without $\star$,

add 9 for card C,

subtract 1 for card A,

getting $9 - 1 = 8$. Note $8 = 1 \cdot 3^2 + 0 \cdot 3 - 1 \cdot 3^0$.

The only problem that can arise occurs when the result of the calculation is negative. In this case, we simply add 27 (3^3) to this negative number. For example, 22 appears on cards A and B without $\star$, and on card C with $\star$.

Add 1 for card A,

add 3 for card B,

subtract 9 for card C,

getting -5. We add 27 to -5, getting 22. (Note $22 = 1 \cdot 3^3 - 1 \cdot 3^2 + 1 \cdot 3^1 + 1 \cdot 3^0$.)

5.25. Hint: Refer to the solution of Exercise 5.24.

5.28. Hint: If $(11111)_b = b^4 + b^3 + b^2 + b + 1 = x^2$, is x larger or smaller than $b^2 + \dfrac{b}{2}$?

What about $b^2 + \dfrac{b+1}{2}$? What must x be equal to?

5.29. Hint: What could the value of $792 \div 297$ possibly be?

5.31. (a) Solution: $(121)_b = b^2 + 2b + 1$
$$= (b + 1)^2$$
$$= [(11)_b]^2, \text{ for each } b > 2.$$

5.32. (a) Solution:

$$2(b - 1) = b + (b - 2) = [1(b - 2)]_b,$$

where $b - 2$ is the units digit and 1 is the b digit.
Similarly,

$(b - 1)^2 = b^2 - 2b + 1 = b(b - 2) + 1$

$= [(b - 2)1]_b.$

5.35. Hint: Construct a table of squares from 1 to 1000.

5.39. Solution: Since $O \cdot T$ ends in T and $N \cdot T$ ends in T, the possibilities are:

$T = 0,$

$T = 5$ with O and N odd,

or T is even and O and N are 1 and 6 or 6 and 1.

Neither N or O can equal 1 since $N \cdot LET \neq LET$ and $O \cdot LET \neq LET$. Also $T \neq 0$ since $O + T = E$, hence $T = 5$ and O and N are odd. Since $N \cdot LET = NOT$, $L = 1$.

If $E = 0$, then $O \cdot LET$ would start with O; but it doesn't. Therefore $E \geq 2$. But $125 \cdot 9 =$ four digits, so neither N nor O can be 9. Therefore N and O are 3 and 7 or 7 and 3.

Since $7 \cdot 145 =$ four digits, $E = 2$ or 3. Since $O + T$ ends in E and O and T are odd, E is even. Therefore $E = 2$ and $LET = 125$.

Since $N \cdot LET$ starts with N and $O \cdot LET$ does not start with O, $N = 3$ and $O = 7$.

Therefore

$$\begin{array}{r} 125 \\ 37 \\ \hline 875 \\ 375 \\ \hline 4625 \end{array}$$

5.40. Hint: First find C. What is the relationship between D and E?

5.43. Hint: For reference purposes, represent the missing digits by letters. (Here, distinct letters might represent the same digit.)

$$\begin{array}{r} a\,b\,3\,c \\ \times\ \ d\,e\,f \\ \hline g\ h\,i\ 0\,j \\ k\,m\,n\,7\,p \\ q\,2\,3\ r \\ \hline s\,4\,t\ u\,v\,5\,j \end{array}$$

First find p; then find c and e, then d, then b, and then a.

5.44. Hint: What could X possibly be? Note that $X^2 +$ the carryover from the previous digit (if there is any such carryover) = _X.

5.45. Hint: $XXX = X \cdot (111)$ and $111 = 3 \cdot 37$.

5.46. Hint: See the hint for Exercise 5.45.

5.47. Hint: How large could CH possibly be? Find EIN from $7(EIN) = \_44\_$.

5.48. Hint: Is the first digit of the quotient greater than, equal to, or less than 8?

5.52. Hint: Label the rows of the division problem as in the solution of Sample Problem 5.4 (page 160). What could row 1 be? What does this imply about the divisor?

5.53. Hint: What can be said about the parity of the last digit of the divisor? What is the relationship between the last two digits of the quotient?

5.56. Hint: Note that the product of the divisor and the last digit of the quotient ends in 000.

5.57. Hint: If the square of a three-digit number has only five digits, what can you say about the size of the first digit of the three-digit number? (Remember that you are working in base six.)

5.58. Solution: In whatever base you are working,

$ABCABC = (1001)ABC.$

$(1001)_{seven} = 344_{ten}.$

$(ABC)_{seven} = 49A + 7B + C.$

Therefore $(344)(49A + 7B + C)$ is a perfect square, and so

$4(86)(49A + 7B + C) = x^2.$

Since 86 contains no square factors, $49A + 7B + C = 86$ times a perfect square (see Practice Problems 4.D, problem 5, page 111). Since A, B, and C are each at most 6, $49A + 7B + C$ is at most 342.

$86 \cdot 2^2 = 86 \cdot 4 = 344$ is too big.

Therefore, $49A + 7B + C = 86 \cdot 1 = 86$. We write 86 in base 7:

$86 = 1 \cdot 49 + 5 \cdot 7 + 2$

$= (152)_{seven};$

$A = 1, B = 5,$ and $C = 2.$

5.59. Hint: See the solution of Exercise 5.58.

5.61. Hint: Consider even bases and odd bases separately.

CHAPTER 6

Exercises

6.3. Hint: Use Theorem 6.5.

6.4. Hint: Use Theorem 6.5.

6.5. Hint: The chessboard may be represented by a graph whose vertices correspond to cells of the chessboard. Two vertices are connected by an edge if it is possible for the knight to move (in one move) between the corresponding cells of the board.

6.6. Hint: Represent the chessboard by a graph. A trip is possible for part (a). No such trip is possible for part (b).

6.7. Solution: There are 6 odd vertices, A, B, C, E, F, and G. Since we are looking for an Eulerian path from A to B, we must eliminate "the oddness" of the vertices C, E, F, and G. We want to do this in the most efficient way (by adding paths of shortest possible length).

Clearly, if we add a path from C to F (or retrace CF) and add a path from G to E (or retrace EG), we will accomplish this aim. The result is this graph:

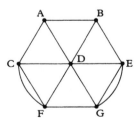

An Eulerian path can be found in the manner described in the chapter. One such path is ACFDCFGEDGEBDAB. The total distance traveled with this route is 1400 meters.

6.8. Hint: Refer to the solution of Exercise 6.7, but note that here not all edges are of the same length.

6.10. Hint: Consider the graph with 7 vertices labeled 0, 1, 2, 3, 4, 5, 6, with each pair of vertices connected by an edge. Each edge can be thought of as representing a domino. For example, the edge connecting vertex 3 to vertex 6 corresponds to the domino three–six. With regard to this graph, to what does a complete chain of dominoes correspond? What if one domino is missing?

6.11. Solution: Suppose we look at a sequence of digits—say, 12132—on the circle. This gives rise to the set of triples 121, 213, 132. Notice that the pair 21 is the end of 121 and the beginning of 213. Similarly, 13 is the end of 213 and the beginning of 132. So think of a triple as a merging of two pairs (for example, 213 represents the merging of 21 with 13). This suggests that we construct a directed graph as follows:

Make nine vertices that are labeled by the two-digit numbers 11, 12, 13, 21, 22, 23, 31, 32, 33. Draw an edge from vertex ij to vertex km, if j = k (for example, draw an edge from 12 to 23, 12 to 22, and 12 to 21). Note that each edge corresponds to a triple. For example, the edge from 12 to 23 corresponds to the triple 123. The resulting graph is pictured below.

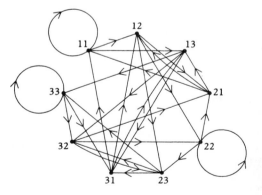

Each vertex has in-degree 3 and out-degree 3. Therefore there exists an Eulerian dicircuit which may be found as suggested in the text.

One such dicircuit is $11 \to 11 \to 12 \to 21 \to 11 \to 13 \to 31 \to 12 \to 23 \to 32 \to 22 \to 21 \to 12 \to 22 \to 22 \to 23 \to 33 \to 32 \to 21 \to 13 \to 33 \to 33 \to 31 \to 13 \to 32 \to 23 \to 31 \to 11$. This result is illustrated at the top of the next column.

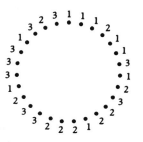

6.12.   Hint: See the solution of Exercise 6.11.

6.13.   Hint: Which vertices are of degree 2? What does that tell you?

6.14.   Hint: Consider vertices of degree 2.

6.15.   Hint: Consider vertices of degree 2.

6.17.   Solution: Note that, in this illustration, it is possible to move from a circled vertex only to an uncircled one, and vice versa. But there are 12 circled

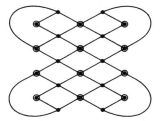

and only 10 uncircled vertices. So even if a runner starts at a circled vertex, by the time he has visited the tenth uncircled vertex, two circled vertices will not yet have been visited. It is impossible to visit both of these without revisiting at least one uncircled vertex. When he tries, he is arrested.

6.18.   Solution: Since we are interested in passing from one state of the filters to another, it seems reasonable to try to represent the problem graph theoretically.

We can represent the state (on, off) of the three filters by a triple of the form (r, b, g), where $r = 0$ if the red filter is off and 1 if it is on, $b = 0$ if the blue filter is off and 1 if it is on, and $g = 0$ if the green filter is off and 1 if it is on. For example, $(1, 1, 0)$ represents the state where the red and blue filters are on and the green filter is off. We construct a graph as shown below, with 8

vertices labeled by the triples representing the possible states of the filters. Two vertices are connected by an edge if it is possible to go from the filter state corresponding to one vertex to the filter state corresponding to the other by changing one filter. (This means that the two triples differ in only one of the three positions.)

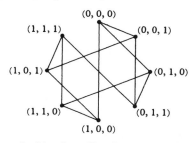

We are looking for a Hamiltonian circuit starting and ending at $(0, 0, 0)$.

By trial and error, we find

$$(0, 0, 0) \rightarrow (0, 0, 1) \rightarrow (0, 1, 1) \rightarrow (0, 1, 0) \rightarrow (1, 1, 0)$$
$$\rightarrow (1, 1, 1) \rightarrow (1, 0, 1) \rightarrow (1, 0, 0) \rightarrow (0, 0, 0).$$

That is, the technician first puts on the green filter, then the blue, then removes the green, then adds the red, then adds the green, then removes the blue, then removes the green, then removes the red.

It is interesting to note that if we considered the triples as coordinates in the 3-dimension rectangular coordinate system, then the 3-dimensional graph is just a cube, as illustrated here.

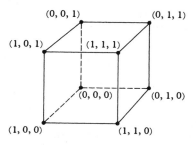

The problem reduces to finding a Hamiltonian circuit on the cube.

6.19.   (b) Hint: If the board is colored in the usual manner, how many white cells are there? How many black? See the solution of Exercise 6.17.

6.20.   Hint: Represent the chessboard by a graph where the cells are represented by the vertices. Two vertices 'are connected by an edge if the corresponding cells of the chessboard are diagonally adjacent. What is the degree of a vertex corresponding to a corner cell? Of a vertex corresponding to a cell on the edge of the board?

6.21.   Hint: Make a graph and consider vertices of degree 2.

6.22.   Hint: Make a graph and consider vertices of degree 2.

6.23.   Hint: *Method 1*   Make a graph with the ten vertices corresponding to Arthur and his nine knights. (It is simplest to arrange the vertices around a circle.) Two vertices are connected if no statement in the problem rules out the possibility of the corresponding people sitting next to each other. Then find a Hamiltonian circuit.

Method 2   The graph could be obtained by first making a graph with vertices as defined above and connecting two vertices corresponding to two people who cannot sit next to one another in one color, say red. Once this is done, join each two vertices that have no common edge by a black edge. This black-lined graph is called the **complement** of the first graph and will be the same as the graph of method 1 above.

6.25.   Hint: This problem is essentially the same as the cannibal-missionary problem. In this problem, however, there is no restriction on who can row.

They can cross the river in 11 trips (a trip in either direction is counted as one trip).

6.26.   Hint: The crossing can be made with 11 trips (a trip in either direction is counted as one trip).

6.28.   (a) Solution: After each round trip, the number of people on the far shore can be increased by at most one. Therefore, if a crossing were possible, at some point there would have to be three people on the far shore (five on the original shore). At some other point, there would have to be four people on the far shore. By the conditions of the problem, any time only three people are on the far shore, they would have to be women. Therefore, to have four people on the far shore, the fourth also has to be a woman. Who would row the boat back and how would the men get across?

(b) Comment: The problem can be solved by making a graph. To describe each state, we need the following information: How many of each sex are on the far shore; on the near shore; on the island. As a result, there are a large number of states (vertices) to be considered. So the graph-theoretical approach seems to be unwieldy. A reasoned trial and error approach seems simpler.

6.30.   Comment: Although this problem can be set up graph-theoretically, graph theory is not really needed.

Hint: Although Hugo cannot handle the 60 lb chest, he is able to pick up the key with his mouth and drop it out of the window. Also note that it may be necessary for one who has gone down to go up again later so that someone else can come down.

6.31.   Hint: John and Val may be with Jon's bag if Jon has both their bags.

6.32.   Solution: Let n be the number of books. If there are more books than there are pages in any one book, then no book can have more than $n - 1$ pages. Construct a bipartite graph: One set of n vertices corresponding to the n books in the library, the other set of $n - 1$ vertices corresponding to the numbers from 1 to $n - 1$. Draw an edge between a book vertex and a number vertex, m, if the book contains m pages. How many edges does the graph contain? Since there are n books, there are n edges. But since there are only $n - 1$ "number vertices," if none were of degree greater than 1, then the sum of the degrees of the number vertices would be at most $n - 1$. This is impossible; hence, at least one number vertex is incident to at least two edges—that is, two books have the same number of pages as this figure illustrates:

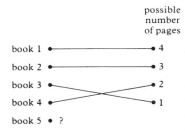

possible
number
of pages

Note that what we have really proved here is an important mathematical result known as the "Cubby Hole" or "Pigeon Hole" Principle. It says that if we are placing objects in boxes and there are more objects than boxes, then one box must receive at least two objects.

6.34. Hint: Note that the interior intersection points are not vertices (we call them apparent vertices), so that we are not really changing the graph if we unravel it in the following manner. Lift up vertex 1 and the connecting edges to a position between 5 and 6, as illustrated.

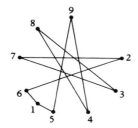

Move 3 to the left, between 7 and 8, etc. When you finish unraveling, you eventually obtain this:

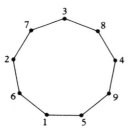

6.35. Hint: Replace the checkerboard by a graph where the vertices correspond to cells and the edges correspond to knight's moves. Then unravel as in the hint for Exercise 6.34.

6.36. Hint: Unravel the graph as in the hint for Exercise 6.34, to the extent possible.

6.37. Hint: Note that the path joining the A's cuts the board into two pieces (a right section and a left). Where must the path connecting the A's go with respect to the E's, C's, B's, and D's?

6.38. Solution: What does this problem have to do with graphs and planarity? If the houses and utilities are considered as vertices, then if it were possible to make all connections, we would obtain a complete bipartite graph on two sets of three vertices. We

have seen in the chapter that such a graph cannot be drawn in the plane. Therefore, there is no solution in this case.

6.39. Solution: Consider the first person. Either he (she) is acquainted with at least three other people or he (she) is not acquainted with at least three other people. Consider the case where person #1 is acquainted with three other people—call them #2, #3, and #4. (The other case is analogous.) If any two of these three people are acquainted, then #1 and these two people are mutually acquainted. Otherwise, people #2, #3, and #4 are mutually unacquainted.

The problem may also be viewed graph-theoretically as follows. Consider a graph with 6 vertices (on a hexagon) corresponding to the 6 people. Connect 2 vertices by a red edge if the corresponding people are acquainted, and by a black edge if they are not acquainted. We obtain a two-color coloring of the complete graph on 6 vertices. The problem then becomes: Show that the graph contains a triangle, all of whose edges are the same color. (Such a triangle is said to be **monochromatic**.) This fact gives rise to an interesting game called Sim, which is presented in the exercises of Chapter 7 (see Exercise 7.68).

6.40. Hint: Construct a complete graph on 5 vertices where the vertices correspond to the 5 men. Connect 2 vertices with a red edge if the men were sitting next to one another initially. When the men resume playing, no one sits next to a person that he sat next to before; therefore, these edges may be eliminated. What remains of the complete graph on 5 vertices is called the complement of the graph (in red) which has been eliminated.

6.41. Hint: See the hint for Exercise 6.40. Find a Hamiltonian circuit that is a subgraph of the complement of the graph corresponding to the original seating arrangement.

6.43. Hint: After three queens are moved, you must still have one queen in each row and one in each column. You can therefore proceed by trial, considering the possibility that the queen in the top row can be moved. If you move this queen to the top row left column, then the queen in the fifth row must be moved. Where can it go? Etc.

Although this problem is not handled graph-theoretically, we include it to introduce two graph-theoretical concepts:

A set of vertices in a graph are said to be **independent** if no two are adjacent to each other (that is, no one edge directly connects the two of them).

A set of vertices is said to **dominate** a graph if every vertex of the graph is adjacent to a vertex in the set.

In this problem, if the chessboard is translated into a graph in the usual manner (where edges correspond to queen's moves), then the set of vertices corresponding to the cells where the queens are currently located forms an independent, dominating subset of the graph.

6.44. Hint: Eliminate those squares that are dominated by the four queens.

6.45. (a) Hint: See the hint to Exercise 6.43 and use the "Cubby Hole" principle (see the solution of Exercise 6.32).

(b) Hint: Which cell in the fifth row from the top could possibly be used?

6.46. Hint: Consider the chessboard with its usual coloring (red, black). If a knight is on a red cell, what can be said about the cells it attacks?

Don't forget to show that the number you obtain is indeed maximal. This can be done by showing that you can pair the cells of the board so that at most one cell of each pair can be occupied by a knight.

CHAPTER 7

Exercises

7.5. Hint: In this game, a position involves two pieces of information—how many sticks remain (denoted by r), and how many sticks were taken by the opponent on the previous move (denoted by t). We use the following notation:

(r, t) denotes a position of the game.

For example, $(8, 2)$ means that 8 sticks remain and 2 sticks were taken on the previous move.

$(r, \neq t)$ means that r sticks remain and that the number of sticks taken on the previous move was not t.

(r) means that there are r sticks remaining and that the number of sticks taken on the previous move is immaterial.

We can now work backward to find winning and losing positions. Clearly, (0) is a winning position. $(1, 1)$ is also winning, since the opponent cannot move; but $(1, \neq 1)$ is losing. We can continue in this manner displaying our results as in the chart below.

winning	(0)	(1, 1)			(3, 3)	(4, 4)	(5, 5)
losing		(1, $\neq$ 1)	(2)	(3, $\neq$ 3)	(4, $\neq$ 4)	(5, $\neq$ 5)	

winning	(6, 3)	(7)				(11, 4)
losing	(6, $\neq$ 3)		(8)	(9)	(10)	(11, $\neq$ 4)

etc.

7.7. Hint: In this game, a position involves two pieces of information—the number of sticks remaining, and the parity (oddness or evenness) of the number of sticks you possess. Work backward.

7.9. Hint: Work backward.

7.10. Hint: Do Exercise 7.9 first.

7.11. Hint: Do Exercise 7.9 first.

7.13. Hint: If, in the winning positions of the regular version of Exercise 7.12, the number of sticks in each pile is expressed in the binary system (as in Nim), what can be said about the column sums? Can a similar observation be applied to this problem?

7.14. Hint: See the hint to Exercise 7.13.

7.16. Hint for parts (a) and (c): Consider n odd and even separately, and try to apply a pairing strategy.

7.17. Hint: What is the connection between this problem and Exercise 7.16?

7.18. Hint: After how many moves will the game be over?

7.19. (a) Solution: A position of the game consists of the number of piles and the number of sticks in each pile. A position of the game with k piles and a_i sticks in the ith pile will be represented by $(a_1, a_2, \ldots, a_k)$.

For example, (10, 3) means 2 piles, one with 10 sticks and the other with 3; and (8) means one pile with 8 sticks. If it turns out in analyzing the game that (m) and (n) are winning positions, then (m, n) is a winning position. Similarly, if (m) is winning and (n) is losing, then the position (m, n) is losing. [If (m) and (n) are both losing, we can't immediately tell whether (m, n) is winning or losing.] (1) and (2) are winning positions since there is no move for the opponent. For this reason, piles of 1 and 2 may be disregarded. We proceed to analyze the game by working backward.

(3) is a losing position, since the opponent will leave (1, 2).

(4) is winning, since the opponent will have to leave (3, 1).

(5) is losing, since the opponent can leave (4, 1).

(6) is losing, since the opponent can leave (4, 2).

(7) is winning, since the opponent must leave (6, 1), (5, 2), or (4, 3), all of which are losing since they consist of one winning and one losing pile.

(8) and (9) are losing positions, since the opponent can leave (7, 1) or (7, 2).

(10) is winning since the opponent must leave (9, 1), (8, 2), (7, 3), or (6, 4), all of which are losing since each contains one losing and one winning subpile.

(11) and (12) are losing since the opponent can leave (10, 1) or (10, 2), both winning.

(13) is losing; the opponent can win by leaving (8, 5), provided he or she then plays properly. (This is not quite obvious and requires further argument.)

(14) is losing; the opponent can leave (10, 4) which is winning because it is made up of two winning positions.

Thus, A wins by leaving two piles: one with 10 and one with 4.

7.21. Solution: To put this problem in the framework of the matchstick games, consider the difference needed to reach 22. That is, the game starts at 22 and proceeds toward 0. With this in mind, it would seem from our analysis of Sample Problem 7.1 in this chapter that the winning positions are the multiples of 5. However, here a position consists not only of the total reached but also of what cards remain. Thus, 5 will be a losing position if no ace remains and a 4 does, because your opponent

will take the 4 and you will then lose.

On the other hand, don't totally discard your previous knowledge: 5 is still a winning position if one of each denomination remains; 10 still wins if at least two of each denomination remain; 15 still wins if at least three of each denomination remain.

Thus, A cannot start by taking 4, because B will reach a winning position (15) by taking 3. Similarly, A cannot start with 3.

If A starts with 2, B can win by taking 3. To analyze this case we would have to consider a number of subcases; but we need not do so because we will show that A can win by starting with 1.

So assume A starts by taking 1. B cannot take 2 or else A will take 4 leaving 15 and winning because at least 3 of each denomination are left.

Similarly, B cannot take 4 or A will take 2 and win.

If B takes 1, A can take 4, leaving 16. B is now forced to take 1 (otherwise A will be able to leave 10 with two cards of each denomination remaining). A can now take 4, leaving 11. Again B is forced to take 1, and again A takes 4, leaving 6. At this point, B loses no matter what she does. Since no aces remain, B must leave 2, 3, or 4, after which A can bring the total to 0.

Thus, B's only reasonable course of action is to take 3 on her first move, leaving 18. At this point A takes 1, again leaving 17. B cannot take 3 or 4 since A can leave 10 with at least two cards of each denomination remaining. Regardless of whether B takes 1 or 2, A should take 4, leaving either 11 or 12 (see the figure below).

$$A_1 \qquad B_1 \qquad A_2 \qquad B_2 \qquad A_3$$

$$22 \xrightarrow{\ 1\ } 21 \xrightarrow{\ 3\ } 18 \xrightarrow{\ 1\ } 17 {\Large<} \begin{array}{l} {\scriptstyle 2}\diagup 15 \xrightarrow{\ 4\ } 11 \\[4pt] {\scriptstyle 1}\diagdown 16 \xrightarrow{\ 4\ } 12 \end{array}$$

From here, B cannot allow A to reach 5, so must take 1 from 11 or 1 or 2 from 12. Again A takes 4. This leaves 6 or 7 —as illustrated below.

$$A_1 \qquad B_1 \qquad A_2 \qquad B_2 \qquad A_3 \qquad B_3 \qquad A_4$$

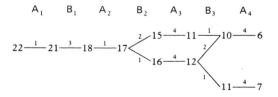

If 6 is left, then three of the aces have been used; and if 7 remains, then all of the aces have been used. If B now brings the total to less than 5, then A wins immediately. If B brings the total to 5, then all the aces are gone, and A can win by taking 4.

7.22. Hint: It is easier to work with the difference between 26 and the previous total. A position includes two pieces of information: the difference between 26 and the total; and the number showing on the top face of the die.

Work backward as in previous problems. Note that a pattern is established once six consecutive entries exactly match six earlier consecutive entries.

7.28. Hint: Consider the triple (m, n, p) where m represents the number of spaces between A and B on the top row, n on the middle row, and p on the bottom row. In this light, the game is related to Nim.

7.29. Solution: (a) By symmetry, there are three possible opening moves for A: center, corner, or middle of a side. Call them cases I, II, and III. The first step in attacking this problem is to play the game a few times to get a feel for it. Do you get the feeling that starting at the corner or middle of the side seems to result in a large variety of possible moves to be considered, whereas starting at the center immediately limits the possibilities that must be considered? Since we are just interested in finding a winning strategy for A (if one exists), let us consider the simplest case first.

Case I A starts by placing an X in the center.

B is now forced to place an O, otherwise A will win on the next move. By symmetry there are only two possibilities—O in a corner or the center of a side.

In either case, A has several alternatives. To keep the analysis simple, consider moves that limit B's possibilities as much as possible. (If this doesn't lead to a winning strategy for A, we will have to try the other alternatives.)

Case Ia In the case that B's first move is the corner, a quick check of possible moves for A reveals one that leaves B no safe reply—A places an O diagonally opposite from B's O, as the figure at the top of the next column shows.

Case Ib If B's first move is the middle of the side, then the move for A that most limits B's moves is to place an O symmetrically opposite to the O on the board, as in the figure below. B now, by symmetry,

has only one move—an O in one of the remaining side boxes. A now forces a win by filling in the fourth side box with an O, as in the figure below:

Thus, A has a winning strategy: Place an X in the middle box and play O's symmetrically opposite to B's, until B presents A with an opportunity to win.

Note that, since case I has led to a winning strategy for A, it is no longer necessary for us to consider cases II and III. On the other hand, if some possible continuations in case I would have led to a win for B, it would have been necessary for us to consider case II (A starts in a corner) and maybe even case III (A starts in the middle of a side).

(b) The misere form of this game is much harder to analyze. A can ensure at least a draw by starting in the center and then playing symmetrically opposite to B by using the symbol not used by B. An example is shown below:

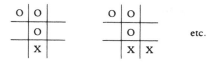

If A's move would complete three in a row, then B would have already completed three in a row on her previous move. On the other hand, the question of whether or not A has a winning strategy is more difficult. In order for A to force a win, he must leave a position such that when B moves she will be forced to make three in a row with either an O or an X. Since B makes the second, fourth, and sixth moves in the game, the only possible positions in which B will be forced to complete three in a row are those shown in the figure below, and their symmetric counterparts (including switching X's and O's).

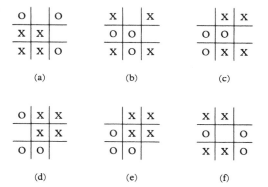

Can A always leave one of these positions, or can B prevent this from happening? B can prevent it, but to see this, many cases must be considered. Let us consider just one example here: If A starts by placing an X in the center, then B can place an O on the side. (Only positions (c), (d), and (e) are now possible.) And if A continues by placing an O in the corner, then B can place an X in an adjacent corner, as

illustrated here. No matter what happens at this point, we cannot end up with any of the above end positions.

7.30. Solution: We use the following notation: X_i indicates A's ith move and O_i indicates B's ith move. There are two possible opening moves for A—bottom corner or bottom middle.

(a) *Case I* A opens in a bottom corner. B has three possible responses, illustrated here:

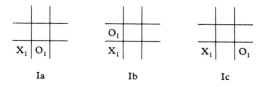

Ia   Ib   Ic

Case Ia Part **a** of the figure below shows that A can force a win. (Note that O_2 and O_4 are forced moves, and that O_3 is essentially forced, otherwise A will win immediately thereafter.) Since case Ia doesn't lead to a win for B, we must consider case Ib.

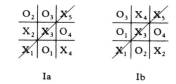

Ia   Ib

Case Ib Again A can force a win, as in figure **b** above. Here all of B's moves are forced. We now try case Ic.

Case Ic A can force at least a draw by playing above B's move and then leaving one of the positions below.

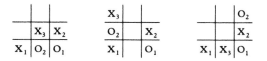

On the other hand, no matter what A does, B can play directly above A's move and continue to do so until B gets the center square (this must happen eventually). With best possible play by A (you supply the details), we get the figure shown below, where A can place his X above the O and the game is a draw.

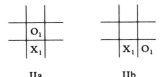

Thus, case I results in a draw if both players play well.

We now must consider case II.

Case II A starts in the bottom middle. B has two possible responses, as this figure shows:

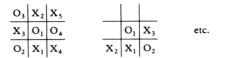

   IIa   IIb

Case IIa is a theoretical draw. An example is illustrated below.

O_3	X_2	X_5
X_3	O_1	O_4
O_2	X_1	X_4

| | O_1 | X_3 | etc.
|---|---|---|
| X_2 | X_1 | O_2 |

Case IIb leads to a win for A, as shown below.

X_4	O_2	$\cancel{X_5}$
O_3	$\cancel{X_2}$	O_4
$\cancel{X_3}$	X_1	O_1

Note that B's moves are essentially forced. Thus, case II also leads to a draw since B can respond as in case IIa.

Since both cases lead to a draw with proper play, the game is a theoretical draw (both players have drawing strategies but neither has a winning strategy).

(b) In the misère form of this game, A wins by forcing B to complete the middle row. That is, A starts anywhere, and whenever B plays the middle row, A plays on top of B; if B plays on the bottom row, A takes the remaining space on the bottom row.

7.37. Hint: By symmetry, the possible first moves for A are as in the figure in the next column.

In all but one of these cases, B has a move to force a win. In the one remaining case, A has a winning strategy.

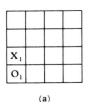

(a)

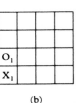

(b)

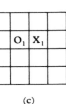

(c)

(d)

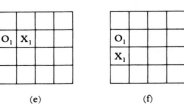

(e)   (f)

7.39. Hint: This game is different from Sample Problem 7.2 in the chapter in that the X's (checkers) must be in touching boxes. As a result, not all positions that were equivalent before are equivalent now.

7.40. (b) Solution: A can ensure a draw by starting in the center and then using central symmetry to B's moves. (For example, if B moves in the upper right-hand corner of the top level, A moves diametrically opposite in the lower lefthand corner of the bottom level. Etc.) Thus, A has a drawing strategy. [See the solution of Exercise 7.29(b).]

However, the game cannot end in a draw. To see this in the case where A adopts the above strategy, first observe that there are essentially four different types of cells in a 3 × 3 × 3 Tic Tac Toe cube. These are illustrated on page 369. There is the middle cell (labeled M), the eight corner cells (labeled C), the twelve non-corner cells that are on an edge of the cube (labeled E), and the six cells that are centers of faces of the cube (labeled F).

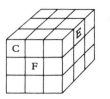

M = cell in the center

C	E	C
E	F	E
C	E	C

bottom
level

E	F	E
F	M	F
E	F	E

middle
level

C	E	C
E	F	E
C	E	C

top
level

Assume that A has adopted the indicated strategy and that the game has ended in a draw with all the cells filled in. Then, by the nature of the strategy, there must be four X's and four O's in the corner cells (C), because each time B took a corner cell, A took one too.

If all four of the corner X's appear in the same horizontal, vertical, or diagonal plane, then some portion of the cube looks like the figure below.

X		X
X		X

There is no way that this plane can be completed without someone being a winner.

If three of these corner X's appear in the same horizontal or vertical plane, then by rotating the cube if necessary, there is no loss in generality in assuming that the plane in question is the bottom plane and that the position is as shown here:

X		X
X		O

bottom
level

	X	

middle
level

X		O
O		O

top
level

Since the position is supposedly drawn, there is only one way in which the bottom and top levels can be completed:

X	O	X
O	O	X
X	X	O

bottom
level

	X	
	?	

middle
level

X	O	O
O	X	X
O	X	O

top
level

But now there is no way to complete the middle level; the cell with the question mark cannot receive an X or an O without one player completing three in a row, as this figure shows:

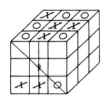

Hence, this case leads to a contradiction.

The only remaining possibility for the four corner X's is as shown below (or an equivalent position).

O		X
	?	
X		O

bottom
level

	X	

middle
level

X		O
O		X

top
level

But now there is no way to fill in the cell that has the question mark. Hence, all cases lead to contradictions and so, if the first player follows the indicated strategy, the game cannot end in a draw. But as the indicated strategy is a drawing strategy (ensures at least a draw), the first player cannot lose and therefore must win by following this strategy.

7.41. (b) Hint: See the solution of Exercise 7.30(b).

7.50. (a) Hint: From a position such as that in part **a** of the figure below, A cannot be cut off from the top of the board. Similarly, from a position such as that in part **b**, B is forced to move to the cell indicated by the arrow, if she wants to cut A off from the top of the board. From a position such as that in part **c** of the figure, A cannot be cut off from the top

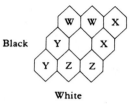

plays the pairing strategy until unable to do so, in which case she makes another free move. Etc.

At the end of the game, by following this strategy, B will never have a center cell, more than one W cell, more than one X cell, more than one Y cell, or more than one Z cell. So B cannot have a path connecting two sides of the board. Since the game cannot end in a draw, A must have a path connecting the top to the bottom of the board. So B wins.

(b) (ii) Solution: In the 4 × 4 case, A can win by labeling the board as in this figure and by playing as

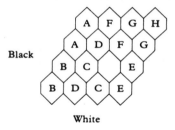

follows: He starts with H and adopts the pairing strategy discussed above for (b) (i). He never plays in the unlabeled cell. A also never occupies both cells with the same letter, so there is no way for A to have a path from the bottom to the top of the board.

(b) (iii) and (iv) Comment: It is known that, for general odd n, B wins the $n \times n$ game and, for general even n, A wins the $n \times n$ game. We do not know the winning strategy.

7.52. Hint: If at the end of the game, an odd number of dots has been used, then an even number of line segments has been drawn, and B made the last move. The reverse is true if an even number of dots has been used.

We are not certain of the general strategy of this game; however, it seems to us that the dot counting idea mentioned above should be applicable.

of the board, since if B moves to the cell indicated by the arrow, then A can move to the cell indicated by *, ensuring a connection with the top of the board.

(b) (i) Solution: B wins by labeling the cells as shown in the figure below and then playing as follows: Whenever A moves to a lettered box, B moves to the other box with the same letter. If A moves to the center, B moves to any box and then

7.54. Hint: (i) Use symmetry.

(ii) For each possible opening move by A, show that B has a successful countermove.

7.56. Hint: Consider how many moves A should try to make in order to win the game. Note that A may be able to influence the number of moves in the game by choosing among patterns such as those shown in the figure below.

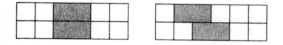

Also consider the 2 × 4 game and note that there is a relationship between what happens in the 2 × n game and what happens in the 2 × (n + 4) game for most values of n.

7.57. (c) (i) Solution: The method is to take cases for A's first move. By symmetry, A has four possible opening moves. Rather than show the complete analyses, we take one opening move that does in fact lead to a win for A.

Let A start as in the figure below. Then, by

symmetry, B has four possible responses. These are shown in parts **a**, **b**, **c**, and **d** of the next figure.

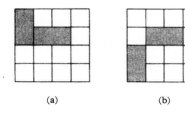

(a)   (b)

(c)   (d)

In each case, A can reply as shown below:

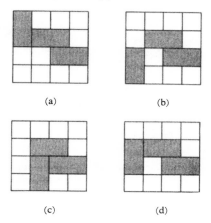

(a)   (b)

(c)   (d)

In each case, no matter what move B makes, A at his next turn can eliminate one of B's possible moves, and so will win the game.

(c) (ii) Solution: Again we consider a winning opening move for A, as shown in the first figure below.

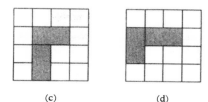

If B doesn't prevent A from moving as in parts **a** or **b** of the next figure, then the game will end after

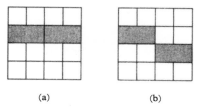

(a)   (b)

B's fourth move and B will lose. Therefore, B has only two reasonable moves, as shown below.

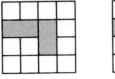

(a)   (b)

In these cases, A responds as follows:

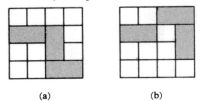

(a)   (b)

In both of these cases, B will have to make the last move.

7.61.   (b) Hint: We present a partial solution. We have found a winning strategy for A which starts with the position in the first figure below.

To this move, B has four possible responses, as shown in parts **a**, **b**, **c**, and **d** below.

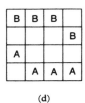

(a)   (b)

(c)   (d)

In a complete solution, all of these cases must be considered. Suppose B moves as in part **a** of the figure, above. A's best response is shown in the single figure below.

Again B has several alternatives. These are shown here as **a**, **b**, and **c**.

(a)   (b)

(c)

We will consider only what happens in the **a** case. Again, you should consider all the other cases. (You can see that to complete this problem, an extensive case analysis is necessary.)

If B moves as in part **a** of the figure, A can now win by moving as below (sacrificing the piece on the right):

B is forced to move the threatened pawn in the third column from the left (either to capture or to push), otherwise A will win on the next move. But either way, A can capture B's pawn in the second column from the left and win on the next move.

7.66. Hint: Does the size of the table matter? Does symmetry play a role?

7.67. Hint: In order to cut down the amount of work needed to analyze this game, we must consider what positions are equivalent to each other.

Consider the positions in parts **a** and **b** of the figure below. Are they really different from each

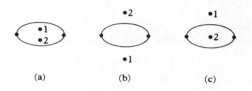

(a)   (b)   (c)

other? It is true that in part **a**, points 1 and 2 are inside the loop, whereas in **b** they are both outside. But any line that can be drawn in part **a** has an analogous line that can be drawn in part **b**. On the other hand, the position in part **c** of the figure is different from those in **a** and **b**. In **c**, it is not possible to draw a line connecting 1 and 2 whereas it is possible in the other figure parts. Thus, in considering the game of Sprouts, the notions of inside and outside are not as important as whether or not points are separated from one another.

Therefore, in the 2-dot game, there are (by symmetry) only two starting moves that the first player can make—see parts **a** and **b** below. Note that the positions in parts **b** and **c** of the figure are equivalent.

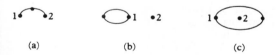

(a)   (b)   (c)

Another consideration in analyzing this game is: How long can the game last? (What is the maximum length of the game?) At the start of the game there are n dots, each having three lives. With each move of the game, two lives of existing dots are eliminated and a new dot having one remaining life is added: There is a net loss of one life. If at any time there is only one life remaining, then it is not possible to make a move and the game is over. Thus, the game can last at most $3n - 1$ moves (starting with $3n$ lives). Note that it is possible for the game to end in fewer than $3n - 1$ moves if we are left with

several dots, each having one life, that are separated from one another. This fact influences the strategies of the two players. For example, the 2-dot game would last at most $3 \cdot 2 - 1 = 5$ moves. Thus, A would make the last move if all 5 moves were available. Therefore, in this game, B would try to isolate a dot, so that the game will end in four moves. A tries to prevent this.

7.68. Hint: (a) (i) For $n = 5$, there are 10 lines that could be drawn; A will draw 5 of these and B the other 5. If A can complete a circuit (see Chapter 6) through 4 of the dots and connect the fifth dot to one of the four (as in the figure below, for example), then B will have to take the remaining five lines and will lose. A's strategy is to try to complete such a figure and B's strategy is to prevent A from doing so.

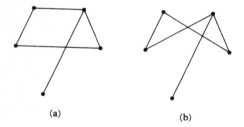

(a)   (b)

(b) Hint: See Exercise 6.39 in Chapter 6.

Comment: The 6-dot version of Sim is the one most frequently played. It is still of interest even though it has been shown by exhaustive computer searches that the second player has a winning strategy. L. Shader ([41], Vol. 51, pp. 60–64) has also found a winning strategy for the second player without the use of a computer.

7.70. Hint: To help in the analysis, label the cells with numbers as shown in this figure:

			1	2	3		
			4	5	6		
	7	8	9	10	11	12	13
	14	15	16	17	18	19	20
	21	22	23	24	25	26	27
			28	29	30		
			31	32	33		

Unless the fox is able to jump a goose, after an odd number of moves by the fox he will be on an even numbered cell; and after an even number of moves, the fox will be on an odd numbered cell.

7.71. Hint: Work backward to find winning positions.

7.72. Hint: Tempo (timing) is important. Oblique Street can allow Fred to attain the correct tempo.

7.76. Hint: You should win 6–3 with proper play.

CHAPTER 8

Exercises

8.2. Hint: Let M_n = the number of moves required if there are n switches in the row and let G_n be the number of moves required to convert n switches from the state in which they are all off to the state in which the nth switch is on and the rest are off. Express G_n in terms of G_{n-1}; find a formula for G_n as a function of n; find a formula for M_n in terms of M_{n-2} and G_{n-1}. Note that it requires the same number of moves to turn n "on" switches off as it does to turn n "off" switches on.

  Comment: This problem is essentially the same as the well-known Chinese Rings Problem (see [3], p. 305).

8.5. (a) Hint: What would the area of the required square be? How big would one of the sides have to be? How could you easily construct a line segment of the required length?

  (b) Hint: What could be the dimensions of the triangle?

  (c) (i) Hint: Label the regions as in this figure:

Which regions make up the square of side c? Are certain regions congruent?

  (ii) Hint: Label the regions as in the figure below. Show that the four quadrilaterals made up of regions

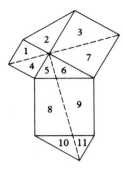

1, 2, and 3, regions 4, 5, 6, and 7, regions 5, 8, and 10, and regions 6, 9, and 11 are all congruent to each other.

8.6. Hint: Figure out how many cells have to be in each section. Then try systematically to describe all possible sections that could have this number of cells. For example, in the 4×4 case—part (b) of the exercise—each section must contain eight cells. There are essentially six different ways of cutting the board up into identical eight cell sections, as the figure below shows:

(1)   (2)

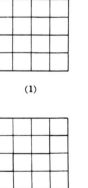

(3)   (4)

(5)   (6)

Each of these may give rise to several different cuttings under rotation or reflection, depending on whether condition (i), (ii), or (iii) is being considered.

For example, if coloring counts, then the two cuttings in the figure below are not the same, whereas they are the same if coloring does not count.

Similarly, if we are allowed to flip pieces over (without regard to color) then the two cuttings that follow are the same, whereas if we are not allowed to flip pieces over, they are not the same.

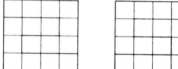

Thus, for each of the six cuts in the first figure, we must count how many different cuts there are depending on (i), (ii), or (iii). For example, in (i), cut (2) in the figure gives rise to the four different solutions shown below.

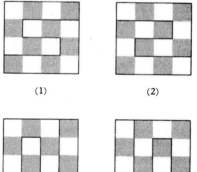

(1)   (2)

(3)   (4)

In (ii), cuts (1) and (4) here are the same as are (2) and (3), giving two solutions arising from this type of cut.

In (iii), all four of the cuttings are the same, giving one solution arising from this type of cut.

8.9. (a) Hint: If the missing cells are in the same row or column, how far apart can they be? If they are in different rows and columns, show that the problem can be reduced to that of covering an odd-by-even board that has diagonally opposite missing corner cells.

(b) Hint: Does the color of the missing cell matter?

(c) Hint: How many cells of each color can each of the tetrominoes have?

(d) Solution: It has become traditional to refer to each of the pentominoes by the letter that it most resembles. These are shown below.

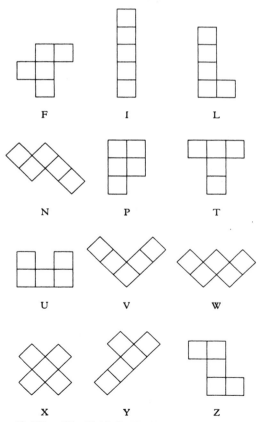

(f) Hint: Use F, N, P, U, V, and X to make one rectangle and the others to make the other.

(g) Hint: For each pentomino, place it in an allowable position—a position in which no fewer than 5 cells are isolated—so that the largest possible number of cells of the pentomino lie on the border of the given region. In this manner, obtain a "maximal edge number" for each pentomino. Add these up. What do you observe?

8.11.   (a) Hint: Where was the remaining center peg, one move before the last? Where was it one move before that?

(b), (c), (d), (h), (i), (k), (l), (m) Hint: Use coloring arguments.

8.13.   (b) Hint: Can this be reduced to the previous case?

(c) Hint: Can this be reduced to previous cases?

8.14.   Hint: Note that there are two T's and two O's.

8.17.   Hint: We will refer to the cubes in the $2 \times 2 \times 2$ model as UFR (for the cube in upper front on the right), UFL (upper front left), UBR (upper back right), etc. Note that L is for left and L is also for lower. Consider UFL in the first figure below. It must have blue on top, green on the left, and red on the front. Let t, x, and z represent the remaining faces of this cube, as indicated in the second figure.

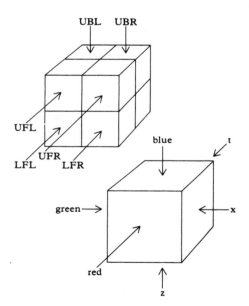

Now consider UFR. It must have blue on top, red on the front, and orange on the right. It also must have x on the left since it is next to UFL. Label the remaining faces u and v as in the figure below. Continue in this manner for each of the other cubes, introducing a new letter each time an unknown face appears but being sure to use the same letter when two faces touch. Note that since x is on UFL, it cannot be blue, red, or green; and, since it is also on UFR, it cannot be orange either. Therefore x must be white or yellow.

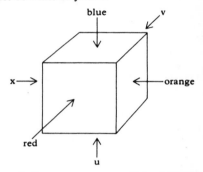

In a similar manner, you will be left with two possibilities for each of the other unknowns. If you choose x to be white, each of the other unknowns will be determined; if you choose x to be yellow, then each of the other unknowns will again be determined. You thus get two solutions, although it turns out that both solutions make use of the same eight cubes.

8.18.   Hint: When a knight makes a move from a cell of one color, what can be said about the color on which he lands? How many moves must he make in all for a reentrant tour?

CHAPTER 9

9.7.   Solution: Label the four coins A, B, C, and D. Label the fifth coin G.

First weighing: weigh A, G against B, C.
If they balance, write $A + G = B + C$.
If A, G is heavier, write $A + G > B + C$. Etc.
The first weighing gives three cases:

Case I $A + G = B + C$.

In this case, D must be the bad coin. A second weighing of D against G tells you whether D is heavy or light.

Case II A + G > B + C.

In this case, D is a good coin, and either A is heavy or B or C is light. A second weighing of B against C determines which coin is bad. That is,

if B = C, then A is bad and heavy.

if B < C, then B is bad and light.

if B > C, then C is bad and light.

Case III A + G < B + C.

This is essentially the same as case II—just interchange the words heavy and light.

9.8.   Hint: Weigh four against four. If they balance, use the solution of Exercise 9.7. If they don't balance, regroup and weigh some of these against others for the second and third weighings.

9.9.   Hint: Label the coins A, B, C, D, and E.

First weighing: A against B.

Second weighing: C against D.

By symmetry, suppose A < B and C < D (otherwise rename so that the first coin weighed in each pair is the lighter coin).

Third weighing: A against C. (Assume A < C; there is no loss of generality in doing this.)

Fourth weighing: E against C.

You now have a partial ordering. In one more weighing you can order four of the coins. Then work with the fifth.

9.10.   Hint: To begin with, there are 120 different possibilities for which coin weighs what amount. If you weigh two coins against each other, there are two possible outcomes, so the best you can do is to eliminate half of the possibilities with each outcome. However, if you weigh two coins against two other coins, there are three possible outcomes (=, <, >). This may enable you to eliminate two-thirds of the cases with each weighing.

You must carefully choose which pair to weigh against which other pair, so that each of the possible outcomes leaves essentially one-third of the cases.

9.11.   (b) Hint: There are 16 oz in 1 lb.

9.12.   (b) Solution: In order to ensure that D ends up facing its original direction, D leaves E and F and goes via 3 to 1; then leaves 1 via 2 to pick up C (so far, one reversal).

D pulls C to 2, leaves it there; passes through 1 to 3 then to 4 to pick up A, B (three more reversals). D pulls A, B to the right and pushes them to 1 and 3 respectively (two reversals). D backs up (very

slightly) to pick up E and F; pulls E, F to the left and backs up, backing C into A, then moves to the left, pulling C, A along (four reversals). D pushes E, F, C, and A to the right on the main track, then heads up 3, pushing B to 1 (two reversals). D then backs up, pulls E, F, C, A to the left and then backs them up to pick up B (three reversals). Finally D pulls E, F, C, A, B to the main track (one reversal), leaves C, A, B, and continues with E and F on its way.

If Connie is willing to leave the stalled cars on the triangular tracks, then only eight reversals are necessary.

9.14.   Hint: What positions near the end could lead to the desired position?

9.16.   Solution: A neat method of handling this type of problem—one involving the expressions "All A's are B's", "Some A's are B's", "No A's are B's", and "x is an A"—is the method of Euler circles.

We denote the set of all A's pictorially as the region bounded by circle A. Similarly, the region bounded by circle B denotes the set of all B's. We can then translate the above statements into Euler diagrams as follows:

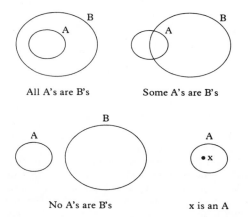

All A's are B's   Some A's are B's

No A's are B's   x is an A

If we now make all composite Euler diagrams which satisfy the premises of a given argument, we can check to see if the conclusion holds in every instance.

In this exercise,

1. All gangs are gengs.

2. Some gengs are gings.

3. No gengs are gongs.

The only possible Euler diagram for premises 1 and 3 is shown below.

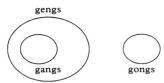

Since this diagram implies conclusion (a) regardless of premise 2, conclusion (a) must hold.

However, the following diagrams also satisfy premise 2 together with 1 and 3.

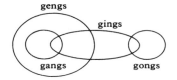

Case I

Conclusion (b) does not hold in this case.

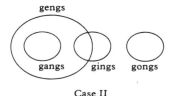

Case II

Conclusions (c) and (d) do not hold in this case.

Therefore, conclusions (b), (c), and (d) do not follow from the premises.

To use this method in general in order to show that a conclusion is valid, we must show that the conclusion is satisfied in every possible Euler diagram which is consistent with the premises. To show that a conclusion is not valid, we must find one Euler diagram which is consistent with the premises but in which the conclusion does not hold.

9.19.   Solution: From statements 1, 2, and 4, all bees are capable of reproducing by parthenogenesis. From statements 3 and 5, all bees are beneficial to the farmer.

But there may be insects other than bees that are also beneficial to the farmer, and these insects

may not be capable of reproducing by parthenogenesis. Hence, we cannot make the desired conclusion.

9.24.   Hint: Let W_n = the number of worms that there were at the end of the nth month. What is the relation between W_n, W_{n-1}, and W_{n-2}?

The sequence of numbers that we get is called the Fibonacci sequence.

APPENDIX B

B.1.   Solution: Let

$$S_n: 1 + 2 + 3 + \cdots + n = \frac{n(n+1)}{2}$$

$$S_1: 1 = \frac{1(2)}{2} = 1, \quad \text{so } S_1 \text{ is true.}$$

Assume S_k is true.

$$1 + 2 + \cdots + k = \frac{k(k+1)}{2} = \frac{k^2 + k}{2}.$$

Add $k + 1$ to both sides.

$$1 + 2 + \cdots + k + (k+1) = \frac{k^2 + k}{2} + (k+1)$$

$$= \frac{k^2 + k + 2k + 2}{2}$$

$$= \frac{k^2 + 3k + 2}{2}$$

$$1 + 2 + \cdots + (k+1) = \frac{(k+1)(k+2)}{2}.$$

This last equation is just S_{k+1}.

Therefore, S_k implies S_{k+1} and so, by induction, since S_1 is true, S_n is true for all $n \geq 1$.

B.3.   Solution: Let

S_n: Any map formed by n straight line segments (each beginning on one edge of the rectangle and ending on another edge) can be colored in two colors so that bordering regions are of different colors.

Then

S_1: Any map formed by 1 straight line segment in the manner defined above can be colored in two colors.

S_1 is clearly true since there are only two regions, as the figure below exemplifies.

Assume S_k is true. We wish to show that S_{k+1} is true. We illustrate our argument with figures for $k = 4$. Suppose we have a map formed within a rectangle by $k + 1$ straight lines in the manner described above. (For example, see the figure below.)

$$k = 4, \quad k + 1 = 5$$

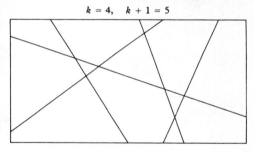

We must show that it can be colored in two colors. To do this, we delete any one of the $k + 1$ lines. This leaves a k-line map.

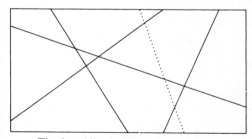

The dotted line is the one being deleted.

Since S_k is assumed to be true, this map may be colored in two colors.

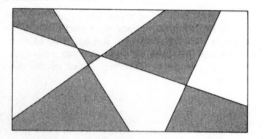

Now reinsert the line that was previously deleted.

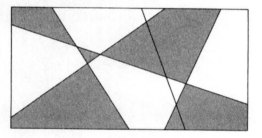

Reverse all the colors on one side of this line, leaving the colors on the other side of the line unchanged.

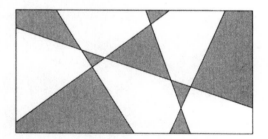

We now have colored the original map in two colors so that bordering regions are in different colors.

To check this, suppose that we have two bordering regions. Their common border is either part of one of the k lines or part of the $k + 1$st line. In the former case, the two regions lie on the same side of the $k + 1$st line, and therefore either remained unchanged in color or both were reversed in the final coloring. Since they received different colors in the k-line coloring, they are still of different color in the final coloring.

In the case that the common border is part of the $k + 1$st line, the two regions were part of the same region in the k-line coloring, and only one of them was changed when the $k + 1$st line was reinserted. They therefore have different colors in the final coloring.

Thus, S_{k+1} is true. We have just shown that if S_k is true, then S_{k+1} is true. Therefore, by mathematical induction, since S_1 is true, S_n is true for all $n \geq 1$.

Appendix C Hints and Solutions

C.1   Hint: What could the sum of each pile possibly be?

C.2   Hint: Break this problem up into cases. This may be done by counting all the triangles of size 1, all made up of four basic triangles, etc.

C.7   Hint: How many arrangements of five names are there? How many of these are in alphabetical order?

C.10   Solution: Suppose red is one of the six colors. Then one face must be red. Turn the cube so that the red face is on top. This leaves five possibilities for the color of the bottom face. Once the color of the bottom face is chosen, there are four colors remaining. Say blue is one of these colors, without disturbing the top and the bottom faces of the cube we may rotate the cube so that the blue face is now in front. There are now three possibilities for the color of the back face (opposite to the blue one); that leaves two possibilities for the color of the face on the right. See figure 46.

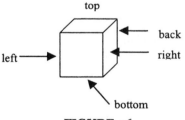

FIGURE 46

The color of the face on the left is then forced (only one left).

Hence by the Multiplication Principle, the are $5 \cdot 2 \cdot 3 = 30$ possible colorings.

C.13   Hint: In how many ways can the tablecloths be placed on the top row? Once this row is placed, in how many ways can the second row be placed? What about the third row?

C.14   Hint: Once the occupant of D101 is known, how many possibilities are there for D102? Continue in this manner but be careful.

C.15   Hint: We are essentially asking the following question. In how many ways can we fill in the array in Figure 47 with the names Manny, Brooke, William and Queenie so that no names appear more than once in any row or any column?

	Night 1	night 2	night 3	night 4
North South East West				

FIGURE 47

How many ways are there to fill in the North row? Once the North row is filled in, how many ways of filling in the East row depends on how the North and South rows are filled in.

Therefore we must subdivide the seating arrangements of the South row into two cases.

C.17   Partial Solution: Break the counting procedure up into disjoint cases:
    Case 1:   No color is repeated – (This case has been handled in Problem C.10).
    Case 2:   One color appears twice and the remaining colors are different.
    Case 2a: The repeated color occurs on opposite faces.
    Case 2b: The repeated color occurs on adjacent faces.
    Case 3:   One color appears three times and the remaining colors are different.
    Case 3a: Two of the faces of the repeated color are opposite one another.
    Case 3b: No two of the faces of the repeated color are opposite one another.
    Case 4:   One color appears four times and the remaining colors are distinct.
    Case 4a: The repeated color appears on two pairs of opposite faces.
    Case 4b: The repeated color does not appear on two pairs of opposite faces.
    Case 5:   One color appears five times and the other color is different.
    Case 6:   One color appears six times.
    Case 7:   Two colors each appear twice and the remaining two colors are different.

Case 7a: Each repeated color appears on two opposite faces.

Case 7b: One repeated color appears on opposite faces but the other does not.

Case 7c: Neither repeated color appears on opposite faces. But the remaining two colors are opposite each other.

Case 7d: Otherwise.

Case 8: Three colors each appear twice.

Case 8a: All pairs of repeated colors appear on opposite faces.

Case 8b: One repeated color appears on a pair of opposite faces, but the others do not.

Case 8c: No repeated color appears on a pair of opposite faces.

Case 9: One color appears three times, another appears twice, the remaining color is different.

Case 9a: The doubly repeated color appears on opposite faces.

Case 9b: The triply repeated color appears on a pair of opposite faces, but the doubly repeated color does not.

Case 9c: Otherwise.

Case 10: Two colors each appear three times.

Case 10a: A pair of opposite faces is the same color.

Case 10b: No pair of opposite faces are the same color.

Case 11: One color appears four times and another color appears twice.

Case 11a: The color which appears twice is on opposite faces.

Case 11b: The color which appears twice does not appear on opposite faces.

In each case, first count the number of ways in which the colors can be chosen, then count the number of ways in which the chosen colors can be arranged and then use the Multiplication Principle.

For example, in case 7a there are $\binom{6}{2}$ ways of choosing the two repeated colors and then there are $\binom{4}{2}$ ways of selecting the remaining two colors. Once the colors are chosen, there is only one way of arranging them. That is, suppose red and black are the two repeated colors in 7a we can turn the cube so that red is on the front and back faces, black is on the left and right

faces. Whichever of the remaining colors is on top is immaterial because the cube can be turned over. Therefore, case 7a gives rise to

$$\binom{6}{2}\binom{4}{2} = 90 \text{ possibilities.}$$

Now in case 7b the analysis is a little different, as the two repeated colors are not equivalent. Namely there are 6 possibilities for the repeated color to appear on opposite faces, this leaves 5 possibilities for the other repeated color, and $\binom{4}{2}$ possibilities for the remaining colors. Again there is only one way to arrange these colors. (For example in Figure 48, cube B is obtained from cube A by switching front and back and rotating by 180° about an axis from front to back.)

This gives $6 \cdot 5 \cdot \binom{4}{2} = 180 \text{ possibilities.}$

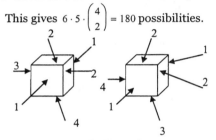

FIGURE 48

In 7c there are $\binom{6}{2}$ ways of selecting the repeated colors. There are $\binom{4}{2}$ ways of choosing the remaining colors. Again there is only one way to arrange them.

So there are $\binom{6}{2}\binom{4}{2} = 90 \text{ possibilities.}$

In case 7d again there are $\binom{6}{2} \cdot \binom{4}{2}$ ways of choosing the colors, but this time once the colors are chosen there is more than one way of arranging them since we lack the symmetry of the other cases. Suppose the repeated colors are red and black and the other two colors are blue and green. Since the red faces are not opposite each other, a red must be opposite a black. We may turn the cube so that the red is in front and black is in back. Without loss of generality put the other red face on top. Since the blue face is not opposite the green face, the remaining black face cannot be on the bottom. So there are two possibilities (right or left) for the position of the remaining black face. Once this position is determined there are two possibilities for the placement of the blue and green faces. Therefore there are

$$\binom{6}{2}\binom{4}{2} = 90 \text{ ways of choosing the colors;}$$

and there are $2 \cdot 2 = 4$ ways of arranging the chosen colors on the cube. Thus case 7d gives rise to $90 \cdot 4 = 360$ possibilities.

In all, case 7 gives us

$$90 + 180 + 90 + 360 = 720 \text{ possible}$$
cubes.

Do the same for each of the other cases and use additivity to obtain the final result.

C.18   b) Hint: Consider disjoint cases.

C.19   a) Solution: There are 5 possibilities for each digit. Therefore, by the Multiplication Principle, there are $4^5 = 625$ four digit numbers.

b) Approach 1: Pair the numbers as follows: Two numbers are paired if the sum of the digits in a corresponding position is 10(e.g. 3757 is paired with 7353). The only number which cannot be paired with another number is 5555. Since from part (a), there are 625 numbers in all, 624 numbers are paired, giving 312 pairs. The sum of each pair of numbers is found to be 11110. Therefore, the sum of all the numbers is 312(11110) + 5555 = 3471875.

Approach 2: Because of the symmetry involved, each position of each digit appears equally often. How often? Since there are 625 numbers and five distinct digits each of which occurs in a position equally often, each of the digits 1, 3, 5, 7, 9 will occur 625/5 = 125 times in each position (that is, there are 125 numbers with a 1 in the last position: 125 with a 3, etc.) The units position adds to 125(1+3+5+7+9); the tens position sums to 10·125(1+3+5+7+9); the hundreds position sums to 100·125(1+3+5+7+9); the thousands position sums to 1000·125(1+3+5+7+9). Altogether we get

(1+10+100+1000)(125)(25)=3471875.

c) All numbers with 1 in the first (leftmost) position precede all numbers with 3 in that position, which precede all numbers with 5 in that position, etc. As there are 125 numbers which have a 1 in this position (see approach 2 above), 125 which

have a 3 in this position, etc. There are 250 numbers which start with a 1 or a 3, and 375 which start with 1, 3 or 5. The 314th number must therefore start with a 5; in fact it is the 64th = (314-250)th number in the list of numbers starting with 5. We now look at the second digit. Of the 125 numbers starting with 5, 25 have 1 as the second digit; 25 have 3; etc.

Since 50<64<75, the second digit of the number we want must be 5. In fact the desired number is the 14th = (64-50)th number in the list of numbers starting with 55.

Continuing in this manner we look at the third digit of the numbers starting with 55. There are 5 which have 1 as the third digit, etc.

Since 10<14<15, 5 is the third digit and the fourth digit is the 4th = (14-10)th number with first three digits 555.

So the 314th number is 5557.

C.20. Hint: See the solution of C.19, but use caution: The number of numbers with last digit 1 is not the same as the number of numbers with next to last digit 1, since 1 digit numbers are also included, etc. Therefore, this problem should be broken down into cases.

C.21. Hint: b) The digits are 0, 1, 2, 3, 4, 5, 6, 7, 8, 9. However the first digit cannot be 0. Suppose we allow 0 to be a first digit (e.g., as in 0132). How many numbers will we then have? We can add up the sum of all the numbers including these in a manner similar to Exercise C.19. Then we subtract the sum of the numbers starting with 0.

C.23. Hint: Break up into cases depending on the position in which the first red checker appears.

C.24. Solution: a) In order for Della to be ahead when the fifth vote was counted, either all four of her votes are among the first five counted or three of her votes are among the first five counted. In all there are $\binom{10}{4}$ for the positions in which Dellā's voted could be counted. Of these there are $\binom{5}{4}\binom{5}{0}$ in which all four votes were counted among the first five votes. (The

$\binom{5}{0}$ indicates that none of her votes were in the last five positions.) Similarly, there are $\binom{5}{3}\binom{5}{1}$ ways in which exactly three are among the first two. (The $\binom{5}{1}$ indicates that her remaining vote could be in any one of the last five positions.)

Thus the probability that Della was ahead when the fifth vote was counted =

$$\frac{\binom{5}{4}\binom{5}{0} + \binom{5}{3}\binom{5}{1}}{\binom{10}{4}} = 11/42.$$

b) If she was ever ahead, then she was ahead for the first time after the first, third, fifth, seventh, or ninth vote was counted. I.e., if she was ahead at the 2nd vote, she must have been ahead at the first, and if she was ahead at the 4th vote, she must have also been ahead at the third, etc.

The probability that she was ahead after the first vote was counted is $\dfrac{\binom{9}{3}}{\binom{10}{4}} = \dfrac{4}{10}$.

The only way she could be ahead at the third vote but not the first is if the first three votes were VDD. There are $\binom{7}{2}$ positions in which Della could receive her remaining two votes, so that the probability that she is ahead after the third vote but not after the first is $\dfrac{\binom{7}{2}}{\binom{10}{4}} = \dfrac{1}{10}$.

There are two ways that Della could be ahead after the fifth vote, but not after the first or third votes, namely if the votes went VDVDD or VVDDD. With either of these ways there are $\binom{5}{1}$ positions for Della's 4th vote. So the probability that Della was

ahead after the fifth vote but not sooner is

$$\frac{2\binom{5}{1}}{\binom{10}{4}} = \frac{1}{21}.$$

Similarly, the probability that Della was ahead after the 7th vote was counted, but not sooner $= \dfrac{5\binom{3}{0}}{\binom{10}{4}} = \dfrac{1}{42}$.

Finally, the probability that Della was ahead after the 9th vote was counted = 0, since by that time Victor has at least 5 votes.

Since the above cases are disjoint the probability that Della was ever ahead is

$$\frac{4}{10} + \frac{1}{10} + \frac{1}{21} + \frac{1}{42} + 0 = \frac{4}{7}.$$

C.25. Hint: Consider cases: 1) five different kinds of trees; 2) two trees of one kind, other three are all different; 3) two pairs of trees of same kind.

C.26. Hint: Consider all possible placements of 2 kings on the board and subtract the number of excluded or illegal positions.

C.27. a) Hint: Consider cases depending on size of the square. (I.e., count the number of squares of each size)

b) Hint: A rectangle with horizontal and vertical sides is determined by its upper left hand and lower right hand corner.

c) Hint: Count the number of squares of each size. Make sure that you have included all possible sizes and orientations. E.g., the two squares in Figure 49 are the same size but have different orientations.

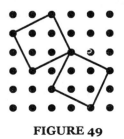

FIGURE 49

C.28. Hint: See the hint for Exercise C.27b).

C.29. a) Hint: What numbers can not be made?

b) Hint: Label the dice (d_1, d_2, d_3) and list the favorable outcomes.

C.30. c) Hint: Since we are only interested in the positions of people relative to one another, in a circular arrangement we may hold one person fixed and consider the seating of the other people in relation to this person.

C.32. Hint: Consider the positions of the yellow beads relative to each other. I.e., they may be touching, there may be one, two or three spaces between them. Count the number of necklaces in each case. Be aware of symmetry or lack thereof.

C.33. Hint: First consider the possible distributions of colors, then the possible arrangements for each particular color distribution.

C.34. b) Solution: There are $\binom{52}{13}$ ways of selecting a 13-card hand. We now consider cases as to whether there are exactly eight, exactly nine, ..., exactly thirteen cards in a particular suit.

We can select 8 cards in the given suit (and 5 cards from the other suits) in $\binom{13}{8}\binom{39}{5}$ ways.

Similarly we can select exactly 9 cards from the selected suit in $\binom{13}{9}\binom{39}{4}$ ways. Etc.

Since there are 4 ways of selecting the suit with eight or more cards there are in all

$$4\left[\binom{13}{8}\binom{39}{5}+\binom{13}{9}\binom{39}{4}+\binom{13}{10}\binom{39}{3}+\binom{13}{11}\binom{39}{2}\right.$$
$$\left.+\binom{13}{12}\binom{39}{1}+\binom{13}{13}\binom{39}{0}\right]$$

favorable outcomes.

Therefore the probability that at least eight cards in the same suit

$$= 4\left[\binom{13}{8}\binom{39}{5}+\binom{13}{9}\binom{39}{4}+\binom{13}{10}\binom{39}{3}\right.$$
$$\left.+\binom{13}{11}\binom{39}{2}+\binom{13}{12}\binom{39}{1}+\binom{13}{13}\binom{39}{0}\right]\Bigg/\binom{52}{13}$$

d) Solution: Here we must be careful as there may be more than one suit with six or more cards. First, break into disjoint cases where there are exactly six cards of a suit, exactly seven cards, etc. The cases with eight or more cards in a suit have been handled in part b).

We can select seven cards in a particular suit and six in the remaining three suits in $\binom{13}{7}\binom{39}{6}$ ways.

We can select six cards in a particular suit and seven in the remaining three suits in $\binom{13}{6}\binom{39}{7}$ ways. As there are four ways of choosing a particular suit it would seem that the number of favorable cases is

$$4\left[\binom{13}{6}\binom{39}{7}+\binom{13}{7}\binom{39}{6}+\binom{13}{8}\binom{39}{5}+\cdots+\binom{13}{13}\binom{39}{0}\right]$$

This is not quite correct, as we have counted some hands twice. For example, a hand which contains 6 spades, 6 hearts, and 1 diamond is included when hands containing 6 spades are counted as well as when we count hands containing 6 hearts. A similar overcount occurs for a hand containing 6 spades and 7 hearts. Etc.

We therefore must subtract off the number of hands that have been counted twice, i.e., the number of hands that contain six or more cards in two or more suits. This happens in two types of hands – hands with seven cards in one suit and six of another and hands with six cards in each of two suits and one card in a third suit.

In the 7, 6 case, there are four ways of selecting the suit with seven cards and then three ways of selecting the six card suit. The seven cards can be chosen in $\binom{13}{7}$ ways; the six in $\binom{13}{6}$ ways giving $4\cdot3\binom{13}{7}\binom{13}{6}$ ways of getting a 7, 6 hand. In the 6, 6 case the two six card suits can be chosen in $\binom{4}{2}$ ways (since they are interchangeable); once the 3 suits are determined, the cards can be chosen in $\binom{13}{6}\binom{13}{6}\binom{26}{1}$ ways giving

$\binom{4}{2}\binom{13}{6}\binom{13}{6}\binom{26}{1}$ ways of getting a 6, 6, 1 hand.

Thus, the total number of favorable outcomes is

$$4\left[\binom{13}{6}\binom{39}{7}+\binom{13}{7}\binom{39}{6}+\binom{13}{8}\binom{39}{5}\right.$$
$$\left.+\dots+\binom{13}{13}\binom{39}{0}\right]-4\cdot 3\binom{13}{7}\binom{13}{6}$$
$$-\binom{4}{2}\binom{13}{6}\binom{13}{6}\binom{26}{1}$$

To compute the probability this number must be divided by $\binom{52}{13}$.

g) Solution: There are 4 ways to choose the suit with four cards, $\binom{13}{4}$ ways to choose these four cards, $\binom{13}{3}\binom{13}{3}\binom{13}{3}$ ways to choose three cards from each of the other suits. The probability of this case =

$$\frac{4\binom{13}{4}\binom{13}{3}\binom{13}{3}\binom{13}{3}}{\binom{52}{13}}.$$

C.35. c) Solution: There are $\binom{52}{5}$ possible five card hands. A five card straight can begin with one of ten cards, A, 2, ..., 10. Each of the five cards in a straight can be of any suit, so the total number of favorable outcomes is $10\cdot 4\cdot 4\cdot 4\cdot 4\cdot 4=10\cdot 4^5$ and the probability of this event is $\dfrac{10\cdot 4^5}{\binom{52}{5}}$. (Note that straight flushes are being included here.)

f) Solution: There are 13 possibilities for the face value of the card repeated 3 times, which leaves 12 possibilities for the face value of the card repeated twice. The three suits of the triply repeated card can be chosen in $\binom{4}{3}$ ways and the two suits of the doubly

repeated card can be chosen in $\binom{4}{2}$ ways, giving the probability of this event =

$$\frac{13\cdot 12\cdot\binom{4}{3}\binom{4}{2}}{\binom{52}{5}}.$$

g) (As in e) there are $13\binom{4}{3}$ ways of choosing the three of a kind. There are $\binom{12}{2}$ ways of choosing the face values of the other two cards. The suit of each of these can be selected in $\binom{4}{1}$ ways.

This gives the desired probability as

$$\frac{13\binom{4}{3}\binom{12}{2}\binom{4}{1}\binom{4}{1}}{\binom{52}{5}}.$$

j) Hint: Count the total number of hands with cards of different face values, subtract the number of straights and the number of flushes but add back the number of straight flushes because these have been subtracted twice.

C.36. a) Hint: Since the five cards you have been dealt are no longer in the deck, only 47 cards remain. If you keep only the two kings, you must draw three cards so that there are $\binom{47}{3}$ possible outcomes. You must also take note that certain cards are no longer in the deck.

For example, if you keep the two kings you can obtain a full house in the following ways:
Case 1: Draw three Q's, three J's or three 10's;
Case 2: Draw three cards of some other face value;
Case 3: Draw one K and two Q's, two J's or two 10's;
Case 4: Draw one K and two cards of some other face value.

The difference between Cases 1 and 2 is that only three Q's, J's, 10's remain in the deck whereas four of each other face values (Ace through 9) are in the deck.
Similarly for Cases 3 and 4.

C.37. Solution: Call the first player B and the second player C. B can beat you only by holding 2 more A's. (B can't have a straight since all the 5's are accounted for.) C can beat you only if he has either two more 9's, two more 8's, two more 6's, a 7 and a 10, or two more hearts.

Consider the ways in which you will not be the winner. Break these up into the following disjoint cases:

Case 1: B beats you.

Case 2: You beat B, but C beats you holding the AH as one of his hidden cards.

Case 3: You beat B, but C beats you not holding the AH.

We compute the probability of each case.

In Case 1 there are $\binom{37}{2}$ possible cards that B could hold. Of these only $\binom{3}{2}$ are favorable. So the probability of Case 1 =

$$\frac{\binom{3}{2}}{\binom{37}{2}} = \frac{6}{37 \cdot 36}.$$

In Case 2, C's fifth card must also be a heart.

As there are only 6 possibilities for this fifth heart, there are $\binom{6}{1}$ ways in which C can beat you. In addition B can hold any two of the remaining 35 cards except for both of the remaining aces. Thus there are $\binom{35}{2} - 1$ possibilities for B's hidden cards. Using the Multiplication Principle, this case gives rise to $\binom{6}{1}\left[\binom{35}{2} - 1\right]$ favorable outcomes. And so the probability of Case 2 =

$$\frac{\binom{6}{1}\left[\binom{35}{2} - 1\right]}{\binom{37}{2}\binom{35}{2}}$$

$$= \frac{24 \cdot 594}{37 \cdot 36 \cdot 35 \cdot 34}.$$

In Case 3, C cannot hold any aces. There are

$$\underbrace{\binom{3}{2}}_{\text{two 9's}} + \underbrace{\binom{3}{2}}_{\text{two 8's}} + \underbrace{\binom{2}{2}}_{\text{two 6's}} + \underbrace{\binom{3}{1}\binom{3}{1}}_{\text{7 and 10}} + \underbrace{\binom{6}{2}}_{\substack{\text{two H's} \\ \text{not AH}}} = 31$$

possible hands that C could hold. For each of these B can hold any two of the remaining 35 cards except for 2 A's. Again by the Multiplication Principle, the number of favorable outcomes $= 31\left[\binom{35}{2} - \binom{3}{2}\right]$.

The probability of Case 3 =

$$\frac{31\left[\binom{35}{2} - \binom{3}{2}\right]}{\binom{37}{2}\binom{35}{2}} = \frac{4 \cdot 31[592]}{37 \cdot 36 \cdot 35 \cdot 34}.$$

The probability that you are not a winner is

$$\frac{6}{37 \cdot 36} + \frac{24 \cdot 594}{37 \cdot 36 \cdot 35 \cdot 34} + \frac{4 \cdot 31 \cdot 592}{37 \cdot 36 \cdot 35 \cdot 34}$$

$$= \frac{7140 + 13536 + 73408}{37 \cdot 36 \cdot 35 \cdot 34}$$

$$= \frac{94084}{37 \cdot 36 \cdot 35 \cdot 34} = \frac{94084}{1585080} \approx .059.$$

Therefore the probability that you win is approximately $1 - .059 = .941$.

C.38. a) Solution: Denote the people by P, Q, R and the gifts they brought by p, q, r respectively. We make a chart (Figure 50) indicating the various receivers.

P	Q	R	
p	q	r	unfavorable; P received p
p	r	q	unfavorable; P received p
q	r	p	favorable
q	p	r	unfavorable; R received r
r	p	q	favorable
r	q	p	unfavorable; Q received q

FIGURE 50

Note that there are $3! = 6$ possible ways of distributing the gifts and that 2 of these are favorable.

Note the number of favorable outcomes could have been obtained as follows. It is equal to $3!$ – the number of unfavorable outcomes.

An outcome is unfavorable if exactly one person receives his or her own gift or every one receives his (her) own gift (it is not possible for exactly two people to receive their own gifts).

There is 1 way in which everyone receives his or her own gift. Also if exactly one person receives his or her own gift, the other two won't. There are 3 ways to determine which one person receives his (her) own gift. Thus the number of favorable outcomes is

$$6 - 1 - \binom{3}{1} \cdot 1 = 2.$$

b) We could proceed by making a chart as in a) but, as there are 24 cases, the approach is cumbersome. We therefore proceed as follows:

Let $D(n)$ = the number of ways in which n people can exchange gifts so that no one receives his or her own gift. In this problem we are looking for $D(4)$. Note that in part a) we found $D(3) = 2$ and it is clear that $D(2) = 1$.

To find $D(4)$, note there are $4!$ ways of distributing the gifts among 4 people. So that

$D(4) = 4!$ – number of unfavorable ways of distributing the gifts. A distribution of 4 gifts is unfavorable if exactly one person receives his or her own gift or if exactly two people receive their own gifts or if everyone receives his or her own gifts.

Exactly one person can receive his or her own gift in $\binom{4}{1} D(3)$ ways (that is, $\binom{4}{1}$ possibilities as to which person receives his or her own gift, and, for each of these, there are $D(3)$ ways to distribute the remaining gifts among the other 3 people so that no one receives the gift that he or she brought).

Similarly, exactly 2 people can receive their own gifts in $\binom{4}{2} D(2)$ ways. There is 1 way in which every one can receive his or her own gift. Thus

$$D(4) = 4! - \binom{4}{1} D(3) - \binom{4}{2} D(2) - 1$$

$$= 24 - 4 \cdot 2 - 6 \cdot 1 - 1 = 9.$$

c) As in b) $D(5) =$

$$5! - \binom{5}{1} D(4) - \binom{5}{2} D(3) \binom{5}{3} D(2) - 1 = 44.$$

d) Analogously $D(6) = 265$

By taking another approach $D(n)$ can be seen in general to be

$$D(n) = n! \left[1 - \frac{1}{2!} + \frac{1}{3!} - \frac{1}{4!} + \dots \pm \frac{1}{n!} \right].$$

This result is obtained in the following manner: The total number of outcomes minus the number of outcomes for which at least one person receives his or her own gift plus the number of outcomes for which at least two people receive their own gifts minus the number of outcomes for which at least three people receive their own gifts plus, etc. (Think about why we have to alternately add and subtract these terms.)

e) We want the probability that Ross and Sanford receive each other's gift given that no one receives the gift that he or she bought. If Ross and Sanford exchange gifts then the other four can be distributed in $D(4)$ ways. Therefore the desired probability is equal to

$$\frac{D(4)}{D(6)} = \frac{9}{265} \approx .033.$$

C.39. Hint: What is the connection with Exercise C.38?

C.43. Solution: If the conditions of part b) of Exercise C.14 are realized, then Jack Smith is either sitting in the same row as Joan Brown and John Jones or else he is sitting in the same row as Joan Jones and John Brown. Both cases are equally likely to happen, so the probability that Jack Smith and Joan Brown are sitting in the same row is $\frac{1}{2}$. Given that they are in the same row, there are 6 possible arrangements only two of which are favorable. Hence the desired probability is

P_r[they are in the same row] $\cdot P_r$[she is to his immediate right/ P_r[they are in the same row] $= \frac{1}{2} \cdot \frac{1}{3} = \frac{1}{6}$.

C.44. Solution: a) Since there are 49 cards unaccounted for, 23 of which are six or less, the desired probability is 23/49.

b) In this case 14 cards are still unaccounted for, 4 of which are favorable. Hence, the probability that if you draw a card the total will not exceed 21 is $\frac{4}{14} = \frac{2}{7}$.

c) If you stand pat at 15, then, since the dealer has a 9 showing, he will beat you if his hole card is 8 or more. This occurs in 7 of the 14 cases, so the probability that he will already beat you is $\frac{1}{2}$. If his hole card is 7 or less, he will have to draw another card. If he has a 7, (probability $\frac{1}{14}$), then the probability that he will not exceed 21 if he draws is $\frac{2}{13}$. Thus the probability of this case is $\frac{3}{14} \cdot \frac{2}{13} = \frac{3}{91}$.

If he has a 6 (probability $\frac{2}{14}$), then the probability that he will not exceed 21 if he draws is $\frac{3}{13}$. Thus the probability of this case is $\frac{2}{14} \cdot \frac{3}{13} = \frac{3}{91}$.

The remaining cases – he has a 5 or he has a 3 – give probabilities of $\frac{1}{14} \cdot \frac{6}{13} = \frac{3}{91}$ and $\frac{1}{14} \cdot \frac{8}{13} = \frac{4}{91}$ respectively.

Thus, the probability that the dealer will beat you if you stand pat is $\frac{1}{2} + \frac{3}{91} + \frac{3}{91} + \frac{3}{91} + \frac{4}{91} = \frac{117}{182}$.

d) If you do draw, then the probabilities for the card(s) the dealer holds and draws will be affected by the card you draw. We therefore consider cases as to what card you might draw. (You should only draw one card since any two of the cards left would bring your total past 21.) If you draw a card larger than a six (probability $\frac{10}{14}$) you automatically lose.

If you draw a three (probability $\frac{1}{14}$), your total will be eighteen.

There will be 13 cards still unaccounted for. Of these, the dealer will win if his hidden card is a ten or a picture card or if his hidden card is a seven and he draws a five, or his hidden card is a six and he draws a five or a six, or if hid hidden card is a five and he draws a six or

a seven. Therefore if you draw a three, the dealer will win with probability

$$\frac{5}{13} + \frac{3 \cdot 1}{13 \cdot 12} + \frac{2 \cdot 2}{13 \cdot 12} + \frac{1 \cdot 5}{13 \cdot 12} = \frac{72}{156}.$$

Therefore

$$\Pr[\text{you win if you draw a three}] = \frac{84}{156}$$

$\Pr[\text{you draw a three and win}]$

$$= \frac{1}{14} \cdot \frac{84}{156} = \frac{84}{2184}$$

$\Pr[\text{you draw a three and lose}]$

$$= \frac{1}{14} \cdot \frac{8}{156} = \frac{72}{2184}$$

Similarly, you could draw a five with probability $\frac{1}{14}$. In this case the dealer will win with probability $\frac{2 \cdot 1}{13 \cdot 12} = \frac{2}{156}$; the dealer will tie you with probability $\frac{2 \cdot 1}{13 \cdot 12} = \frac{2}{156}$; and you will win with probability $1 - \frac{2}{156} - \frac{2}{156} = \frac{152}{156}$.

Therefore,

$\Pr[\text{you draw a five and win}]$

$$= \frac{1}{14} \cdot \frac{152}{156} = \frac{152}{2184}.$$

$\Pr[\text{you draw a five and lose}]$

$$= \frac{1}{14} \cdot \frac{2}{156} = \frac{2}{2184}.$$

$\Pr[\text{you draw a five and tie}]$

$$= \frac{1}{14} \cdot \frac{2}{156} = \frac{2}{2184}.$$

In a similar manner

$$\Pr[\text{you draw a six and win}] = \frac{300}{2184}$$

$$\Pr[\text{you draw a six and lose}] = 0$$

$$\Pr[\text{you draw a six and tie}] = \frac{12}{2184}$$

Therefore if you draw a card the probability that you win is

$$= \frac{84}{2184} + \frac{152}{2184} + \frac{300}{2184} = \frac{536}{2184} \approx .245;$$

the probability you lose is

$$= \frac{10}{14} + \frac{72}{2184} + \frac{2}{2184} = \frac{1634}{2184} \approx .748;$$

and the probability that you tie is

$$= \frac{2}{2184} + \frac{12}{2184} = \frac{14}{2184} \approx .006 .$$

Since the probability that you win if you draw (.245) is less than the probability that you win if you stand pat (.357), and the probability that you lose if you draw (.748) is greater than the probability that you lose if you stand pat (.643), you should stand pat.

C.45. Hint: See solution to Exercise C.45.

C.47. Solution: a) Since each of the 7 remaining people can put a "one" or a "two," there are 2^7 possible outcomes. Of these $\binom{7}{4}$ result in 4 ones and 3 twos, and $\binom{7}{4}$ result in 4 twos and 3 ones. This gives

$$\frac{2\binom{7}{4}}{2^7} = \frac{35}{64} \approx .55.$$

b) Given that the teams are selected, there are $\binom{7}{3}$ possibilities for Dan's teammates. Therefore the desired probability is

$$\frac{\binom{6}{2}}{\binom{7}{3}} = \frac{3}{7}.$$

c) There are four possible scenarios which should be considered:
　　i) Dan and Eliot are both on the same team as the designated player;
　　ii) Dan and Eliot are both on the opposite team as the designated player;
　　iii) Dan is on the same team as the designated player but Eliot is not;
　　iv) Eliot is on the same team as the designated player but Dan is not.
In each case, there are five players remaining.
In case i), one of the five must be on the team with Dan and Eliot, so this case can occur in $\binom{5}{1} = 5$ ways; in case ii), two of the five must be on the team with Dan and Eliot, so this case can occur in $\binom{5}{2} = 10$ ways; in case iii), two of the five must be on the team with Dan and the designated player, so this

case occurs $\binom{5}{2} = 10$ ways; and in case iv), two of the five must be of the team with Eliot and the designated player, so this case occurs in $\binom{5}{2} = 10$ ways. Since cases i) and ii) are favorable, the required probability is

$$\frac{5 + 10}{5 + 10 + 10 + 10} = \frac{15}{35} = \frac{3}{7}.$$

C.49. Hint: What is the probability that no two of them have the same birthday?

C.54. Hint: Play the movie in reverse.

C.62. Solution: Clearly, 1 must be placed in the upper left-hand corner and 8 must be placed in the lower right. Figure 51 shows all possible placements of 2, 3 and 4.

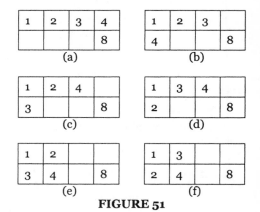

FIGURE 51

In (a), the placement of 5, 6, and 7 is forced. (a) gives rise to 1 case (b), (c) and (d) each give rise to three possibilities in each either the 5, 6 or 7 can be on the top row. In (e) and (f) the placement of 5 is forced; therefore each gives rise to two possibilities depending on the placement of the 6 and 7.

　　　1 + 3 + 3 + 3 + 2 + 2 = 14
　　　(a) (b) (c) (d) (e) (f)

C.63. Hint: Consider the possible patterns for the placement of 1, 2, 3 and 4. Separately consider the possible placements of 6, 7, 8 and 9. How can these patterns be numbered?

C.64. Hint: Label each hot dog by the number indicating the order in which it will be eaten. That is, the first hot dog to be eaten will receive a 1; the second, a 2, etc. In how many ways can these numbers be assigned to the hot dogs in the leftmost column? Once this column has been assigned numbers, how many ways can numbers be assigned to the middle column, etc?

C.66. Solution: There is only 1 way to go from A to B; there are only two ways to go from A to C (A→C or A→B→C); there is only one way to go from A to D.

Therefore there are 1 + 2 + 1 = 4 ways to go from A to E since we must get to E directly from either B, C or D.

Continuing in the same manner we can label each vertex of the diagram with the number of ways in which that vertex can be reached from A. The label of each vertex is the sum of the labels of the vertices which could immediately precede the vertex in question.

C.67. Hint: See the solution of Exercise C.66.

C.68. Hint: Label each letter by the number of ways it can be reached. See the solution of Exercise C.66.

C.69. Hint: First, count the number of routes to I in a manner similar to that of Exercise C.68.

C.70. Hint: See the solution of Exercise C.66 and the hint to Exercise C.68.

Answers to Selected Problems

CHAPTER 1

Exercises

1.1. Sue's mother (Jo) owns the garter snake.

1.4. Allen is the third baseman; Bill is the center fielder; Chuck is the catcher; Ed is the shortstop; Harry is the pitcher; Jerry is the second baseman; Mike is the right fielder; Paul is the first baseman; Sam is the left fielder.

1.6. Tuesday.

1.7. Freddy came in last in the high jump.

1.8. Alice beat Bob, 9–1; Alice beat Carol, 7–3; Alice beat Ted, 10–0; Bob beat Ted, 6–4; Bob tied Carol, 5–5; and Carol beat Ted, 8–2.

1.10. Partial answer: Patti beat Nancy, 5–0; Patti beat Manny, 5–0; Patti beat Tom, 5–2.

1.11. Cleo was the big winner. They are seated in alphabetical order clockwise.

1.12. Leopold and Eli.

1.13. Mr. Lawyer—the plumber.

1.16. John ordered three bologna and received one salami and two bologna.

1.17. They are labeled $.40 and $.45; they contain $.75 between them.

1.18. Bobbie is the cheerleader; Frankie, the class president; Gerry, the principal's child; Jo Jo, the basketball player; Ronnie, the volleyball player; Sal, the valedictorian.

1.19. (a) Helen and Max play for East; Sylvia and Lee for West; Emily and Ted for North; Vicki and Paul for South; Becky and Irv for Central.
(b) Thursday: South beat Central, North beat East, and South beat North; Friday: South beat Central in the first match; then Central beat South to win the championship.

1.20.

Competitor	School	Entry number	Prior standing	High jump
Jimenez	Mamaraneck	5	4	4
Harris	Jupiter	2	1	5
Boone	West Roch.	1	3	2
Manners	Marmaduke	3	2	3
Twofeathers	Holbrook	4	5	1

1.21. The Norwegian drinks water; the Japanese owns the zebra.

1.22. "Kumquats" was at The Sarcophagus; "Labradorian" at The Thumbscrew; "Mahatma" at The Reptilian; "Nonconformist" at The Purgatory; and "Ottoman" at The Quagmire.

Who went with whom to what show can then be determined from the chart below:

	Florence	Glenda	Helen	Inez	Joan
Albert	O	N	M	K	L
	Thur	Sat nite	Wed	Fri	Sat mat
Barney	K	M	O	L	N
	Sat mat	Thur	Fri	Sat nite	Wed
Chuck	N	K	L	M	O
	Fri	Wed	Thur	Sat mat	Sat nite
Danny	L	O	K	N	M
	Wed	Sat mat	Sat nite	Thur	Fri
Ernie	M	L	N	O	K
	Sat nite	Fri	Sat mat	Wed	Thur

For example, the upper left box reads: Albert and Florence went to see the "Omnipotent Ottoman" on Thursday.

CHAPTER 2

Practice Problems

2.A 1. (a), (d), and (e) are statements.

3. (a) This sentence is true for some values of x ($x = 3$) and false for others. But, as x is not a specific number, we cannot say that the sentence is true or that it is false.

(b) If it were a statement, it would have a truth value. But both possible truth values lead to contradictions. That is, if the sentence is true, then it must be false (since it claims to be false), but if it is false, then it is a true statement.

2.B 1. (a) Since both the antecedent and the consequent are true, the statement is true.

(b) Since the antecedent (Jarvis got an F on the final) is false, the statement is true.

(c) Since the antecedent (Jarvis got an F on the

final) is false, the statement is true (regardless of whether or not Jarvis passed the course).

(d) Since the antecedent (Jarvis passed the course) is true and the consequent (Jarvis got an A on the final) is false, the statement is false.

(e) Since the antecedent (Jarvis failed the course) is false, the statement is true.

3. (a) $f \rightarrow t$. If you recently fired a gun, then the paraffin test is positive.

(b) $(\sim r) \rightarrow g$. If it doesn't rain tomorrow, then I will go to the beach. Or, equivalently $(\sim g) \rightarrow r$; If I don't go to the beach tomorrow then it will be raining.

(c) $h \rightarrow b$. If I hear you say so, then I believe it.

(d) $a \rightarrow l$. If you ask me nicely, then I'll leave.

(e) $s \rightarrow c$. If Tex sees a cowboy movie, then he cries.

5. (a) $p \rightarrow q$. This is true if Alice and Bob both passed math, if they both failed math, or if Alice failed but Bob passed.

(b) $q \rightarrow p$. This is true if Alice and Bob both passed math, if they both failed, or if Bob failed but Alice passed.

(c) $p \wedge q$. This is true if Alice passed math and Bob also passed math.

7. (a) Jack is a good golfer if and only if he is not a good tennis player.

(b) Jack is a good golfer, but he is not a good tennis player.

(c) Jack is a good tennis player.

(d) Jack is not a good golfer.

(e) If Jack is not a good tennis player, then he is a good golfer.

2.C 1. (a) We can conclude that no one was home.

(b) We can conclude that Archie was not early.

2.D 1. (a) $q \rightarrow (\sim p)$.   (b) $s \rightarrow [(\sim p) \vee r]$.

(c) $r \rightarrow p$.   (d) $(r \leftrightarrow p) \rightarrow [s \wedge (\sim q)]$.

3. (a) If the poet is dreaming, then Titania is not the queen of the fairies.

(b) Either the poet is not dreaming, or else Titania is not the queen of the fairies.

(c) It is not possible that both "the poet is dreaming" and "Titania is queen of the fairies" are true.

(d) If it is not true that the poet is either dreaming or jesting, then if Oberon is the king of the fairies then Titania is the queen of the fairies.

5. (a) No one is home.

(b) Archie is not early and Janice is not absent.

(c) No one is home.

7. (a) Either three krimmls are not worth one glunk or the Martian canals are not empty.

(b) Three krimmls are not worth one glunk.

(c) I am not a Martian.

(d) Either the Martian canals are not empty or I am not a Martian.

(e) We cannot conclude anything else. (The statement is true regardless of whether Xcag Zemph is the ruler of Mars or not and regardless of whether or not the canals are empty.)

2.E 1. (a) If Jonas does not have at least ten cents, then he does not have two nickels in his wallet.

(b) If someone could enter, then the door is not locked.

(c) If I won't hear some music, then I won't go to the concert.

(d) If a triangle is not equilateral, then its three angles are not equal.

3. (b) and (c) are equivalent; (a) and (e) are equivalent.

2.F 1. (a) Premises:

P_1: Either Rachelle is brilliant or she has a wonderful personality.

P_2: If Rachelle has a wonderful personality, then she has an active social life.

Conclusion: Either Rachelle is brilliant or she has an active social life.

$$b \vee p$$
$$p \rightarrow a$$
$$\therefore \ b \vee a$$

(b) Premises:

P_1: If Bessie is a cow, then she moos.

P_2: Bessie is not a cow.

Conclusion: Bessie does not moo.

$$c \rightarrow m$$
$$\sim c$$
$$\therefore \ \sim m$$

(c) Premises:

P_1: Either Lucretia is forceful or she is creative.

P_2: Lucretia is forceful.

Conclusion: Lucretia is not creative.

$$f \vee c$$
$$f$$
$$\therefore \ \sim c$$

(d) Premises:

P_1: If Lee Wong is a student, then he is bright.

P_2: If Lee Wong is bright, then he studies logic.

Conclusion: If Lee Wong is a student, then he studies logic.

$$s \rightarrow b$$
$$b \rightarrow l$$
$$\therefore \ s \rightarrow l$$

3. (a) Valid.   (b) Valid.

(c) Invalid (p and q may both be true).

(d) Invalid (p and q may both be false).

5. (a) Valid.

(b) Invalid. Bessie might be a bull, or even a person who moos. In this case, both premises would be true but the conclusion false.

(c) Invalid. If Lucretia were forceful and creative, then both premises would be true but the conclusion would be false.

(d) Valid.

7. (a) Invalid.   (b) Valid.

Exercises

2.2.   The first was truthful and the second was a liar.

2.4.   The fourth troll is a Truthful and the others are Liars.

2.5.   The tall troll is Lowax; the short one, Waldar; and the medium one, Gaut.

2.7.   Winken is a Liar; Blinken is a Truthful; and Finken is an Alternator who is lying.

2.10.   The first robot is a Lawbake and the second is a Mendible. Captain Cooke was using telic coins.

2.13.   Sykes did it. Tolliver was also present.

2.15.   Fifi.

2.17.   He is under 5 feet tall, has white fur, and 5 toes on each foot.

2.18.   Dr. Mandlebaum is the hematologist; Coates,

the endocrinologist; Mavis, the cardiologist; and Rowe, the gastroenterologist.

2.19.   Danny received his present from Jason; Ethan received his from Danny; Mark received his from Ethan; Mindy received hers from Mark; and Jason received his from Mindy.

2.20.   Earl is the mason, Moe is the carpenter, Luis is the plumber, and Randy is the architect.

2.22.   Amy bought comic books; Betty bought a coloring book; Carmen bought the ball; and Dee bought candy.

2.24.   Martians do not have three heads, they are green, they can fly.

2.25.   His mechanic. He also should replace his spark plugs and distributor cap.

2.27.   A blue shirt and white socks (and a brown suit—if he is wearing any suit at all).

2.29.   No, Luigi's plane could not fly more than 25 feet high.

2.30.   Yes, Pamela's porridge is putrid.

2.32.   Attila did not die at the age of seventy-nine; the emotional development of primates does not parallel that of reptiles; we cannot tell whether or not Attila was reincarnated as a snake.

CHAPTER 3

Practice Problems

3.A   1.   (a) n; $n + 15$.   (b) n; $2n - 3$.

(c) n; $n(n - 2)$.   (d) d; $\frac{1}{5}d$.   (e) a; $\frac{1}{a}$.

3.B   1.   (a) $n = n^2 - 3$.   (b) $n^2 = n + 12$.
(c) $n + 1 \leq \frac{1}{4}n^2$.   (d) $f \geq s + m$.

Exercises

3.2. $91.90.

3.4.   54 pieces.

3.5.   Tom should get $6.30 and Don $2.10.

3.8.   200 yards × 900 yards, and 450 yards × 400 yards.

3.10. $\dfrac{1 + \sqrt{5}}{2}$.

3.11. $x = 5$.

3.12. $.50.

3.13. $109.40.

3.14. $5.40.

3.15.   Ada weighs 299; Brendan, 348; Corinne, 327; Darryl, 272; Eva, 311; and Floyd, 370.

3.17.   Minski spent $17 and his wife spent $34; Pinski spent $11 and his wife spent $40; Dubinski spent $36 and his wife spent $15.

3.18.   The 50 bonor note.

3.19.   78.

3.21.   64.

3.22.   76 people (the baby wasn't a person when it left Fort McConnell).

3.25.   Chin Lee is 28, Sue Ling is 42.

3.26.   Mutt is 40, Jeff is 50.

3.28.   The butcher is 48, the baker is 42, and the candlestick maker is 60.

3.31.   The hare can't win. Just as he is finishing the first half of the race, the tortoise is crossing the finish line.

3.32.   3 seconds.

3.33.   10:55 AM.

3.34.   168 miles.

3.36.   16 nautical miles.

3.37.   55 miles.

3.38.   (a) 350 yards.   (b) The cyclist who starts at the southern end.   (c) At the southern tip of the track.

3.39.   40 miles per hour.

3.41.   2 miles per hour.

3.42.   60.

3.43.   45.

3.44.   Yes (see Hints and Solutions section).

3.45.   18 stations; 16 minutes.

3.46. $3\frac{1}{3}$ miles per hour.

3.48.   (a) $25\frac{5}{7}$ minutes.

3.49.   21 minutes.

3.50. $32\frac{2}{3}$.

3.51.   They save 1 hour $22\frac{1}{2}$ minutes.

3.52.   (a) 50 days.   (c) Forever.

3.53.   I am a male doctor.

3.55.   Neither. The amount of oil in the vinegar was equal to the amount of vinegar in the oil.

3.56.   (a) $3\frac{1}{33}$ gallons.

3.58.   Beulah (260 lb) is married to Bobby (260); Barbra (160) is married to Whelan (320); Belinda (200) is married to Wally (300). Bubba is single and weighs 280.

3.59.   5.

3.61.   (a) The ratio of the length of the right arm to the left is 6 to 5.   (b) 20 oz.   (c) 10 oz.

3.62. $\frac{1386}{97}$.

3.66.   36.

3.68.   There are 28 beggars and I have 220 cents.

3.69.   7.

3.71.   8 and 5.

3.72.   He turned the entire packet over behind his back.

3.73.   31.

3.74.   When the card was replaced, he counted off 7 cards to make sure that the selected card was the eighth card of his half.

3.75.   He peeked at the original bottom card of the deck and predicted that card.

3.77.   He counted to the $32 - 9 = 23$rd card.

3.79.   The four of diamonds was the top card of the deck at the beginning of the trick.

CHAPTER 4

Practice Problems

4.A   1.   (a) Since $k = 6(x + 2y)$, then by part (ii) of Theorem 4.1, $2 \mid k$, $3 \mid k$, and $6 \mid k$.

3. $a + b = c$. Suppose $d \mid a$ and $d \mid b$. Then $a = de$ and $b = df$, for some integers e and f.

$$c = a + b = de + df = d(e + f).$$

Since $e + f$ is an integer, $d \mid c$.

Similarly, if $d \mid a$ and $d \mid c$, then $a = de$ and $c = df$ for some integers e and f.

$$b = c - a = df - de = d(f - e).$$

Hence, $d \mid b$.

The remaining case is similar.

4.C   1.   (a) $15120 = 2^4 \cdot 3^3 \cdot 5 \cdot 7$.

(b) $2183 = 37 \cdot 59$.

(c) $409 = 409$.

(d) $72814 = 2 \cdot 7^2 \cdot 743$.

4.D   1.   (a) $15120 = 2^4 \cdot 3^3 \cdot 5 \cdot 7$ is neither a perfect square nor a perfect cube.

(b) $46656 = 2^6 \cdot 3^6$ is both a perfect square and a perfect cube.

(c) $1728 = 2^6 \cdot 3^3$ is a perfect cube but is not a perfect square.

3.   (a) $\sqrt{46656} = 2^3 \cdot 3^3 = 8 \cdot 27 = 216$.

(b) $\sqrt[3]{46656} = 2^2 \cdot 3^2 = 36$.

$\sqrt[3]{1728} = 2^2 \cdot 3 = 12$.

5.   (a) x must equal 11 times a perfect square.

(b) y must be a perfect square.

7.   A number $m = p_1^{r_1} p_2^{r_2} \cdots p_k^{r_k}$ is a perfect nth power if each exponent of each prime is divisible by n—that is, $n \mid r_i$ for all $i = 1, 2, \ldots, k$.

4.E   1.   Fifty.

3.   (a) 1, 2, 4, 7, 8, 14, 28, 49, 56, 98, 196, 392.

(b) 1 and 353.

(c) 1, 2, 4, 7, 8, 14, 16, 28, 56, 112.

5.   (a) 105.   (b) It is odd.   (c) It is even.

4.F   1.   No; $\gcd(3, 6) = 3$ but $3 \nmid 5$.

3. k is divisible by 3.

4.G   1.   (a) (i) 4;   (ii) 2;   (iii) 1;   (iv) 3.

(b) 37 and 54 are relatively prime.

3.   (a) 3^4.

(b) p^m, where m is the smaller of r and s.

(c) $3^2 \cdot 5^6$.

(d) $p_1^{m_1} p_2^{m_2} \cdots p_k^{m_k}$.

(e) $2^3 \cdot 3^2 \cdot 5 \cdot 7 = 2520$.

4.H 1. quotient $= 36$, remainder $= 0$.

3. $q = 12, r = 3$.

5. $q = -19, r = 2$.

4.I 1. $\bar{0} = \{\ldots, -8, -4, 0, 4, 8, 12, \ldots\}$

$\bar{1} = \{\ldots, -7, -3, 1, 5, 9, 13, \ldots\}$

$\bar{2} = \{\ldots, -6, -2, 2, 6, 10, 14, \ldots\}$

$\bar{3} = \{\ldots, -5, -1, 3, 7, 11, 15, \ldots\}$.

3. $12; 35; n$.

4.J 1. (a), (b), (d), and (f) are true.

3. 2.

5. 20.

7. $x = 3k + 1, k = 0, \pm 1, \pm 2, \ldots$.

4.K 1. $x + y \equiv 11 \equiv 3 \pmod 8$.

$xy \equiv 30 \equiv 6 \pmod 8$.

3. (a) $4x \equiv 4 \pmod 5$ or $-x \equiv -1 \pmod 5$.

(b) $2x \equiv 2 \pmod 3$.

5. (a) $y \equiv 1 \pmod 7$.

(b) $-2x + 2y \equiv 2 \pmod 7$.

7. 1.

9. $x = 2$.

4.L 1. (a) 5. (b) 1. (c) 3.

4.M 1. 8.

3. 5.

4.N 1. (a) $x \equiv 5 \pmod{81}$.

(b) $x \equiv 5 \pmod{12}$.

4.O 1. (a) One (mod 8).

(b) Three (mod 9), but only one (mod 3).

(c) None.

(d) Two (mod 98), but only one (mod 49).

3. (a) $x \equiv 11 \pmod{12}$.

(b) There aren't any solutions.

(c) $x \equiv 2, 6, 10$ or $14 \pmod{16}$; or, equivalently, $x \equiv 2 \pmod 4$.

(d) $x \equiv 1 \pmod 3$.

4.P 1. (a) $x \equiv 3 \pmod{12}$.

(b) $x \equiv 11 \pmod{12}$.

(c) $x \equiv 38 \pmod{59}$.

4.Q 1. $x = 66, y = 59$; and $x = 85, y = 76$.

3. $x = -4 + 53k, y = -3 + 37k$, for $k = 0, \pm 1, \pm 2, \ldots$.

Exercises

4.2. 118.

4.4. \$10,737,418.24.

4.5. 648.

4.6. 12.

4.7. There are 5 boys and 2 girls; the parents have been married for 11 years.

4.9. Quincy owned 15 chickens; his sister, Trixie, owned 5. Ralph owned 14 chickens; his sister, Vera, owned 7. Pedro owned 13 chickens; his sister, Sandy, owned 12.

4.10. $a = 144, b = 1728, c = 169, d = 2197$.

4.12. The ages are 6, 6, and 1. (Note the two oldest are twins.)

4.15. \$34.41.

4.18. Two.

4.19. \$1.22.

4.20. She sold 5 kitchen clocks, 3 cuckoo clocks, and 2 grandfather clocks.

4.24. The magician would have said, "Count up to N" (where N is 12 more than the second number).

4.26. By placing the proper card in the tenth position before the trick began. (Use algebra and digital roots.)

4.29. 2.

4.30. 61; Sean got the best deal. The owner of the gum ball machine got the worst.

4.31. \$4.50.

4.33. (a)

Divisor	Smallest	Largest
2	123456798	987654312
3	123456789	987654321
4	123457896	987654312
5	123467895	987643215
6	123456798	987654312
7	123456879	987654213

Divisor	Smallest	Largest
8	123457896	987654312
9	123456789	987654321
10	There are none	There are none
11	123475869	987652413
12	123457896	987654312
13	123456879	987654213
14	123457698	987653142
15	123467895	987643215
16	123457968	987654312

·	0	1	2	3	4	5	6	7
0	0	0	0	0	0	0	0	0
1	0	1	2	3	4	5	6	7
2	0	2	4	6	10	12	14	16
3	0	3	6	11	14	17	22	25
4	0	4	10	14	20	24	30	34
5	0	5	12	17	24	31	36	43
6	0	6	14	22	30	36	44	52
7	0	7	16	25	34	43	52	61

3. (a) $(11211)_{three}$.   (b) $(1112)_{three}$.
5. (a) $(11101110)_{two}$.   (b) $(100011)_{two}$.
7. (a) $(1112112)_{three}$.   (b) $(111101)_{three}$.

5.D 1. (a) C = 1.
(b) F is larger than B. (AB < 100, so F · AB < F · 100.)
(c) B = 3 and F = 5 or B = 6 and F = 8.
(d) If B = 3 and F = 5, then A = 6 or 7; if B = 6 and F = 8, then A = 7.
(e) B = 3, F = 5, A = 7. 1095 ÷ 73 = 15.

4.35. Six.

4.36. (c) Tuesday.   (d) 1920,   1948,   1976.
(e) None. (f) Sunday, Wednesday, and Friday.
(g) 1905, 1911, 1916, 1922, 1928, etc. (mod 28).
(h) 171.

CHAPTER 5

Practice Problems

5.A 1. (a) 13.   (b) 344.
(c) $3.833 \ldots = 3\frac{5}{6}$.   (d) 1299.

5.B 1. (a) $(100000100111)_{two}$.
(b) $(2212022)_{three}$.
(c) $(13355)_{six}$.
(d) $(6041)_{seven}$.
(e) $(125E)_{twelve}$.
3. (a) $(10)_{two}$.   (b) $(10)_{three}$.   (c) $(10)_{b}$.

5.C 1.

+	0	1	2	3	4	5	6	7
0	0	1	2	3	4	5	6	7
1	1	2	3	4	5	6	7	10
2	2	3	4	5	6	7	10	11
3	3	4	5	6	7	10	11	12
4	4	5	6	7	10	11	12	13
5	5	6	7	10	11	12	13	14
6	6	7	10	11	12	13	14	15
7	7	10	11	12	13	14	15	16

Exercises

5.1. He became numb Burr, Count Turr, after writing the 9 in 8196.

5.2. (b) Any digit could be the last digit of a cube.

5.4. R. Velt.

5.6. 11, 12, 15, 24, and 36.

5.9. (b) 11, 12, 15, 22, 23, 26, 33, 34, 37, 44, 45, 48, 55, 56, 59, 66, 67, 77, 78, 88, 89, 99.
(c) 11, 14, 22, 25, 33, 36, 44, 47, 55, 58, 66, 69, 77, 88, 99.

5.11. The only such number is 65.

5.13. He wrote 1089 on the index card before the trick began.

5.14. December 4.

5.15. 10.

5.16. 255. Tweedledee should always guess the number in the middle of the remaining permissible range.

5.18. He added the numbers in the upper left hand corners of the indicated cards.

5.19. (b) It reverses the order.

(c) After four deals the cards would be back in their original order.

5.23. (a) 22.

5.25. Between 1 and 13, weights of 1, 3, and 9 grams suffice; between 1 and 40, weights of 1, 3, 9, and 27 will do; in general, one weight for each power of 3 up to the desired total will be needed.

5.27. 2.

5.29. 19.

5.30. If the base is even, n is odd if and only if the last digit (the units digit) of its representation is odd; if the base is odd, n is odd if and only if the sum of the digits in the representation of n is odd.

5.33. MY.

5.34. MAID = 6423.

5.35. A nag.

5.37. 14579 + 85919 = 100498.

5.39. 125 · 37 = 4625.

5.40. 246 · 386 = 94956.

5.45. 37 · 29 = 1073.

5.46. 74 · 59 = 4366.

5.47. CHEIN = 38921; AVERBACH = 76905738.

5.50. 12800874 ÷ 142 = 90147.

5.52. 101010101 ÷ 271 = 372731.

5.53. 3040774 ÷ 178 = 17083.

5.55. The two divisors are 333 and 29 respectively; the two quotients are 300324 and 10356 respectively.

5.56. 631938 ÷ 625 = 1011.1008.

5.58. 152152 is the only one.

5.60. Any base greater than or equal to five.

5.61. Base seven or base eight.

5.63. Base four. 102003 ÷ 33 = 1031.

CHAPTER 6

Practice Problems

6.A 1. (a)

vertex	A	B	C	D	E	F
degree	2	4	3	3	4	4

(b)

vertex	A	B	C	D
degree	3	5	3	3

6.B 1. (a) ABCDEFGHIBJCEJFHJIA

(b) ABCFEDGHIFHEBDA

(c) None exists.

6.C 1. (a) BACFJIHGDBCEFIEHDEB

(b) FIHGDABCFEDBEH

(c) None exists.

6.E 1. (a) ABGDFADB; CHGEBCEH; MLGJHMJL; and KFGILKIF. Four are needed.

(b) ABHGACDFEC; BD; FH; and GE. Four are needed.

3. (a) ABGDFADBCHGEBCEHLKGIHLIKJ-FGHKJHF. Three retracings are necessary.

(b) ABJGACDBDFJFEGEC. Three retracings are necessary.

5. (a) ABGDFADBCHGEBCEHLKGIHLIKJ-FGHKJHFA.

(b) ABJGACDBDFJFEGECA.

6.F 2. (a) Not possible. There are too many odd vertices.

(b) Dicircuit: BCFIHGDABFEDHEB.

(c) Not possible. E has in-degree 4 and out-degree 2.

6.G 1. (a) Four colors.

(b) Four colors.

(c) Two colors.

(d) Three colors.

3. As this figure shows, the number of colors needed are (a) three, (b) five, (c) four.

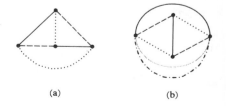

(a)   (b)

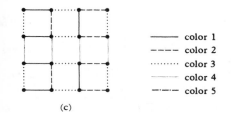

——— color 1
- - - - color 2
········· color 3
·········· color 4
—·—·— color 5

(c)

Exercises

.2.   Colorado.

.5.   (a) No.   (b) $n = 3$.

(c) AGIBHJCLFDKEKDFLCJHBIGA.

.7.   ACFDCFGEDGEBDAB; 1400 meters.

.8.   (a) 3350 meters. Her route could be:
DABDEBCFBFEHDGHFIH.

.11.   11121131232212223321331323.

.13.   There are two different routes plus their
reversals.

.16.   THISISTOODIFFICULT or
THISISODIFFICULTOT.

.17.   No.

.19.   (a) Yes, it is possible.

(b) It is not possible.

(c) It can be done on an $n \times n$ board if n is even
but not if n is odd. More generally, it can be done
on an $m \times n$ board unless m and n are both odd, in
which case it cannot be done.

.28.   (b)

Trip number	Resulting state		
	Near shore	Island	Far shore
1	$h_1h_2h_3h_4w_1w_2$	w_3w_4 (kayak)	
2	$h_1h_2h_3h_4w_1w_2w_3$ (kayak)	w_4	
3	$h_1h_2h_3h_4w_1$	w_4	w_2w_3 (kayak)
4	$h_1h_2h_3h_4w_1w_2$ (kayak)	w_4	w_3
5	$h_1h_2w_1w_2$	w_4	$h_3h_4w_3$ (kayak)
6	$h_1h_2h_4w_1w_2$ (kayak)	w_4	h_3w_3
7	$h_1h_4w_1$	w_4	$h_2h_3w_2w_3$ (kayak)
8	$h_1h_4w_1$	w_3w_4 (kayak)	$h_2h_3w_2$
9	$h_1h_4w_1w_4$ (kayak)	w_3	$h_2h_3w_2$
10	w_1w_4	w_3	$h_1h_2h_3h_4w_2$ (kayak)
11	w_1w_4	w_2w_3 (kayak)	$h_1h_2h_3h_4$
12	w_1w_4		$h_1h_2h_3h_4w_2w_3$ (kayak)
13	$w_1w_2w_4$ (kayak)		$h_1h_2h_3h_4w_3$
14	w_1		$h_1h_2h_3h_4w_2w_3w_4$ (kayak)
15	w_1w_2 (kayak)		$h_1h_2h_3h_4w_3w_4$
16			everyone

6.29.   20.

6.30.   Hugo descends and gets out. The chest is
lowered and Hugo gets in with the chest. Jon
descends as Hugo and the chest rise. Hugo and the
chest are removed at the top, and Val descends—
causing Jon to rise. The first three steps are
repeated and Jon gets out. At this point Jon and
Val are at the bottom and John is in the tower with
Hugo, the chest, and the key. Now Hugo descends
and gets out, then the chest is lowered. John descends
while Val, Hugo, and the chest ascend. (If the basket
will hold only two people at a time, Jon and Val
ascend while John and Hugo descend, and the rest
of this answer is modified slightly.) Val removes the
chest and Hugo, and descends while Jon ascends.
The first three steps are then repeated again. Hugo
gets out at the top and Jon gets out at the bottom,
causing the chest to come tumbling down. Hugo
picks up the key in his mouth and drops it out of
the window. He then descends.

6.35.   It can be done in sixteen moves.

6.36.   Twenty-two moves.

6.37.

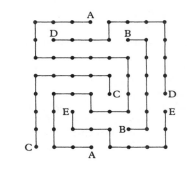

6.38. No.

6.40. Amos, Crawford, Everett, Burl, and then Dirk.

6.43. There are two possible final positions (see the figure below). They may be reached in several different ways.

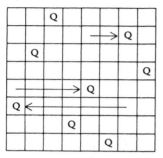

(a)

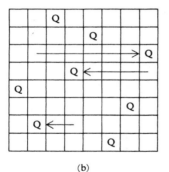

(b)

6.46. 18.

CHAPTER 7

Practice Problems

7.A 1. (a) Chance is involved; the game is of perfect information and is finite.

(b) The game is finite and decisionmaking is chance-free, but it is not a game of perfect information.

(c) Backgammon is a game of perfect information, but it involves chance and is not finite.

(d) Gin rummy involves chance, is not a game of perfect information, and is not finite.

(e) Chinese checkers is a chance-free game of perfect information, but it is not finite.

(f) Tic Tac Toe is a finite, chance-free game of perfect information.

(g) Parcheesi is a game of perfect information, but it is not chance-free and is not finite.

7.B 1. Tic Tac Toe may end in a draw. The argument does show that the first player does have at least a drawing strategy.

2. The argument used for Hex shows that A has at least a drawing strategy, since A's earlier moves can't hurt him later on.

7.F 1. (a), (b), and (d) are equivalent.

3. Placing an X in the upper righthand corner.

7.G 1. The positions in (a) are equivalent; so are the positions in (c). The positions in (b) are not equivalent.

7.H 1. (a) Yes; fill in the middle column.

(b) Yes; fill in the top row.

(c) No.

(d) Yes; complete the second column.

7.I 1. (a) Losing.   (b) Winning.

3. (a) Take 7 from the pile of 8, leaving (1, 6, 7).

(b) Leave (7, 11, 12), (5, 11, 14) or (2, 12, 14).

(c) Give up.

(d) Take 10 from the 11, leaving (1, 2, 4, 7).

Exercises

The answers below either present a list of the winning positions in the game or give the identity (A or B) of the player who wins. Occasionally, when A is the winner, a correct first move is also indicated. (Other first moves might also be correct.)

7.1. (b) (i) Winning positions: $n = 6k$; $k = 0, 1, 2, \ldots$.

(b) (ii) Winning positions: $n = 6k + 2$; $k = 0, 1, \ldots$.

7.3. (i) Winning positions: $n = (m + j)k$; $k = 0, 1, \ldots$.

(ii) Winning positions: $n = (m + j)k + j$; $k = 0, 1, \ldots$.

7.4. (a) (ii) A wins by taking 4.

(b) (ii) Winning positions: $n = 7k + 1$ and $7k + 3$; $k = 0, 1, \ldots$.

7.7. (i) Winning positions: $n = 8k$ or $8k + 1$ if you possess an odd number, and $8k + 4$ or $8k + 5$ if you possess an even number.

7.8. (i) Winning positions: (j, k), $j \leq k$ where $k - j$ is divisible by 4.

(ii) Winning positions: $(4j, 4k + 1)$ or $(4j + 1, 4k)$ or $(4j + 2, 4k + 2)$ or $(4j + 3, 4k + 3)$, j, $k = 0, 1, 2, \ldots$.

7.9. (a) (i) and (ii) A wins by taking 2 sticks from the first pile to leave $(9, 15)$.

7.11. B wins.

7.12. (a) (ii) A wins by leaving $(3, 3, 3)$.

(b) (ii) Winning positions: $(0, 0, 1)$ and (k, k, k) for $k = 2, 3, 4, \ldots$.

7.13. (a) A wins by leaving $(2, 5, 7, 7)$, $(3, 4, 7, 7)$, or $(3, 5, 6, 7)$.

(b) Winning positions: Those for which all column sums are divisible by 3 when the numbers of sticks in each pile are written in the binary system.

7.14. Winning positions: Those for which all column sums are divisible by $m + 1$ when the numbers of sticks in each pile are written in the binary system.

7.15. (a) (i) A wins by wiping out any pile.

(b) (i) Winning positions: $(0,$ even, even$)$, $(0,$ odd, odd$)$; all other positions are losing.

7.16. (a) A wins by taking the middle stick.

(b) A wins by taking one end stick.

(c) A wins by taking either the middle stick if n is odd or the middle two sticks if n is even.

7.19. (a) The first player wins by making piles of 10 and 4 sticks.

(b) The first player wins by making piles of 12 and 3.

7.20. (a) Winning positions: leave two odd piles.

(b) Winning positions: leave one pile containing 1 or 2 sticks and the other pile containing $3k + 1$ for some $k = 0, 1, 2, \ldots$.

7.21. A wins by taking an ace.

7.22. A wins by turning up a 4.

7.24. A wins by selecting 24, 25, 28, or 29 (or any number of the form $9k + 1$, $9k + 2$, $9k + 6$, or $9k + 7$ for $k = 1, 2, \ldots$.

7.25. (a) November 30.   (b) January 20.

7.26. B wins by always staying on the same diagonal as A.

7.27. A wins by moving four cells diagonally and then using a pairing strategy indicated by the lines in the figure below. That is, if B moves to a cell with a line ending in it, A moves to the cell at the other end of that line. A could also win by moving two or three cells diagonally, but it would take longer.

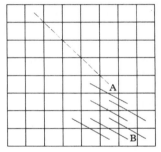

7.28. (a) A wins by moving any one of his counters as far to the right as possible.

(b) If m is odd A wins; otherwise B wins.

7.29. (a) A wins by starting in the center.

(b) A draws by starting in the center and then playing symmetrically opposite to B, using the symbol (X or O) not used by B.

7.30. (a) The game is a draw.

(b) A wins by forcing B to complete the middle row.

7.31. A wins by starting in the middle on the bottom.

7.32. A wins by starting in the bottom row, second column.

7.33. A wins by starting in the middle.

7.34. (a) A wins by starting in one of the four central boxes.

7.35. A wins by starting in the center.

7.36. A wins by placing two adjacent X's in the interior of the third row.

7.37. A wins by starting as in this figure:

7.38. A wins by taking one box.

7.39. (a) and (b) A wins by taking the middle row.

7.40. (a) A wins by starting in the center.
  (b) A wins. See the Hints and Solutions section.

7.41. (a) A wins by starting in the center of the lower level.
  (b) A wins by forcing B to complete three O's in a row on the second level.

7.42. A wins by starting with an X in the middle, with an O below it.

7.45. (a) A wins by starting with the fifth or sixth cell.
  (b) B wins.

7.46. (a) B wins.   (b) B wins.
  (c) A wins by starting in the center.

7.47. (a) B wins.
  (b) A wins by starting with the two central cells.

7.48. (b) A wins by starting in the center and playing by central symmetry until the opportunity to win presents itself.

7.49. (a) B wins.   (b) The game is a draw.
  (c) B wins, 2 to 1.

7.50. (a) (i) A wins by starting in the center.
  (ii) A wins easily by starting in one of the two cells on the middle of the shorter diagonal. (He can also win by starting on one of the end cells of the shorter diagonal, but in this case the game is more complicated.)
  (iii) A wins by starting in the center cell.
  (b) B wins on the $n \times n$ boards for n odd; and A wins on the $n \times n$ boards for n even. However, we

do not know the winning strategy except on the 3×3 and 4×4 boards.

7.51. (a) A wins in all three cases.
  (b) B wins in both cases.

7.52. (a) (i) B wins.
  (ii) A wins by starting in the center and then playing by central symmetry.
  (iii) A wins by starting in the middle of the top row and playing by vertical symmetry.
  (iv) A wins.
  (v) We think B wins.
  (b) (i) A wins.
  (ii) and (iii) B wins.

7.53. (i) If n is odd, A wins; if n is even, B wins.
  (ii) If n is odd, B wins; if n is even, A wins.

7.54. (i) A wins.   (ii) B wins.

7.55. (a) (i) A wins.   (b) (i) B wins.

7.56. (i) A wins except when $n = 1, 5, 9$, or 13.
  (ii) A wins except when $n = 2, 3$, or 6.

7.57. (a) (i) A wins by starting in the middle row.
  (ii) B wins.
  (b) (i) A wins by starting in the center.
  (ii) B wins.
  (c) (i) A wins by starting in the middle of the second row.   (ii) A wins by starting at the left side of the second row.
  (d) (i) B wins.   (ii) B wins.

7.58. (i) B wins.   (ii) B wins.

7.59. (a) (i) B wins.   (ii) A wins.
  (b) (i) A wins.   (ii) A wins.
  (c) (i) B wins.   (ii) B wins.

7.60. (i) B wins.

7.61. (a) B wins.
  (b) A wins by starting on the left.
  (c) B wins for $n = 2, 3, 6, 9$, or 10; A wins in the other cases. For $n = 4$ or 8, A starts by pushing on the left side; for $n = 5$ or 7, A starts by pushing in the center.
  (d) A wins by pushing in the center.

7.63. (i) (a) Draw.   (b) We think it is a draw. Certainly B has at least a drawing strategy.
  (ii) (a) Draw.   (b) A wins.   (c) A wins.

7.64. (a) A wins by starting in the center and using central symmetry until B completes the third vertex of a square.

(b) B wins. (c) A wins. (d) B wins.

7.65. (i) (a) A wins by placing the checker in the lower lefthand corner.

(b) B wins.

(c) If n is even, then B wins; if n is odd, then A can win by proper placement of the checker. (Proper placement depends on the parity of m.)

(ii) (a) A wins by placing the checker in a cell next to the corner.

(b) B wins.

(c) If n is even, then B wins; if n is odd, A wins by proper placement of the checker. (Proper placement depends on the parity of m.)

7.66. A wins by placing the first checker in the exact center of the table.

7.67. (a) B wins.

(b) A wins by drawing a loop starting and ending at one dot so that the remaining two dots are separated from each other.

7.68. (a) (i) The game is a draw. (ii) A wins.
(iii) A wins.

(b) No.

7.71 (a) The dwarfs win. (b) The dwarfs win.

(c) The giant wins.

7.72. (a) Fred wins. (b) Fred wins.

7.75. Either in d–3 or f–2.

7.79. Play 1d–1e. Then don't allow your opponent to isolate any more points.

7.82. (a) Move to C–4. (b) Move to C–4.

CHAPTER 8

Practice Problems

8.A 1. (a) 12; even. (b) 13; odd.
(c) 0; even.

8.B 1. (a) Possible: 3, 5, 4, 2, 1, 3, 5, 4, 2, 1, 4, 5, 3, 4, 1, 2, 5, 3.
(b) Impossible. (c) Impossible.

Exercises

8.2. (a) 5. (b) 10. (c) 21. (d) 42.
(e) $1 + 2^2 + 2^4 + \cdots + 2^{n-1} = \frac{1}{3}(2^{n+1} - 1)$.
(f) $2^1 + 2^3 + 2^5 + \cdots + 2^{n-1} = \frac{1}{3}(2^{n+1} - 2)$.

8.3. (a) 4. (b) 6. (c) 8. (d) 11.
(e) $M_7 = 14$; $M_8 = 17$; $M_9 = 20$; $M_{10} = 24$.

8.4. (a) and (d) See the figure below.

(a)   (d)

8.5. (b) See the figure below.

(b)

8.6. (a) 1 in all cases.
(b) (i) 22. (ii) 11. (iii) 6.
(c) (i) 2. (ii) 2. (iii) 1.
(d) (i) 30. (ii) 30. (iii) 15.
(e) None in all cases.

8.7. (a) (i) None. (iii) 1.
(b) (i) 6. (iii) 4.
(c) 1 in all cases.
(d) (i) 13. (iii) 7.
(e) 1 in all cases.

8.8. Here are the T and the boat (next page):

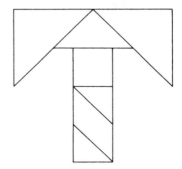

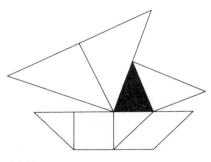

8.9. (a) Yes.

(b) Not necessarily; it depends on the color of the missing cell.

(c) No.

(d) See the Hints and Solutions section.

(e) A region with nine cells will suffice.

(f) Here are the two rectangles:

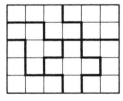

(h) 35.

8.10. (a) The center cube or any cube lying in the middle of an edge of the $3 \times 3 \times 3$ cube.

8.11. (a) The pegs that started in 24, 42, 46, or 64.

(d) The statements are true of all of the holes indicated by X in the figure below.

		X		
X	X		X	X
X	X		X	X
X				X
X	X		X	X
X	X		X	X
		X		

(e) Start by jumping 34–32.

(j) It can be done in four moves starting with 33–53–35–13–31.

(k) No.

(m) 11, 32, 41, 44, or 53. Actually, we have found solutions that would leave the single peg in any of these holes except 32.

8.12. (a) 15 moves are needed.

(b) $n(n + 2)$ moves are required.

(c) 120 moves are required.

(d) 34–33, 32–34, 31–32, 33–31, 43–33, 23–43, 13–23, 33–13, 53–33, 54–53, 34–54, 32–34, 33–32, 35–33, 45–35, 43–45, 23–43, 22–23, 32–22, 12–32, 11–12, 31–11, 33–31, 53–33, 55–53, 54–55, 34–54, 44–34, 43–44, 23–43, 21–23, 31–21, 33–31, 34–33, 32–34, 12–32, 13–12, 33–13, 53–33, 43–53, 23–43, 33–23, 35–33, 34–35, 32–34, 33–32.

8.15. It can be done in 12 moves, with one of the cousins ending up in the foyer and the den vacant.

8.17. One solution is shown in this figure:

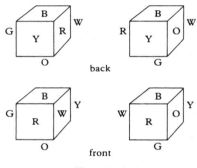

Upper level

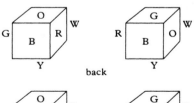

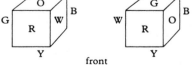

Lower level

8.19. (a) Yes. Start sowing from cup number 3 (just before the bowl).

(b) No.   (c) No.   (d) No.

(e) Yes. Start sowing from cup number 3 (the second cup before the bowl).

(f) No.   (g) No.   (h) No.   (i) No.

CHAPTER 9

9.3. 4, 9, 10, ace, 3, 6, 8, 2, 5, 7.

9.6. First weigh three of the coins against three other coins. This will limit the possibilities to three coins. Then weigh one of these against another.

9.11. (a) Weigh one coin from the first sack, two from the second sack, etc.

9.13. It can be done in fourteen reversals, with each train ending up in its original order.

9.18. No.

9.19. No.

9.20. Yes: All people who believe that knowledge is obtainable through reason would take a chance that a bucket of paint might fall on their heads.

9.22. 499. Any way of doing it will require 499 moves.

9.24. 233.

9.25. 95 millimeters.

9.26. 74.

9.27. (a) 14.   (b) 3.

9.29. The die on the right in the first figure can't be either of the dice in the bottom figure unless a number is repeated on the die.

9.30. There is such a set; it contains 50 numbers.

9.31. (a) One answer is to make knots at the 1 inch, 4 inch, and 6 inch marks on the string.

APPENDIX A

A.1. (a) $x = 3$ or -2.

(b) $x = \frac{1}{3}$.

(c) $x = \frac{7}{8}$.

(d) $x = \dfrac{9 + \sqrt{73}}{4}$ or $\dfrac{9 - \sqrt{73}}{4}$.

(e) $x > -\frac{5}{4}$.

(f) No solution.

(g) $x = 8,\ y = -3$.

(h) $x = 3,\ y = 4$ and $x = 4,\ y = -3$.

(i) $x = 1,\ y = -3,\ z = 5$.

Appendix C Answers

PPS C.A

1. a) (h,h,h), (h,h,t), (h,t,h),(h,t,t), (t,h,h), (t,h,t),
    (t,t,h), (t,t,t).
    b) i) (h,h,h), (h,h,t), (h,t,h), (h,t,t)
       ii) (h,h,h), (h,t,t), (t,h,h), (t,t,t)
       iii) (h,h,h), (h,h,t), (h,t,h), (t,h,h)
3. outcome set one: (use r for right and w for
    wrong)
    a) {(r,r,r,r), (r,r,r,w), (r,r,w,r), (r,r,w,w), (r,w,r,r),
    (r,w,r,w), (r,w,w,r), (r,w,w,w), (w,r,r,r), (w,r,r,w),
    (w,r,w,r), (w,r,w,w), (w,w,r,r), (w,w,r,w), (w,w,w,r),
    (w,w,w,w)}
    b) i) (r,r,r,r)
       ii) (r,r,r,w), (r,r,w,r), (r,w,r,r), (w,r,r,r)
       iii) (r,r,w,w), (r,w,r,w), (r,w,w,r), (w,r,r,w),
    (w,r,w,r), (w,w,r,r)
    outcome set two: {4r, 3r1w, 2r2w, 1r3w, 4w}
    b) i) 4r
       ii) 3r1w
       iii) 2r2w

PPS C.B

1. a) $\frac{1}{2}$
    b) $\frac{1}{2}$
    c) $\frac{1}{2}$
2. $\frac{7}{8}$ $\frac{7}{8}$

PPS C.C

1. a) 1 to 5
    b) 5 to 1

PPS C.D

1. a) Equally likely outcomes: {(h,h,h), (h,h,t),
    (h,t,h), (h,t,t), (t,h,h), (t,h,t), (t,t,h), (t,t,t)}.
    b) 0 heads: $\frac{1}{8}$
       1 head: $\frac{3}{8}$
       2 heads: $\frac{3}{8}$
       3 heads: $\frac{1}{8}$

PPS C.E

1. a) $\frac{26}{52} = \frac{1}{2}$
    b) $\frac{12}{52} = \frac{3}{13}$
    c) $\frac{6}{52} = \frac{3}{26}$
    d) $\frac{20}{52} = \frac{5}{13}$
    e) $\frac{32}{52} = \frac{8}{13}$

PPS C.F

1. a) $4 \cdot 5 \cdot 10 \cdot 4 = 800$
    b) $4 \cdot 5 \cdot 10 \cdot 4 \cdot 4 = 3200$

PPS C.G

1. a) 5040
    b) 40320
    c) 56
    d) 28

PPS C.H

1. a) 720
    b) 30240
    c) 3628800
    d) 10
    e) 1

PPS C.I

1. 479001600
3. 79

PPS C.J

1. a) 120
    b) 120
    c) 1
    d) 1
    e) 1
    f) 252

PPS C.K

1. $\dfrac{1}{19958400}$

PPS C.L

1. 1. a) $(x+y)^4 = x^4 + 4x^3y + 6x^2y^2 + 4xy^3 +$
       b)
    $(x+y)^5 = x^5 + 5x^4y + 10x^3y^3 + 10x^2y^4 + 5xy^5 +$
       c) $(x-y)^3 = x^3 - 3x^2y + 3xy^2 - y^3$
    3. 0

PPS C.M

1. $\frac{4}{25}$

PPS C.N

1. 1. E_1 and E_4 are independent.

PPS C.O

1. 1. $\frac{1}{18}$

PPS C.P

1. a) $\left(\frac{18}{25}\right)^4 = \left(\frac{9}{19}\right)^4$

   b) $1 - \left(\frac{18}{38}\right)^4 - 4\left(\frac{18}{38}\right)^3\left(\frac{16}{38}\right)$ or

   $\left(\frac{16}{38}\right)^4 + 4\left(\frac{16}{38}\right)^3\left(\frac{18}{38}\right) + 6\left(\frac{16}{38}\right)^2\left(\frac{18}{38}\right)^2$

   c)

   $\left(\frac{16}{38}\right)^4 + 4\left(\frac{16}{38}\right)^3\left(\frac{18}{38}\right) + 12\left(\frac{16}{38}\right)^3\left(\frac{2}{38}\right) + 4\left(\frac{16}{38}\right)\left(\frac{2}{38}\right)^3$

PPS C.Q

1. $1 - \left(\frac{1}{2}\right)^{10} = \frac{1023}{1024}$

PPS C.R

1. 5 to 1

PPS C.S

1. $\frac{3}{13} + \frac{30}{52} - \frac{33}{52} = \frac{9}{52} \approx 17.3$ cents

PPS C.T

1. $\frac{3}{8} + \frac{6}{8} + \frac{3}{8} + \frac{0}{8} = \frac{3}{2}$; $1.50

C.1   a) 5 (including one pile with all the cards).
      b) 2.

C.2   78.

C.3   a) $\frac{1}{20}$.
      b) $\frac{19}{20}$.
      c) $\frac{4}{20} = \frac{1}{5}$.

C.4   a) $\frac{688}{4800} \approx .143 > \frac{1}{7}$.
      b) 0.
      c) $\frac{1}{4}$

C.5   Jill selected 5 and 15; Jack's other number is 3.

C.6   a) 60.
      b) $\frac{1}{5}$.

C.7 $\frac{1}{120}$.

C.8 $\frac{1}{30}$.

C.9   24.

C.10 30.

C.11 a) 792.
      b) 945.

C.12 $\frac{2}{9}$.

C.13 12.

C.14 a) 2592.
      b) 72.
      c) $\frac{1}{6}$

C.15 576.

C.16 a) $\dfrac{\binom{12}{5}\binom{8}{3}\binom{6}{2}}{\binom{26}{10}} \approx .13$.

      b) $\dfrac{\binom{12}{10}}{\binom{26}{10}} \approx .000012$.

      c) $\dfrac{\binom{18}{10} + \binom{20}{10} + \binom{14}{10} - \binom{12}{10}}{\binom{26}{10}} \approx .04$.

C.17 2316.

C.18 a) 17576.
      b) 3125.

C.19 a) 625.
      b) 3471875.
      c) 5557.

C.20 a) 780.
      b) 3542650.

C.21 a) 4536.
      b) 24917760.
      c) 1658.

C.22 35.

C.23 14.

C.24 a) $\frac{11}{42}$
      b) $\frac{4}{7}$

C.25 3840.

C.26 a) 3612.
      b) 3136.

c) 3472.
d) 2576.
e) 3696.

C.27   a) 55.
       b) 170.
       c) 50.

C.28 $\dbinom{m}{2}\dbinom{n}{2}.$

C.29   a) 319.

       b) $\dfrac{5}{216} \approx .02.$

C.30   a) $10! = 3628800.$
       b) $2(9!) = 725760.$
       c) $9! = 362880.$
       d) $2(8!) = 80640.$

C.31 $\dfrac{6!}{2} = 360.$

C.32   76.

C.33   83.

C.34   a) $\dfrac{4}{\dbinom{52}{13}} \approx 6.3 \times 10^{-12}.$

       b) $\dfrac{4\left[\dbinom{13}{8}\dbinom{39}{5} + \dbinom{13}{9}\dbinom{39}{4} + \dbinom{13}{10}\dbinom{39}{3}\right]}{\dbinom{52}{13}}$

       $+ \dfrac{4\left[\dbinom{13}{11}\dbinom{39}{2} + \dbinom{13}{12}\dbinom{39}{1} + \dbinom{13}{13}\dbinom{39}{0}\right]}{\dbinom{52}{13}}$

       $\approx .005.$

       c) $\dfrac{4\left[\dbinom{13}{7}\dbinom{39}{6} + \ldots + \dbinom{13}{13}\dbinom{39}{0}\right]}{\dbinom{52}{13}}$

       $\approx .04.$

       d) $\dfrac{4\left[\dbinom{13}{7}\dbinom{39}{6} + \ldots + \dbinom{13}{13}\dbinom{39}{0}\right]}{\dbinom{52}{13}}$

       $- \dfrac{4 \cdot 3\dbinom{13}{7}\dbinom{13}{6} - \dbinom{4}{2}\dbinom{13}{6}\dbinom{13}{6}\dbinom{26}{1}}{\dbinom{52}{13}}$

       $\approx .21.$

       e) $\dfrac{4 \cdot 3\left[\dbinom{13}{12}\dbinom{13}{1} + \dbinom{13}{11}\dbinom{13}{2} + \ldots + \dbinom{13}{7}\dbinom{1}{ }\right]}{\dbinom{52}{13}}$

       $\approx 9.8 \times 10^{-5}.$

       f) $\dfrac{4 \cdot 3\left[\dbinom{13}{6}\dbinom{13}{5}\dbinom{26}{2} + \dbinom{13}{6}\dbinom{13}{4}\dbinom{26}{3}\right]}{\dbinom{52}{13}}$

       $+ \dfrac{\dbinom{4}{2}\left[\dbinom{13}{6}\dbinom{13}{6}\dbinom{26}{1} + \dbinom{13}{5}\dbinom{13}{5}\dbinom{26}{3}\right]}{\dbinom{52}{13}}$

       $\approx .1.$

       g) $\dfrac{4\dbinom{13}{4}\dbinom{13}{3}\dbinom{13}{3}\dbinom{13}{3}}{\dbinom{52}{13}} \approx .11$

       h) $\dfrac{4\dbinom{13}{4}\dbinom{13}{4}\dbinom{13}{4}\dbinom{13}{1}}{\dbinom{52}{13}} \approx .03$

i) $$\dfrac{4\left[\dbinom{13}{4}\dbinom{13}{3}^3 + \dbinom{13}{4}^3\dbinom{13}{1}\right]}{\dbinom{52}{13}}$$

$$+\dfrac{\dbinom{4}{2}\cdot 2\dbinom{13}{4}^2\dbinom{13}{3}\dbinom{13}{2}}{\dbinom{52}{13}}$$

$\approx .15.$

j) $$\dfrac{4\dbinom{13}{5}\cdot 3\dbinom{13}{3}^2\dbinom{13}{2}}{\dbinom{52}{13}} \approx .16.$$

k) $$\dfrac{\dbinom{12}{4}\dbinom{40}{9}+\dbinom{12}{5}\dbinom{40}{8}}{\dbinom{52}{13}}$$

$$+\dfrac{\dbinom{12}{6}\dbinom{40}{7}+\dbinom{12}{12}\dbinom{40}{1}}{\dbinom{52}{13}} \approx .34.$$

l) $$\dfrac{\dbinom{48}{9}}{\dbinom{52}{13}} \approx .003$$

m) $$\dfrac{\dbinom{20}{13}}{\dbinom{52}{13}} \approx 1.2\times 10^{-7}$$

n) $$\dfrac{\dbinom{36}{13}}{\dbinom{52}{13}} \approx .004$$

35 . a) $$\dfrac{4\dbinom{13}{5}}{\dbinom{52}{5}} \approx 2.0\times 10^{-3}$$

b) o.

c) $$\dfrac{10\left(4^5\right)}{\dbinom{52}{5}} \approx 9.9\times 10^{-4}$$

d) $$\dfrac{40}{\dbinom{52}{5}} \approx 1.5\times 10^{-5}$$

e) $$\dfrac{13\dbinom{48}{1}}{\dbinom{52}{5}} \approx 2.4\times 10^{-4}$$

f) $$\dfrac{13\cdot 12\dbinom{4}{3}\dbinom{4}{2}}{\dbinom{52}{5}} \approx 1.4\times 10^{-3}$$

g) $$\dfrac{13\dbinom{4}{3}\dbinom{12}{2}\dbinom{4}{1}^2}{\dbinom{52}{5}} \approx .02$$

h) $$\dfrac{13\dbinom{4}{2}\dbinom{12}{3}\dbinom{4}{1}^3}{\dbinom{52}{5}} \approx .42$$

i) $$\dfrac{\dbinom{13}{2}\dbinom{4}{2}^2\cdot 4\dbinom{11}{1}}{\dbinom{52}{5}} \approx .95$$

j) $$\dfrac{\dbinom{13}{5}4^5 - \left(10\right)4^5 - 4\dbinom{13}{5} + 40}{\dbinom{52}{5}} \approx .50$$

C.36 a) straight flush: o;

four of a kind: $$\dfrac{\dbinom{2}{2}\dbinom{45}{1}}{\dbinom{47}{3}} \approx .003$$

full house:
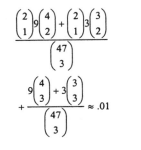

$$\frac{\binom{2}{1}9\binom{4}{2}+\binom{2}{1}3\binom{3}{2}}{\binom{47}{3}}$$

$$+\frac{9\binom{4}{3}+3\binom{3}{3}}{\binom{47}{3}}\approx.01$$

flush: 0
straight: 0
three of a kind:

$$\frac{\binom{2}{1}\binom{9}{2}\binom{4}{1}^2+\binom{2}{1}\binom{9}{1}\binom{3}{1}\binom{4}{1}\binom{3}{1}+\binom{2}{1}\binom{3}{2}\binom{3}{1}^2}{\binom{47}{3}}$$

$$\approx.11$$

two pair:

$$\frac{9\binom{4}{2}8\binom{4}{1}+9\binom{4}{2}3\binom{3}{2}}{\binom{47}{3}}$$

$$+\frac{3\binom{3}{2}9\binom{4}{1}+3\binom{3}{2}2\binom{3}{1}}{\binom{47}{3}}\approx.16$$

one pair:

$$\frac{\binom{4}{1}^3\binom{9}{3}+\binom{9}{2}3\binom{4}{1}^2\binom{3}{1}}{\binom{47}{3}}$$

$$+\frac{9\binom{3}{2}\binom{4}{1}\binom{3}{1}^2+\binom{3}{1}^3}{\binom{47}{3}}\approx.71$$

no pair: 0
   b) straight flush: $\frac{2}{47}$
      four of a kind: 0
      full house: 0
      flush: $\frac{7}{47}$

straight: $\frac{6}{47}$
three of a kind: 0
two pair: 0

one pair: $\frac{11}{47}$

no pair: $\frac{21}{47}$

C.37   .941

C.38   a) 2.
      b) 9.
      c) 44.
      d) 265.
      e) $\dfrac{9}{265}\approx.033.$

C.39 $\dfrac{2293839}{3628800}\approx0.63.$

C.40 $\dfrac{922}{4165}\approx0.22.$

C.41   a) $\dfrac{5}{36}\left(1+\dfrac{25}{36}+\left(\dfrac{25}{36}\right)^2\right)=\dfrac{14105}{(36)^3}\approx0.30.$

      b) $\frac{2821}{6^5}\approx0.36.$

C.42   a) $\frac{1}{8}.$
      b) $\frac{1}{4}.$
      c) $\frac{1}{7}$
      d) $\frac{2}{8}=\frac{1}{4}$

C.43 $\frac{1}{6}$

C.44   a) $\dfrac{23}{49}\approx.47.$

      b) $\dfrac{4}{14}\approx.29.$

      c) $\dfrac{117}{182}\approx.64.$

      d) stand pat.

C.45   a) $\dfrac{22}{49}\approx.45.$

      b) $\dfrac{3}{14}\approx.21.$

      c) $\dfrac{856}{2184}\approx.55.$

      d) stand pat.

.46 a) $\dfrac{\binom{8}{4}}{2^8} \approx .27.$

b) $\dfrac{2\binom{6}{2}}{\binom{8}{4}} = \dfrac{3}{7} \approx .43.$

.47 a) $\dfrac{2\binom{7}{4}}{2^8} \approx .55.$

b) $\dfrac{3}{7} \approx .43.$

c) $\dfrac{\binom{5}{1}+\binom{5}{3}}{\binom{7}{3}} = \dfrac{3}{7} \approx .43.$

.48 a) i) $\dfrac{2\binom{6}{2}}{2^6} = \dfrac{15}{32} \approx .47.$

ii) $\dfrac{\binom{6}{3}}{2^6} = \dfrac{5}{16} \approx .31.$

b) $\dfrac{3}{5}.$

c) $\dfrac{2}{5}.$

d) $\dfrac{22}{50}.$

.49 $1 - \dfrac{365 \cdot 364 \cdot \ldots \cdot 341}{365^{25}} \approx .57.$

(Note it is greater than $\dfrac{1}{2}$.)

C.50 a) $\left(\dfrac{1}{4}\right)^{10} = \dfrac{1}{1048576} \approx 9.53 \times 10^{-7}.$

b) $\left(\dfrac{3}{4}\right)^{10} = \dfrac{59049}{1048576} \approx .05.$

c) $\dfrac{109}{262144} \approx 4.15 \times 10^{-4}.$

C.51 a) .684.
b) .967.
c) .999.

C.52 a) .0001.
b) .3024.
c) .6976.

C.53 a) 32.
b) One route ends in slot 1; five in slot 2; ten in 3; ten in 4; five in 5; and one in 6.
c) $\dfrac{1}{6}.$

C.54 512.

C.55 a) $\dfrac{56}{2187} \approx .026.$

b) $\dfrac{2851}{19683} \approx .145.$

c) $\dfrac{8272}{19683} \approx .420.$

C.56 7 to 5.

C.57 a) $\dfrac{7}{72} \approx 097.$

b) $\dfrac{2}{27} \approx .074.$

c) $-\dfrac{17}{216} \approx -\$.08.$

C.58 $-\$.125 \approx -\$.13.$

C.59 a) Thomas.
b) Neil.
c) Rudolph.
d) Rudolph wins with probability $\dfrac{7}{27}$;

Thomas,. with probability $\dfrac{10}{27}$; and

Neil, with probability $\dfrac{10}{27}$.

e) Rudolph's expectation is $-\$19\dfrac{24}{27} \approx -\$19.89$;

Thomas' is $-\$51\dfrac{22}{27} \approx -\$51.81$; and Neil's is

$-\$71\dfrac{19}{27} \approx -\$71.71$.

C.60 a) \$3.
      b) There is none; no matter what he pays the game is favorable to him. (The expectation is infinite.)

C.61 -\$.29.

C.62 14.

C.63 42.

C.64 1680.

C.65 462.

C.66 a) 1.
      b) 2.
      c) 1.
      d) 4.
      e) 394.

C.67 The corner of G Street and Fourth Avenue.

C.68 60.

C.69 a) 15376.
      b) 400000.

C.70 276.

Index

Pages on which definitions are given are in **boldface** type.

A CATALOG OF SELECTED
DOVER BOOKS
IN ALL FIELDS OF INTEREST

A CATALOG OF SELECTED DOVER
BOOKS IN ALL FIELDS OF INTEREST

CONCERNING THE SPIRITUAL IN ART, Wassily Kandinsky. Pioneering work by father of abstract art. Thoughts on color theory, nature of art. Analysis of earlier masters. 12 illustrations. 80pp. of text. 5⅜ x 8½.   0-486-23411-8

CELTIC ART: The Methods of Construction, George Bain. Simple geometric techniques for making Celtic interlacements, spirals, Kells-type initials, animals, humans, etc. Over 500 illustrations. 160pp. 9 x 12. (Available in U.S. only.)   0-486-22923-8

AN ATLAS OF ANATOMY FOR ARTISTS, Fritz Schider. Most thorough reference work on art anatomy in the world. Hundreds of illustrations, including selections from works by Vesalius, Leonardo, Goya, Ingres, Michelangelo, others. 593 illustrations. 192pp. 7⅛ x 10¼.   0-486-20241-0

CELTIC HAND STROKE-BY-STROKE (Irish Half-Uncial from "The Book of Kells"): An Arthur Baker Calligraphy Manual, Arthur Baker. Complete guide to creating each letter of the alphabet in distinctive Celtic manner. Covers hand position, strokes, pens, inks, paper, more. Illustrated. 48pp. 8¼ x 11.   0-486-24336-2

EASY ORIGAMI, John Montroll. Charming collection of 32 projects (hat, cup, pelican, piano, swan, many more) specially designed for the novice origami hobbyist. Clearly illustrated easy-to-follow instructions insure that even beginning papercrafters will achieve successful results. 48pp. 8¼ x 11.   0-486-27298-2

BLOOMINGDALE'S ILLUSTRATED 1886 CATALOG: Fashions, Dry Goods and Housewares, Bloomingdale Brothers. Famed merchants' extremely rare catalog depicting about 1,700 products: clothing, housewares, firearms, dry goods, jewelry, more. Invaluable for dating, identifying vintage items. Also, copyright-free graphics for artists, designers. Co-published with Henry Ford Museum & Greenfield Village. 160pp. 8¼ x 11.   0-486-25780-0

THE ART OF WORLDLY WISDOM, Baltasar Gracian. "Think with the few and speak with the many," "Friends are a second existence," and "Be able to forget" are among this 1637 volume's 300 pithy maxims. A perfect source of mental and spiritual refreshment, it can be opened at random and appreciated either in brief or at length. 128pp. 5⅜ x 8½.   0-486-44034-6

JOHNSON'S DICTIONARY: A Modern Selection, Samuel Johnson (E. L. McAdam and George Milne, eds.). This modern version reduces the original 1755 edition's 2,300 pages of definitions and literary examples to a more manageable length, retaining the verbal pleasure and historical curiosity of the original. 480pp. 5⁵⁄₁₆ x 8¼.   0-486-44089-3

ADVENTURES OF HUCKLEBERRY FINN, Mark Twain, Illustrated by E. W. Kemble. A work of eternal richness and complexity, a source of ongoing critical debate, and a literary landmark, Twain's 1885 masterpiece about a barefoot boy's journey of self-discovery has enthralled readers around the world. This handsome clothbound reproduction of the first edition features all 174 of the original black-and-white illustrations. 368pp. 5⅜ x 8½.   0-486-44322-1

ANIMALS: 1,419 Copyright-Free Illustrations of Mammals, Birds, Fish, Insects, etc., Jim Harter (ed.). Clear wood engravings present, in extremely lifelike poses, over 1,000 species of animals. One of the most extensive pictorial sourcebooks of its kind. Captions. Index. 284pp. 9 x 12.   0-486-23766-4

1001 QUESTIONS ANSWERED ABOUT THE SEASHORE, N. J. Berrill and Jacquelyn Berrill. Queries answered about dolphins, sea snails, sponges, starfish, fishes, shore birds, many others. Covers appearance, breeding, growth, feeding, much more. 305pp. 5¼ x 8¼.   0-486-23366-9

ATTRACTING BIRDS TO YOUR YARD, William J. Weber. Easy-to-follow guide offers advice on how to attract the greatest diversity of birds: birdhouses, feeders, water and waterers, much more. 96pp. 5³⁄₁₆ x 8¼.   0-486-28927-3

MEDICINAL AND OTHER USES OF NORTH AMERICAN PLANTS: A Historical Survey with Special Reference to the Eastern Indian Tribes, Charlotte Erichsen-Brown. Chronological historical citations document 500 years of usage of plants, trees, shrubs native to eastern Canada, northeastern U.S. Also complete identifying information. 343 illustrations. 544pp. 6½ x 9¼.   0-486-25951-X

STORYBOOK MAZES, Dave Phillips. 23 stories and mazes on two-page spreads: Wizard of Oz, Treasure Island, Robin Hood, etc. Solutions. 64pp. 8¼ x 11.
                                                               0-486-23628-5

AMERICAN NEGRO SONGS: 230 Folk Songs and Spirituals, Religious and Secular, John W. Work. This authoritative study traces the African influences of songs sung and played by black Americans at work, in church, and as entertainment. The author discusses the lyric significance of such songs as "Swing Low, Sweet Chariot," "John Henry," and others and offers the words and music for 230 songs. Bibliography. Index of Song Titles. 272pp. 6½ x 9¼.   0-486-40271-1

MOVIE-STAR PORTRAITS OF THE FORTIES, John Kobal (ed.). 163 glamor, studio photos of 106 stars of the 1940s: Rita Hayworth, Ava Gardner, Marlon Brando, Clark Gable, many more. 176pp. 8⅜ x 11¼.   0-486-23546-7

YEKL and THE IMPORTED BRIDEGROOM AND OTHER STORIES OF YIDDISH NEW YORK, Abraham Cahan. Film Hester Street based on *Yekl* (1896). Novel, other stories among first about Jewish immigrants on N.Y.'s East Side. 240pp. 5⅜ x 8½.   0-486-22427-9

SELECTED POEMS, Walt Whitman. Generous sampling from *Leaves of Grass*. Twenty-four poems include "I Hear America Singing," "Song of the Open Road," "I Sing the Body Electric," "When Lilacs Last in the Dooryard Bloom'd," "O Captain! My Captain!"—all reprinted from an authoritative edition. Lists of titles and first lines. 128pp. 5³⁄₁₆ x 8¼.   0-486-26878-0

SONGS OF EXPERIENCE: Facsimile Reproduction with 26 Plates in Full Color, William Blake. 26 full-color plates from a rare 1826 edition. Includes "The Tyger," "London," "Holy Thursday," and other poems. Printed text of poems. 48pp. 5¼ x 7.
                                                               0-486-24636-1

THE BEST TALES OF HOFFMANN, E. T. A. Hoffmann. 10 of Hoffmann's most important stories: "Nutcracker and the King of Mice," "The Golden Flowerpot," etc. 458pp. 5⅜ x 8½.   0-486-21793-0

THE BOOK OF TEA, Kakuzo Okakura. Minor classic of the Orient: entertaining, charming explanation, interpretation of traditional Japanese culture in terms of tea ceremony. 94pp. 5⅜ x 8½.   0-486-20070-1

DRIED FLOWERS: How to Prepare Them, Sarah Whitlock and Martha Rankin. Complete instructions on how to use silica gel, meal and borax, perlite aggregate, sand and borax, glycerine and water to create attractive permanent flower arrangements. 12 illustrations. 32pp. 5⅜ x 8½. 0-486-21802-3

EASY-TO-MAKE BIRD FEEDERS FOR WOODWORKERS, Scott D. Campbell. Detailed, simple-to-use guide for designing, constructing, caring for and using feeders. Text, illustrations for 12 classic and contemporary designs. 96pp. 5⅜ x 8½. 0-486-25847-5

THE COMPLETE BOOK OF BIRDHOUSE CONSTRUCTION FOR WOODWORKERS, Scott D. Campbell. Detailed instructions, illustrations, tables. Also data on bird habitat and instinct patterns. Bibliography. 3 tables. 63 illustrations in 15 figures. 48pp. 5¼ x 8½. 0-486-24407-5

SCOTTISH WONDER TALES FROM MYTH AND LEGEND, Donald A. Mackenzie. 16 lively tales tell of giants rumbling down mountainsides, of a magic wand that turns stone pillars into warriors, of gods and goddesses, evil hags, powerful forces and more. 240pp. 5⅜ x 8½. 0-486-29677-6

THE HISTORY OF UNDERCLOTHES, C. Willett Cunnington and Phyllis Cunnington. Fascinating, well-documented survey covering six centuries of English undergarments, enhanced with over 100 illustrations: 12th-century laced-up bodice, footed long drawers (1795), 19th-century bustles, 19th-century corsets for men, Victorian "bust improvers," much more. 272pp. 5⅜ x 8¼. 0-486-27124-2

ARTS AND CRAFTS FURNITURE: The Complete Brooks Catalog of 1912, Brooks Manufacturing Co. Photos and detailed descriptions of more than 150 now very collectible furniture designs from the Arts and Crafts movement depict davenports, settees, buffets, desks, tables, chairs, bedsteads, dressers and more, all built of solid, quarter-sawed oak. Invaluable for students and enthusiasts of antiques, Americana and the decorative arts. 80pp. 6½ x 9¼. 0-486-27471-3

WILBUR AND ORVILLE: A Biography of the Wright Brothers, Fred Howard. Definitive, crisply written study tells the full story of the brothers' lives and work. A vividly written biography, unparalleled in scope and color, that also captures the spirit of an extraordinary era. 560pp. 6⅛ x 9¼. 0-486-40297-5

THE ARTS OF THE SAILOR: Knotting, Splicing and Ropework, Hervey Garrett Smith. Indispensable shipboard reference covers tools, basic knots and useful hitches; handsewing and canvas work, more. Over 100 illustrations. Delightful reading for sea lovers. 256pp. 5⅜ x 8½. 0-486-26440-8

FRANK LLOYD WRIGHT'S FALLINGWATER: The House and Its History, Second, Revised Edition, Donald Hoffmann. A total revision–both in text and illustrations–of the standard document on Fallingwater, the boldest, most personal architectural statement of Wright's mature years, updated with valuable new material from the recently opened Frank Lloyd Wright Archives. "Fascinating"–*The New York Times*. 116 illustrations. 128pp. 9¼ x 10¾. 0-486-27430-6

PHOTOGRAPHIC SKETCHBOOK OF THE CIVIL WAR, Alexander Gardner. 100 photos taken on field during the Civil War. Famous shots of Manassas Harper's Ferry, Lincoln, Richmond, slave pens, etc. 244pp. 10⅝ x 8¼. 0-486-22731-6

FIVE ACRES AND INDEPENDENCE, Maurice G. Kains. Great back-to-the-land classic explains basics of self-sufficient farming. The one book to get. 95 illustrations. 397pp. 5⅜ x 8½. 0-486-20974-1

LIGHT AND SHADE: A Classic Approach to Three-Dimensional Drawing, Mrs. Mary P. Merrifield. Handy reference clearly demonstrates principles of light and shade by revealing effects of common daylight, sunshine, and candle or artificial light on geometrical solids. 13 plates. 64pp. 5⅜ x 8½. 0-486-44143-1

ASTROLOGY AND ASTRONOMY: A Pictorial Archive of Signs and Symbols, Ernst and Johanna Lehner. Treasure trove of stories, lore, and myth, accompanied by more than 300 rare illustrations of planets, the Milky Way, signs of the zodiac, comets, meteors, and other astronomical phenomena. 192pp. 8⅜ x 11. 0-486-43981-X

JEWELRY MAKING: Techniques for Metal, Tim McCreight. Easy-to-follow instructions and carefully executed illustrations describe tools and techniques, use of gems and enamels, wire inlay, casting, and other topics. 72 line illustrations and diagrams. 176pp. 8¼ x 10⅞. 0-486-44043-5

MAKING BIRDHOUSES: Easy and Advanced Projects, Gladstone Califf. Easy-to-follow instructions include diagrams for everything from a one-room house for bluebirds to a forty-two-room structure for purple martins. 56 plates; 4 figures. 80pp. 8¾ x 6⅜. 0-486-44183-0

LITTLE BOOK OF LOG CABINS: How to Build and Furnish Them, William S. Wicks. Handy how-to manual, with instructions and illustrations for building cabins in the Adirondack style, fireplaces, stairways, furniture, beamed ceilings, and more. 102 line drawings. 96pp. 8¾ x 6⅜. 0-486-44259-4

THE SEASONS OF AMERICA PAST, Eric Sloane. From "sugaring time" and strawberry picking to Indian summer and fall harvest, a whole year's activities described in charming prose and enhanced with 79 of the author's own illustrations. 160pp. 8¼ x 11. 0-486-44220-9

THE METROPOLIS OF TOMORROW, Hugh Ferriss. Generous, prophetic vision of the metropolis of the future, as perceived in 1929. Powerful illustrations of towering structures, wide avenues, and rooftop parks–all features in many of today's modern cities. 59 illustrations. 144pp. 8¼ x 11. 0-486-43727-2

THE PATH TO ROME, Hilaire Belloc. This 1902 memoir abounds in lively vignettes from a vanished time, recounting a pilgrimage on foot across the Alps and Apennines in order to "see all Europe which the Christian Faith has saved." 77 of the author's original line drawings complement his sparkling prose. 272pp. 5⅜ x 8½. 0-486-44001-X

THE HISTORY OF RASSELAS: Prince of Abissinia, Samuel Johnson. Distinguished English writer attacks eighteenth-century optimism and man's unrealistic estimates of what life has to offer. 112pp. 5⅜ x 8½. 0-486-44094-X

A VOYAGE TO ARCTURUS, David Lindsay. A brilliant flight of pure fancy, where wild creatures crowd the fantastic landscape and demented torturers dominate victims with their bizarre mental powers. 272pp. 5⅜ x 8½. 0-486-44198-9

Paperbound unless otherwise indicated. Available at your book dealer, online at **www.doverpublications.com**, or by writing to Dept. GI, Dover Publications, Inc., 31 East 2nd Street, Mineola, NY 11501. For current price information or for free catalogs (please indicate field of interest), write to Dover Publications or log on to **www.doverpublications.com** and see every Dover book in print. Dover publishes more than 500 books each year on science, elementary and advanced mathematics, biology, music, art, literary history, social sciences, and other areas.